Birkhäuser

Static & Dynamic Game Theory: Foundations & Applications

Series Editor

Tamer Başar, University of Illinois, Urbana-Champaign, IL, USA

Editorial Advisory Board

Daron Acemoglu, MIT, Cambridge, MA, USA
Pierre Bernhard, INRIA, Sophia-Antipolis, France
Maurizio Falcone, Università degli Studi di Roma "La Sapienza", Roma, Italy
Alexander Kurzhanski, University of California, Berkeley, CA, USA
Ariel Rubinstein, Tel Aviv University, Ramat Aviv, Israel; New York University, NY, USA
William H. Sandholm, University of Wisconsin, Madison, WI, USA
Yoav Shoham, Stanford University, CA, USA
Georges Zaccour, GERAD, HEC Montréal, Canada

More information about this series at http://www.springer.com/series/10200

Daniel T. Jessie · Donald G. Saari

Coordinate Systems for Games

Simplifying the "me" and "we" Interactions

 Birkhäuser

Daniel T. Jessie
Department of Mathematics, Institute
for Mathematical Behavioral Sciences
University of California
Irvine, CA, USA

Donald G. Saari
Department of Economics, Department
of Mathematics, Institute for Mathematical
Behavioral Sciences
University of California
Irvine, CA, USA

ISSN 2363-8516   ISSN 2363-8524   (electronic)
Static & Dynamic Game Theory: Foundations & Applications
ISBN 978-3-030-35846-4   ISBN 978-3-030-35847-1   (eBook)
https://doi.org/10.1007/978-3-030-35847-1

Mathematics Subject Classification (2010): 91-XX, 91Cxx, 91Fxx, 91-00, 91-02

This book is published under the imprint Birkhäuser, www.birkhauser-science.com by the registered
company Springer Nature Switzerland AG
The registered company address is: Gewerbestrasse 11, 6330 Cham, Switzerland

Dan Jessie dedicates the book to his wife,
with love,
Kiki

Don Saari dedicates this book, in loving
memory, to his departed wife
Lillian

Preface

As the title asserts, the purpose of this book is to create coordinate systems for the space of games. It is fair to wonder, "What is this? Why is it of any interest?"

Start with the reality that game theory has an embarrassment of riches with its long and growing list of fascinating games. How can they be handled? A common strategy, as indicated in Chap. 1, is to assign some of them intriguing names such as the Prisoner's Dilemma, Battle of the Sexes, Stag Hunt, Hawk–Dove, Ultimatum, and Centipede (all of which are described in the book), and then get to know particular choices quite well. But this concentration on individual "parts" is an ad hoc process that leaves open questions about the "whole." How are the games related? Why are some so complex? Which ones can be analyzed in a similar manner? Can information learned from one game transfer to others? How can more complex games be created, and how should they be analyzed? Stated in a manner to suggest answers, a natural goal should be to create a global structure; a structure that can unify all games, can facilitate and simplify their analysis where a common procedure is applicable to *all* of them, and can handle newly developed issues.

To indicate how to do so, consider vectors—such as the direction and strength of the wind on a stormy threatening evening, or that of a rock rolling down a hill, or the initial direction and speed of a baseball hit by a bat. As true with game theory, there are far too many interesting vectors to know all of them with any familiarity. Some, yes, but not all. This problem of the plenty is efficiently handled by becoming intimate with just certain choices: With two dimensions, the selection can be the unit vectors in the horizontal and vertical directions (the x and y axes). Thanks to the resulting $x - y$ coordinate system, all other vectors can be comfortably described in terms of the ones that we know so well.

This technique of designing a coordinate system to advance our understanding is standard throughout mathematics, where choices reflect what is important. The commonly used $x - y$ system, for instance, is clumsy when dealing with circles or ellipses, such as plotting the orbit of a satellite circling the Earth: A more effective choice is to use polar coordinates, which describe the circle being considered and a point's position on that circle. When identifying a location on a sphere, such as on the surface of the Earth, it is convenient to use longitude and latitude. Problems in linear algebra are radically simplified by describing inputs in terms of eigenvectors. In dynamics, we have learned to analyze what happens in terms of motion tending toward and away from equilibria (stable and unstable manifolds).

The same approach, which enjoys all of the many powerful advantages, is developed here for game theory: Our main theme is to create a convenient coordinate system for games. But the selection of an appropriate system reflects intended objectives, which means that the choice depends on what game theoretic or solution features are to be emphasized.

Our priority is to create an easily used coordinate system that stresses individual strategic behavior, which is the widely studied Nash structure (Chap. 1). There are other options, such as coordinating individual choices. An example is finding an optimal route to drive to work: As widely experienced, too many drivers making the same choice leads to annoying, time-wasting traffic jams. This emphasis on coordinating solutions gives rise to the much used potential games (Chaps. 3, 7). Another possibility is to focus on Pareto behavior (solutions where any change hurts at least one of the players); this leads to what has been called a CoCo (cooperative, competitive) feature (Chaps. 3, 7).[1]

Different priorities define quite different systems, so it is interesting, even surprising, that these three dissimilar choices can be derived in the *same* manner (Chap. 7). Indeed, a goal of Chap. 7 is to describe a common methodology that can be used to handle other concerns from the social, managerial, and behavioral sciences.

Gains of having a coordinate system for games parallel those strong advantages that accompany standard coordinate systems.

- Coordinate systems help to geometrically identify similarities and differences among points and vectors. Similarly, such a system for games connects all of them in an easily used manner as illustrated in Chap. 2 with our Nash decomposition. In Chap. 3, it is shown how to connect and relate *all possible* 2×2 normal form games (each of two players has two strategies) with just the geometry of a simple three-dimensional cube. (For purists, it is the torus T^3.) Geometric relationships among games, such as Hawk–Dove, Stag Hunt, and Battle of the Sexes, now follow.

- Readers who have struggled to create appropriate games for an article, the classroom, or lab experiments know how time consuming and difficult this can be. But coordinates can simplify the creation of new examples. After all, often the goal is to determine how much is needed in each direction. As an illustration, imagine how difficult it can be to create a rich collection of games with *identical* individual strategic properties but with a wide variety of other traits that are designed to entice players to emphasize different features. Starting in Sect. 2.2 and continuing through the book, it is shown how doing so becomes easy with these coordinates.

---

[1]There are other approaches, such as to characterize the topology of the space of all 2×2 games [10, 20], but the coordinate system approach is more general and easier to use.

- Analysis! A suitably constructed coordinate system simplifies analyzing and discovering new results. An example involving weather forecasts on the evening news is that freeze warnings apply to all locations at a certain elevation—the east–west and north–south positions need not matter. The point is that an appropriately designed coordinate system decouples those attributes that cause different types of behavior. This feature, which accompanies our system for games, clarifies concepts such as establishing that the difference between pay-off and risk dominance is, essentially, a "coordinate directional difference" (Sect. 2.5.7). Being able to separate attributes simplifies discovering and developing new results. For instance, it turns out that the standard "best response" approach affects only certain coordinate directions of a game; this immediately explains (Sect. 3.4) why complexity measures based on this approach must fail to reflect a game's true difficulties.

Book Outline

Chapter 1. Beyond serving as a hasty introduction for those without game theory experience and a quick review for all others, Chapter 1 introduces an inherent dynamic (loosely borrowing notions from dynamical systems such as repellers, attractors, and hyperbolic points) that conveniently captures strategic aspects within games. This dynamic, which includes "out of equilibrium" behavior, is of particular value (in later chapters) when analyzing games where players have multiple strategies, when modeling games that are intended to have specified features, and, definitely, when searching for mixed strategies.

Chapter 2. The construction starts in Chap. 2 by decomposing games into a (uniquely defined) coordinate system that emphasizes what individuals can individually attain; this is the Nash structure described in Chap. 1. As an appropriately designed system decouples behaviors, it can be correctly guessed that certain directions (in the associated coordinate system) contain *all aspects* of a game, and *only these features*, that capture what individuals can attain on their own. This is the "me" (or Nash) component. For a mental picture, treat this me-information as defining the x-axis in a plane. With this mental sketch, whatever is in the y-direction *cannot* have anything to do with individual conduct. So, expect all information involving actions *among* the players, such as coordinating behavior, externalities, cooperation, mutual advantage, and so forth, to be orthogonal to a game's me-component. These interactions among players, whether positive or negative, constitute the game's we-portion. Everything is surprisingly easy to compute.

The system is illustrated by applying it to well-known games (Sect. 2.5). In this manner, the source of their complexities becomes immediately and explicitly exposed; it is the differing tensions between the me and we coordinates. As a sample of other contributions, these coordinates provide new tools to better understand approaches such as tit-for-tat and grim trigger in repeated games (Sect. 2.5.2).

Chapter 3. A measure of whether a coordinate system is appropriately designed is if it simplifies discovering new results; this is the theme of Chap. 3. A mantra of game theory, for instance, is "best response," which is where a player reacts to a given situation by selecting a personally optimal option. Thus a "best response" is a "me-action," which immediately signals that any theory based on best response *cannot* handle complexities of a game generated by we-actions such as coordination, cooperation, externalities, and so forth (Sect. 3.4). Elsewhere in this chapter, games are identified with vectors, which simplifies the analysis and introduces unexpected properties. It now becomes natural, for example, to compute how much of one game is embedded in another, which helps to explain similarities and differences in behavior and strategies.

Coordinates are typically used to compare objects. This feature is described in Sect. 3.6 where we construct a map of all 2×2 games to expose how they are related. One of several consequences is that by displaying the geometric positioning of *all* "weak Nash" points, the map identifies the source of their properties. Another use is that with the map, it is easy to compute the proportion of all games having, say, two, or maybe zero, Nash cells. This map also helps to extend the Monderer–Shapley result, which proves that the mixed strategy equilibrium of an 2×2 game is precisely that of either a particular zero-sum or a particular identical play game. The new result proves that the *full Nash structure* (i.e., beyond mixed strategies) of almost any 2×2 game is identical to either that of a particular zero-sum or a particular identical play game. Which game? It is determined by a game's map position.

Chapter 4. The first three chapters emphasize 2×2 games (each of two players has two strategies). Starting in Chap. 4, players can have any finite number of choices. The associated coordinates remain essentially the same, but properties of games can change because, well, added strategies and coordinates introduce higher dimensions. As an analogy, a way to imprison an object living in flat-land (the two-dimensional plane) is to draw a circle around it. But by adding the dimension of "up," the circle no longer suffices. For similar reasons, certain results about 2×2 games do not survive once the players have added options.

A difficulty with adding strategies is that it increases a game's complexity, which can hinder the design of games as well as concealing where certain features (e.g., mixed equilibria) are located. Then, certain necessary but annoying algebraic computations become messier. Techniques are created to simplify these issues.

Chapter 5. It is inconvenient to represent some games in the (normal form) format of the first four chapters. Instead, an insightful tree diagram (extensive form) approach describes the game by indicating what follows what. While a game's representation differs the decomposition remains consistent (Chap. 5). Of interest, using coordinates to express this added complexity exposes unexpected features.

Chapter 6. Adding players can unleash fascinating concerns. With two players, each interacts with the other, but with three or more, coalitions can form: It may be to the advantage of Ann and Barb, for instance, to gang up on poor Connie. As shown in this chapter, hidden symmetry structures must be understood to fully analyze all such strategic behaviors (and to create any number of new examples).

Topics include an analysis of coalition-free Nash points and a description of new features that arise with the standard tit-for-tat or grim trigger approaches for multi-player repeated games.

Chapter 7. Here is a question: How can one discover appropriate decompositions and their associated coordinate systems? A valued approach is *representation theory*, which, beyond game theory, can simplify our understanding of a variety of concerns coming from the social, managerial, and behavioral sciences. (This mathematical approach is used in physics, geology, and chemistry, but, so far, not in the social sciences.)[2] And so an objective of Chap. 7 is to provide an intuitive introduction of this methodology while deriving our Nash decomposition as well as the decompositions for potential games and CoCo solutions.

Chapter 8. This chapter provides a very short summary of the book.

Irvine, CA, USA   Daniel T. Jessie
June 2019   Donald G. Saari

Acknowledgements There are so many people to thank for their thoughts and comments, which reaches back to 2011 when we developed this approach. Of value are the critiques and suggestions from participants of our weekly research seminar (called the "Don Squad") and colleagues here at the University of California, Irvine, as well as comments from participants at various professional meetings and colloquia where these results were described. Special warm thanks go to Ryan Kendall, who not only contributed to our thoughts but who has been promoting these results while using them (with Dan Jessie) to design a rich trove of games for lab experiments. Specific thanks for comments concerning a draft of this book go to Jean-Paul Carvalho, Santiago Guisasola, Ryan Kendall, Tomas McIntee (who made the interesting suggestion of moving Chap. 7 up to Chap. 2, which we might have done had the book been targeted strictly to mathematicians), Louis Narens, Hannu Nurmi, Katri Sieberg, and Junying (June) Zhao. Also, our thanks for suggestions from five anonymous reviewers.

---

[2]More accurately, one of us (DGS) has used this approach, starting with [23, 24], to discover all possible ranking paradoxes and properties of positional voting methods and of certain decision approaches from psychology. The approach was used, but nothing was said about the mathematical source. So here, for what appears to be the first time in the social and behavioral science literature, the underlying mathematics is described.

Contents

Chapter 1
Introduction

1.1 Review

Most readers probably know something about game theory, so selective fundamentals are described as a quick primer for those who are not familiar with the topic and as a hasty review for all others. What differs from a standard introduction is that certain basic structures of games are outlined by borrowing concepts from dynamical systems, and it offers an intuitive commentary of the book's peculiar "me and we" subtitle, which refers to personal opportunity versus possible mutual benefits.

A way to start is with an experience one of us (DGS) had many years ago when invited to speak to a group of business decision-makers. The intent of the lecture was to build upon the obvious premise:

> When making a decision, be sure to consider an opponent's actions.

This principle is first learned as a child when introduced to the game of tic-tac-toe with the mysterious placement of those X's and O's and then enhanced after advancing to the red and black plastic discs in a checker game. To review the apparent, the presentation started with the following game[1]:

$$\mathcal{G}_{Lecture} = \begin{array}{c} \\ T \\ B \end{array} \begin{array}{|cc|cc|} \multicolumn{2}{c}{L} & \multicolumn{2}{c}{R} \\ \hline \$10,000 & -\$20 & -\$100 & \$250 \\ \$100 & \$50 & \$150 & \$100 \\ \hline \end{array} \qquad (1.1)$$

To remind the reader how to interpret the above, Row (the row player) selects either the T (Top) or B (Bottom) row, while Column (the column player) chooses either the L (Left) or R (Right) column. If, for instance, the choices are B and L, then the players receive the outcomes in the BL cell that are listed in the (Row, Column) order. This game's BL entry is ($100, $50), so Row receives $100 and Column earns $50.

---

[1] This example also is in the book [30, Chap. 5].

© Springer Nature Switzerland AG 2019

D. T. Jessie and D. G. Saari, *Coordinate Systems for Games*, Static & Dynamic Game Theory: Foundations & Applications, https://doi.org/10.1007/978-3-030-35847-1_1

The lecture participants were told that they were row players; they had to select either T or B. Without exception, *everyone* chooses Top. When questioned, the universal response was to cash in on the $10,000 bonanza. But this choice creates a problem: By Row choosing T, Column must select between *losing* $20 by playing L or *earning* $250 by playing R. Column's rational decision is to avoid losing; she chooses R, which leads to the TR outcome that transforms Row's dreams of enjoying a $10,000 vacation into an ugly $100 loss.

The lesson learned is to *Respect thy opponent.* Lacking contrary evidence, expect she or he to select a personal best response to your choice. With Eq. 1.1,

- If Row selects T, then Column chooses between L, with a loss, or R, with a nice gain: Column's best response is R leading to the TR outcome, which hurts Row.
- If Row selects B, then Column selects between L, with a gain of $50, or R with a prize of $100. Expect the choice to be BR giving Row the outcome of $150.

Conversely, Row makes the best response to the Column's choice.

- If Column selects L, then Row has an opportunity of enjoying that bonanza, so the choice is T, which causes a loss for Column.
- If Column selects R, then Row chooses between T with its loss of $100, or B with its $150 reward. Expect Row's choice to create the BR outcome, which yields a reasonable $100 for Column.

Combining both players' reactions, a sensible outcome is BR with rewards of $150 to Row and $100 to Column. Each player could have earned more had particular opportunities emerged. But, when interacting with a thoughtful opponent, it can be counterproductive to anticipate that such selections are obtainable. With BR, neither player can do better on their own, which leads to the concept of a Nash equilibrium.

Definition 1.1 With $n \geq 2$ players, a Nash equilibrium is where each player's strategy yields a personal maximum based on what is available with the other players' choices.

This definition includes all considerations that could be used to determine a maximum. It even allows an agent's outcome to be the same no matter what strategy is selected because, by default, this constant outcome is the maximum available choice.

This requirement where each player enjoys a personal maximum means that

at a Nash equilibrium, no player can unilaterally change strategies to obtain a personally better payoff (i.e., outcome).

To illustrate with BR in Eq. 1.1, Column's selection of R limits Row's choice to B and T, where Row suffers a loss from the TR cell. Similarly, with Row's selection of B, Column's other option of L is a poorer outcome. Neither player can unilaterally do better, so BR is a Nash equilibrium.

Conversely, if a selection is *not* a Nash equilibrium, some player has not exercised the best response: This player is guaranteed a personally better conclusion by changing strategy. Illustrating with BL in Eq. 1.1, because Column selects L, Row can snag that $10,000 windfall by changing from B to T. Similarly, by Row playing B, Column could ensure a personally improved outcome by playing R. This unilateral emphasis on personal rewards reflects a game's "me" attribute.

1.2 An Inherent Dynamic

As true with $\mathcal{G}_{Lecture}$ (Eq. 1.1), a central purpose of games is to model interactions to better understand what can happen. The games in Sect. 1.3, for instance, investigate events ranging from something as common as a couple deciding what to do over the weekend to negotiations between countries facing the threat of nuclear annihilation.

The modeling typically starts by listing each agent's available strategies. Then, appropriate cell entries are assigned to reflect a player's preferences of one setting over another. A common but frustrating "construction season" example is where two drivers approach a "Merge to the left" sign. Each driver has the same two strategies: Merge Left, or stay to the Right. You know the mental gymnastics, "If I merge, that is fine, but if I stay to the right until the last second, I can pass that other car. On the other hand, if she ..."

With each driver's two strategies, this conflict can be expressed in a game that has four cells with entries determined by potential rewards: "Let's see, if we both merge, we have an orderly setting. But if she merges and I don't, then I gain an advantage at her expense. Unfortunately, should neither of us merge until the end, each of us will bear a cost." Assuming her mental calculus is similar, the entries in a game could be of the form

$$\mathcal{G}_{Traffic} = \begin{array}{c} \\ Merge \\ Stay\,Right \end{array} \begin{array}{c} Merge \qquad Stay\,Right \\ \begin{array}{|cc|cc|} \hline 16 & 16 & 4 & 20 \\ \hline 20 & 4 & 8 & 8 \\ \hline \end{array} \end{array} \qquad (1.2)$$

Here is a legitimate worry: *Where did these numbers, these payoffs, come from?* Will lessons learned by analyzing this game be limited to these specific values, or, as we should hope, do they reveal basic behavioral principles? This concern is central to game theory because often, as with $\mathcal{G}_{Traffic}$, the numbers are artificially invented even though the objective is to explain general, complex interactions. This means there is a tacit expectation that discovered conclusions reflect the *game's structure*, rather than the specific numerical entries. This implicit belief is that a specified game represents what happens with a *class of games* sharing the same structure. The first requirement is robustness; slight changes in the selected payoffs should not alter what is learned nor the game's central structure.

What structure? What is, for instance, the basic form of $\mathcal{G}_{Traffic}$ that captures the essence of a relevant class of games? For this fundamental but tacit assumption to be useful, a game's essential components must be identified. This is a theme of this book, which starts next by the use of dynamics. Then, a game's basic components are extracted with a "coordinate system" developed in Chap. 2 (with a mathematical explanation in Chap. 7).

Fig. 1.1 Dynamics

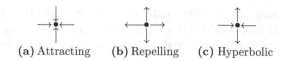

(a) Attracting   (b) Repelling   (c) Hyperbolic

1.2.1   Nash Dynamics

The game $\mathcal{G}_{Traffic}$ displays interesting behavior: If she (Row) merges, it is to my (Column's) advantage to change from "merge" to "stay on the right." Rather than specific numerical values, this action depends on which payoff value is larger. This sense of capturing how a player unilaterally seeks better outcomes generates a dynamic. Traditionally, only Nash points are sought, but doing so ignores much of what else can happen. After all, Nash equilibria focus on the game's equilibrium structure, but considering what else is admissible provides a rich insight into the game's disequilibrium behavior.

The "Nash dynamic" structure developed here loosely mimics the behavior of dynamical systems[2] where a qualitative appreciation of a system's underlying motion is gained from the locations and characteristics of the equilibria. Dynamics include three general types as depicted in Fig. 1.1:

- An attractor (Fig. 1.1a): The point is attractive in that, all nearby motions move toward it.
- A repeller (Fig. 1.1b): This point repels interest; all nearby motions move away.
- Hyperbolic points (Fig. 1.1c): Motion in some directions moves away, and in other directions, moves toward.

Interestingly, in the generic setting (i.e., robust choices where arbitrarily small changes in payoff values do not change the analysis; i.e., ignore games where no player has an advantage to move between cells), the Nash dynamics endows each cell in a game with one of these characteristics. A "Nash cell," which is a Nash equilibrium, assumes the role of an attractor; each agent cannot unilaterally obtain a better outcome, so motion is toward this cell. The next definition reflects, respectively, aspects of a repeller and a hyperbolic point. (The definition generalizes to any number of strategies.)

Definition 1.2 For an $n \geq 2$-person game $\mathcal{G}$ where each player has two strategies, a Nash cell is where the entries define a Nash equilibrium. A cell is "repelling" if each agent can unilaterally move to a personally improved cell. A cell is "hyperbolic of order (k, l)," where the positive integers satisfy $k + l = n$, if k agents cannot unilaterally move to a personally better cell but l of them can. A hyperbolic cell identifies which players can defect; e.g., with two-person games, they become "hyperbolic-column" or "hyperbolic-row".

---

[2]No knowledge of dynamical systems is needed. Instead, treat the borrowed "names" as a convenient way to indicate how many and which agents have an incentive to move.

According to Definition 1.2, for 2×2 games (each of two agents has two strategies), at a Nash cell, nobody wishes to defect. At a hyperbolic cell, precisely one player can defect, and at a repelling cell, both players can unilaterally do so. Illustrating with the $\mathcal{G}_{Traffic}$ (Eq. 1.2) cells,

- Surprisingly, the counterproductive "Stay Right–Stay Right" (BR) is an attractor or Nash cell.
- Although Merge–Merge (TL) is a cooperative choice, it is a repelling cell. Either driver can be rewarded by unilaterally defecting from cooperation!
- Merge–Stay Right (TR) is a hyperbolic-row cell. Only Row can benefit with a unilateral change.
- Stay Right–Merge (BL) is a hyperbolic-column cell. Only Column can benefit by unilaterally changing.

Incidentally, this discussion of motion appears to contradict the standard structure of one-shot games where it is assumed that, independently and simultaneously, each player selects a strategy. Fine, interpret moves as individual "if she, then he ..." brainstorming.

1.2.2   Structure of 2×2 Games

These definitions with their associated consequences form the first step toward identifying a game's essential structural aspects. Comparing cell entry values is what determines the cell type. In this manner, $\mathcal{G}_{Traffic}$'s *structural information* becomes clearer. For instance, drivers typically accept that the correct action is to merge, which is captured by the TL cell's cooperative appeal. But, TL is a repeller; beyond providing opportunities to each player, the rewards can encourage them to renege.

Of interest is that these terms characterize the generic structures of *all* 2×2 games, which provides structural information for modeling and interpretations.

Theorem 1.1 *Generically (remember, this is "robustness" condition where very slight changes in payoffs do not change the analysis), for a 2×2 game $\mathcal{G}$:*

1. *There are the same number of hyperbolic-column cells as there are hyperbolic-row cells. If there are two of a particular type, they cannot be adjacent.*
2. *With no Nash cells, all four cells are hyperbolic.*
3. *With one Nash cell, there is one repelling and two hyperbolic cells; the repelling cell could be adjacent to the Nash cell, or diametrically opposite.*
4. *If there are two Nash cells, they are diametrically opposite each other, and there are two repelling cells that also are diametrically opposite each other.*
5. *There are no games with three or four Nash cells.*

There is a dual conclusion in terms of repelling cells:

Corollary 1.1 *Generically, for a 2 × 2 game G,*

1. *if the game has no repelling cells, all four cells are hyperbolic,*
2. *if there is one repelling cell, there is one Nash cell and two hyperbolic cells, the repelling and Nash cells can either be adjacent or diametrically opposite,*
3. *if there are two repelling cells, there are two Nash cells that are diametrically opposite, and*
4. *there are no games with three or four repelling cells.*

Outline of proof: The proof is particularly simple by using the material developed in Sect. 2.4.1. For now, basic ideas are outlined and illustrated in Fig. 1.2.

No Nash cell. Start with any cell in the above figure, say A. By not being Nash, some player can unilaterally move from A to obtain a personally preferred cell; suppose it is Column. (Everything is essentially the same if it is Row, but with a reversed motion.) By being in the top row, Column selects between cell A or B. Column already is at A, so the assumption that she has a personally better choice requires her to move to cell B (as indicated by the arrow at the top of Fig. 1.2a).

B is not a Nash cell, so some agents can move to a personally better cell. The top arrow shows that it cannot be Column (Column prefers B to A), so it must be Row; i.e., B is hyperbolic-row. He moves from B to cell D (the arrow on the right side). Again, D is not Nash, so a player can change to a personally preferred cell; as Row moved from B to D (arrow along the right side), the advantage goes to Column; she moves to C (arrow on the bottom). The same argument applies where Row changes from C to the A cell. As the arrows indicate, each cell is hyperbolic providing a cyclic movement. The vertical arrows identify hyperbolic-row cells; the horizontal arrows identify hyperbolic-column cells.

One Nash cell. Suppose B is the single Nash cell. By being Nash, the two arrows on the top and right of Fig. 1.2b must point toward B. The top arrow means that Column finds B to be an improvement over A. Now, either Row finds C preferable to A (Fig. 1.2b) or Row finds A preferred to C (Fig. 1.2c). Figure 1.2b setting identifies A as a repelling cell. Because C cannot be a Nash (B is the only Nash cell), it must be that Column prefers D to C (so C is hyperbolic-column) as given by the bottom

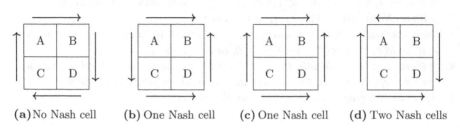

(**a**) No Nash cell   (**b**) One Nash cell   (**c**) One Nash cell   (**d**) Two Nash cells

Fig. 1.2 Dynamical structure of cells

arrow. Thus, Fig. 1.2b has two hyperbolic cells (C and D), and a repelling cell (A) adjacent to the Nash cell (B).

Turning to Fig. 1.2c, the choice made of an arrow from C to A identifies A as an hyperbolic cell. It remains to find whether Column prefers C to D, or D to C. For the first, the bottom arrow would point from D to C, which makes D a repelling cell, which is a mirror image of Fig. 1.b. Thus, the only new choice is for Column to prefer D to C (bottom arrow of Fig. 1.2c), which is in Fig. 1.2c diagram. Here, A is hyperbolic-column, D is hyperbolic-row, and C is a repelling cell diametrically opposite the Nash cell.

Two Nash cells. Nash cells have both arrows pointing toward it, so it is impossible for two Nash cells to be adjacent. Select any two diametrical opposed cells to be the Nash cells, say A and D; drawing the arrows completes the diagram and identifies the remaining two cells as being repelling.

Hyperbolic-column, -row cells. Finally, an arrow leaving a cell identifies who can defect. It follows from each diagram in Fig. 1.2 that the number of hyperbolic-column cells equals the number of hyperbolic-row cells. □

1.3 Standard Games

The following description of 2 × 2 games is based on the Nash (unilateral) "me" game dynamic. It is interesting how this dynamic defines categories (Theorem 1.1, Fig. 1.2) whereby seemingly different games belong to the same class. If games agree on the "me" traits, then all differences must reflect dissimilarities of a hidden mutually beneficial "we" term.

1.3.1 Games with One Nash Cell

For some games, the analysis is almost boringly obvious, such as with the identical play game (i.e., both entries in each cell are the same)

$$\mathcal{G}_{Obvious} = \begin{array}{c} \\ T \\ B \end{array} \begin{array}{cc} L & R \\ \hline \boxed{\begin{array}{cc} 3 \quad\quad 3 & 5 \quad\quad 5 \\ 1 \quad\quad 1 & 7 \quad\quad 7 \end{array}} \end{array} \qquad (1.3)$$

where even highly naive players, who are so myopic that they totally ignore an opponent's potential response, would stumble on the appropriate conclusion. With $\mathcal{G}_{Obvious}$, each player gravitates to the personal optimum of 7. Each prefers BR over any other cell, so BR is a Nash equilibrium. This game reflects the strong "me" creed of "do onto thyself!"

To prove that BR is the only Nash point, notice that BL offers each player a measly 1, which is so repulsive that each prefers *any* other cell; this makes BL a repeller. According to Theorem 1.1, only TL needs to be examined to determine whether it is a Nash cell. But, Column prefers TR over TL, so TL is hyperbolic-column. Consequently (Theorem 1.1), TR must be hyperbolic-row, which completes the analysis.

1.3.1.1 Prisoner's Dilemma

A more challenging game is one that all of us have, in some form and at some time, encountered—the *Prisoner's Dilemma.* The name reflects the cliched plot of a TV crime story where, in separate rooms, two alleged criminals are interrogated about a crime. If the two miscreants cooperate, not with officials but with their partner-in-crime, they do much better and may even get off. The conventional TV dialog has the interrogators offering attractive side-deals to encourage either player to defect on the partner-in-crime; if so, the partner suffers. Should both defect, both suffer.

$$\mathcal{G}_{PD} = \begin{array}{c} \\ \text{C} \\ \text{D} \end{array} \begin{array}{cc} \text{C} & \text{D} \\ \begin{array}{|cc|cc|} \hline 4 & 4 & -4 & 6 \\ \hline 6 & -4 & -2 & -2 \\ \hline \end{array} \end{array} \qquad (1.4)$$

Clearly, the group's desired $\mathcal{G}_{PD}$ (Eq. 1.4) outcome is TL. But TL is *not* Nash. Check; TL is repelling because Column personally prefers the TR outcome and Row personally prefers the BL outcome. As each player can unilaterally defect ensuring a personally preferred outcome—at the other player's expense—TL is a repelling cell. This incentive to defect reflects, of course, why the law officials offer side deals.

The existence of a repelling point in a 2×2 game ensures there is a Nash cell (Theorem 1.1). It is not BL; Column's wretched outcome would encourage Column to move to the right to attain a personally improved outcome, so BL is a hyperbolic-column cell. Similarly, by Row seeking relief from her dismal TR offering, TR is hyperbolic-row. Consequently (Theorem 1.1), even though the remaining BR cell grants each player a miserable negative outcome, it must be the game's Nash outcome! It is.

And so, a Prisoner's Dilemma game has a repelling cell offering both players an improved outcome over the Nash cell. Staying with the theme of understanding the structure of classes of games, a game's entries reflect this behavior if and only if it assumes the form

$$\mathcal{G}_{PD,Gen} = \begin{array}{c} \\ \text{C} \\ \text{D} \end{array} \begin{array}{cc} \text{C} & \text{D} \\ \begin{array}{|cc|cc|} \hline B_1 & B_2 & D_1 & A_2 \\ \hline A_1 & D_2 & C_1 & C_2 \\ \hline \end{array} \end{array} \qquad (1.5)$$

where

$$A_j > B_j > C_j > D_j. j = 1, 2, \text{ or } j = \text{Row, Column.} \qquad (1.6)$$

Be honest; rather than "inspiring," the informative Eq. 1.6 inequalities are boring. An improvement for 2×2 games (which includes Eq. 1.6) is that a Prisoner's Dilemma game is where each player prefers the payoffs in a repelling cell to those in a Nash cell.[3]

A more complete structural description is in the next chapter.

This game captures the sense where cooperation ensures a community gain, but reneging offers personal benefits at society's expense. The reader need not have been a prisoner, nor even a "person of interest," to have experienced this phenomenon. Instead, consider any societal situation requiring cooperation such as $\mathcal{G}_{Traffic}$ (Eq. 1.2), or where social norms or regulations are imposed to ensure a desired cooperative conclusion, but someone defaults. In writing a joint research paper, both are rewarded with the final publication, but a coauthor not sharing responsibilities reaps the reward of the publication without expending effort. If neither do anything, nothing is done. Such a situation can be modeled as a form of a Prisoner's Dilemma.

The "achieving a personal maximum" description makes a Nash equilibrium sound appealing, but beware! After all, as Eqs. 1.4, 1.5 demonstrate, rather than a Nash equilibrium being the desired conclusion, it merely identifies a setting where no player can unilaterally do better—rather than bells ringing, it can reflect an unhappy marriage doomed because moving elsewhere (divorce) is not feasible. Both Eq. 1.4 players prefer the repelling cell's payoffs over that of the Nash cell, but a repelling cell *always* generates incentives for each player to unilaterally move away. A cooperative effort—a "we" coordinated undertaking—is required to sustain mutually beneficial outcomes. This comment identifies several concerns.

1. The sterile Eqs. 1.4 and 1.5 representations fail to convey the fascinating conflict between personal and group interests. A preferred representation would starkly expose the differences; it would explicitly identify cooperative interactions. (This is done with our decomposition.)
2. There is a well developed, general theory for the Nash, or "me" portion of a game. While there exist clever ad hoc descriptions, including coordinated and anti-coordinated actions, a general approach to handle a game's cooperative portion is missing.
3. If cooperative (we) aspects of a game identify a non-Nash cell as a mutually beneficial outcome, the fact the cell is not Nash ensures that cooperative efforts can be unilaterally sabotaged for an agent's personal advantage. How can cooperation be sustained?
4. The story of a joint project where one partner is a free rider *need not* be modeled as a Prisoner's Dilemma. After all, only one person may need to have a completed paper to satisfy tenure and other academic rewards, so only the other person can free ride. Consequently, rather than repelling, the desired cell could be hyperbolic. This raises an interesting modeling question: Is there a simple way to construct *all possible games* reflecting an intended tension between mutual and personal rewards?

---

[3]In a 2×2 game, for *each* player to prefer the repelling cell's offerings, the repelling cell must be diagonally opposite the Nash cell. See Theorem 2.3.

1.3.2 Games with Two Nash Cells

Each of the following games can be described with an Eq. 1.6 type of representation to describe a wide class of games with similar behaviors. But, be honest, the dull Eq. 1.6 inequalities fail to stimulate or explain, why these games encourage defection from a cooperative action.

As an example, each game in this section has the same Nash dynamic structure of two Nash and two repelling cells. While the games share similar individual actions, their differences reflect variations in the "we," or cooperative component, which is made explicit in the decomposition that starts in the next chapter.

1.3.2.1 Battle of the Sexes

Anyone with interpersonal connections, whether dorm roommates or a romantic relationship, has experienced a conflict of preferences. A game capturing this common feature is the *Battle of the Sexes* where the wife wants to go to the opera, while the husband much prefers the football game. In a positive relationship, each gains pleasure by joining the other; neither wants to be alone. All of this is represented in Eq. 1.7 where conflict leads to zero pleasure; but with agreement, one partner enjoys the outcome much more than the other; e.g., the husband, Column, is happier at the football game than at the opera.

Neither spouse wants discord (represented by the zeros in cells of disagreement), so BL and TR are repelling cells from which either partner could unilaterally flee. This repelling cell structure (Theorem 1.1) anoints TL and BR as Nash cells and equilibria.

$$\mathcal{G}_{Sexes} = \begin{array}{c} \\ \text{Opera} \\ \text{Football} \end{array} \begin{array}{cc} \text{Opera} & \text{Football} \\ \begin{array}{|cc|cc|} \hline 6 & 2 & 0 & 0 \\ \hline 0 & 0 & 2 & 6 \\ \hline \end{array} \end{array} \qquad (1.7)$$

Each partner has a preferred activity (the "me" term), but both want to be together, which captures a desire for cooperation (a "we" feature). This cooperative "we" component, which should identify a group desire, is made explicit in our decomposition.

1.3.2.2 Hawk–Dove, or Chicken

Distinctly different from the caring reflected in the Battle of the Sexes is the antagonism that characterizes the Hawk–Dove game. In 1973, John Maynard Smith and George Price introduced this game into the biological literature [33] to examine the conflict between animals where the rewards of a contested resource are balanced with costs from a possible fight. A special case is

$$\mathcal{G}_{Hawk,Dove} = \begin{array}{c} \\ \text{Hawk} \\ \text{Dove} \end{array} \begin{array}{cc} \text{Hawk} & \text{Dove} \\ \boxed{\begin{array}{cc|cc} 0 & 0 & 6 & 2 \\ 2 & 6 & 4 & 4 \end{array}} \end{array} \qquad (1.8)$$

where an expensive fight (TL) leaves both with nothing. Should one player be antagonistic while the other cowers (BL and TR), antagonism is rewarded. Should both back off (BR), rewards are shared.

Obviously, TL is a repelling cell; each player can unilaterally move from nothing to something. The same is true for BR; in the R column, Row can ensure a better conclusion by bullying her way to the TR cell. In the B row, Column can do better with an aggressive move to the BL cell. Consequently (Theorem 1.1), BL and TR are Nash cells.

This game displays a forceful "me" element where each player seeks a personally gratifying conclusion. It also involves an anti-coordination factor where a Nash outcome requires doing the opposite of the opponent, which is impossible to achieve with a strictly unilateral focus. None of this is clear from an initial glance at Eq. 1.8; the decomposition (next chapter) clarifies what happens while establishing cooperative possibilities.

Equation 1.8 also captures aspects of a game called *Chicken*, which, in the 1950s, was actually played by testosterone-driven, intelligence-deprived teenagers. (This game is central to the plot of the 1955 critically acclaimed movie *Rebel Without a Cause.*) In one form, teenagers would speed in cars toward each other (Hawk). A driver who swerves out of the collision path (Dove) loses to be disparaged as a chicken—a coward. Should neither swerve (Hawk, Hawk), the game literally *ends*.

A version of this game is the political brinkmanship between countries as manifested by the Cuban crisis of the early 1960s where the US and the Soviet Union played Chicken. But here, the possibility of a collision was replaced with the danger of nuclear annihilation. A more contemporary example is the 2017 exchange of insults between the leaders of the US and North Korea. In modeling brinkmanship, one might hope for a (Dove, Dove) conclusion, but (Dove, Dove) is not a Nash outcome, so it requires cooperation. This feature is discussed later with "we" component of the game's decomposition.

Both Hawk–Dove and Battle of the Sexes have equivalent Nash dynamics of two Nash cells separated by two repelling cells. The significant difference between these games is the attitude the players have toward each other, which is the game's "we" aspect.

1.3.2.3 Hunting Stag

The philosopher Jean-Jacques Rousseau captured "the whole can differ from the sum of the parts" conflict by comparing gains achieved by societal cooperation versus personal choices where individuals accept poorer but risk-free rewards. (See Skyrms [35].) Two hunters are off to snag a stag, which requires cooperation. One hunter, for instance, could be the chaser spooking the deer in the direction where the other hunter lies in wait. On the other hand, seeking a hare is risk-free. OK, the outcome is a paltry rabbit, but, independent of what a partner may decide to do, a positive conclusion is ensured. After all, on a tough hunt during a chilly day, a stag hunter might be tempted to renege, snare a rabbit or two, and go home to watch the football game on TV.

A special case of this interaction, appropriately called the *Stag Hunt Game*, is

$$\mathcal{G}_{Stag} = \begin{array}{c} & \text{Stag} \quad \text{Hare} \\ \begin{array}{c} \text{Stag} \\ \text{Hare} \end{array} & \begin{array}{|cc|cc|} \hline 7 & 7 & -1 & 3 \\ \hline 3 & -1 & 3 & 3 \\ \hline \end{array} \end{array} \tag{1.9}$$

The two Nash cells are obvious; one is where TL, or Stag–Stag, leads to generous rewards that are gained by both players sharing the stag, while the other is BR where both snare rabbits and return home with (limited) success. Consequently (Theorem 1.1), TR and BL are repelling cells, whereby either player can unilaterally move to a personally more rewarding outcome. (In this manner, the stag hunt has strong "me" similarities with Hawk–Dove and Battle of the Sexes.)

Rousseau's story reflects the conflict between the "me"—what can be obtained on our own—and "we"—requiring coordinated effort. While the story captures this game's selfish and cooperative aspects, without thoughtful analysis, the Eq. 1.9 matrix representation fails to display this tension. Our decomposition separates the competing factors.

1.4 Mixed Strategies

Some games do not have Nash cell. Yet, they have a Nash equilibrium. This situation, which is examined next, introduces tools that are central for much of what follows.

1.4.1 Games Without Nash Cells

If a 2×2 game fails to have a Nash cell, then all four cells are hyperbolic (Theorem 1.1). Consequently, the Nash dynamic (see Fig. 1.2a) on opposite edges have arrows that point in opposing directions. (Each pair is either hyperbolic-row or hyperbolic-column.) This twisting dynamic is central for what follows.

A game with this behavior is

$$\mathcal{G}_{Cyclic} = \begin{array}{c} & \text{L} \quad\quad \text{R} \\ \begin{array}{c} \text{T} \\ \text{B} \end{array} & \begin{array}{|cc|cc|} \hline 4 & 0 & -2 & 6 \\ \hline -2 & 2 & 6 & -2 \\ \hline \end{array} \end{array}. \tag{1.10}$$

To see the cycle, start with the TL cell. With Row playing T, Column can do better by changing from L to R because the TR cell's outcome rewards Column with 6 rather than 0. But with this TR choice, Row examines the two cells in the R column; here Row's best response is B to replace the losing -2 in TR with the better 6 in BR. By Row playing B, it is to Column's advantage to select L to personally move from a loss (with BR) to a gain in the BL choice. But with Column playing L, Row's best response is to play T, which moves to the TL cell—this completes the cycle. The best response dynamic demonstrates that $\mathcal{G}_{Cycles}$ (Eq. 1.10) does not have a Nash cell.

How should a player react? The answer reflects lessons learned from personal experiences ranging from chess to tennis, where predictability creates an exploitable weakness. Should Column tend to play L, Row can select T; should Column show a preference for R, Row should select B: Predictability can be a liability.[4]

To become unpredictable, players could randomly select options with some probability; these random choices, which are called *mixed strategies,* arise in settings of flipping a coin or the random strategic choices made in athletics to keep the opponent guessing. Strategies that occur with probability one are the *pure strategies*

1.4.2 Finding the Probabilities

The Nash dynamic of $\mathcal{G}_{Cyclic}$ (Eq. 1.10) has the cyclic motion of Fig. 1.2a. According to this dynamic, should Column play L, or have a tendency to play L, then Row's best response (arrow on left of Fig. 1.2a) is to play T. But should Column play, or tend to play R, then the twisting dynamic of Fig. 1.2a shows that Row now should play B. At some point, Row experiences a transition from playing T to selecting B; the transition depends on the tendency, or likelihood, of Column playing one choice over the other.

These tendencies of Column and of Row can be captured with probabilities. So, let $q \geq 0$ and $1 - q \geq 0$ be, respectively, the probability that Column selects L and R, while $p \geq 0$ and $1 - p \geq 0$ are, respectively, the probability that Row selects T and B. With Column's given value of q, Row can compute expected winnings by playing T or B. For instance, Column plays L with probability q, so if Row plays T, the expected value of TL is $4q$ and of TR is $-2(1 - q)$. Thus, Row's expected winnings are

$$EV(T) = 4q + (-2)(1 - q) = 6q - 2$$
$$EV(B) = -2q + 6(1 - q) = 6 - 8q \qquad (1.11)$$

where $EV(T) - EV(B) = 14q - 8$, and $EV(T) - EV(B) = 0$ when $q = \frac{4}{7}$. According to Eq. 1.11, whenever $EV(T) > EV(B)$, which is $q > \frac{4}{7}$, Row's best response is to play T. This makes sense, $q > \frac{4}{7}$ means that Column is leaning toward

---

[4] As an example, Edward Thorpe proved that card-counting in blackjack and other games can change the odds to a player's advantage, where what might happen next becomes more predictable. His successful book [36] describing how to do this forced casinos to change the rules of games.

L, so T is Row's best reaction. Conversely, if $EV(T) < EV(B)$ (or $q < \frac{4}{7}$), Row should play B. Treat the probabilistic value of $q = \frac{4}{7}$ as a traffic control; it identifies in which direction a player should move.

Remember, Row seeks the best response, which is to select an appropriate p value. Whatever the p selection, with Column's choice of q, the likelihood of selecting TL is $p \times q$, so Row's expected value of this entry is the probability TL occurs times the winning or $4pq$. Computing the expected value for all four cells, Row's expected payoff is

$$
\begin{aligned}
EV(Row) &= 4pq - 2p(1-q) - 2(1-p)q + 6(1-p)(1-q) \\
&= pEV(T) + (1-p)EV(B) \\
&= p(EV(T) - EV(B)) + EV(B),
\end{aligned} \tag{1.12}
$$

In reacting to Column's choice of q, Row's best response is determined by the bottom line of Eq. 1.12. Namely, if $EV(T) - EV(B) < 0$, then Row's best choice is the smallest possible p value, or $p = 0$: Play B with probability 1. If $EV(T) - EV(B) > 0$, Row's best response is the largest possible p value, or $p = 1$—select T with certainty. If $EV(T) - EV(B) = 0$, then Row's best response is any choice of p. Row's best response curve has

- $p = 0$ for $0 \le q < \frac{4}{7}$,
- all p values for $q = \frac{4}{7}$, and
- $p = 1$ for $\frac{4}{7} < q \le 1$

This best response curve, which is the heavy solid line of Fig. 1.3a, is consistent with the twist of Fig. 1.2a. If Column displays a strong tendency to play L (so q has a large value), Row should play T (let $p = 1$); if Column has a strong tendency to play R (so q has a value closer to 0), Row should play B (or $p = 0$). So Eq. 1.12 identifies when Row should switch from T to B based on Column's tendencies (the q value).

An interesting feature of Fig. 1.2a is that the twisting action (leading to Eq. 1.12) also applies to Column's reactions to Row's actions. Thus, a similar computation determines Column's expected gains based on Row's choice of p.

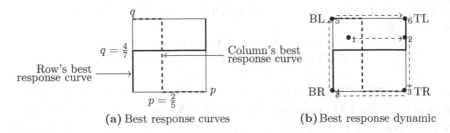

(a) Best response curves   (b) Best response dynamic

Fig. 1.3 Mixed strategy

$$EV(L) = 0p + 2(1 - p) = 2 - 2p$$
$$EV(R) = 6p + (-2)(1 - p) = -2 + 8p. \tag{1.13}$$

where $EV(L) - EV(R) = 4 - 10p$, so $EV(L) - EV(R) = 0$ at $p = \frac{2}{5}$. Column's expected outcome is

$$EV(Column) = 0pq + 6p(1 - q) + 2(1 - p)q - 2(1 - p)(1 - q)$$
$$= qEV(L) + (1 - q)EV(R) \tag{1.14}$$
$$= q(EV(L) - EV(R)) + EV(R)).$$

For all $p < \frac{2}{5}$, $EV(L) - EV(R) > 0$, so Column's best response is the largest q value of $q = 1$, which is to play L. If $p > \frac{2}{5}$, then $EV(L) - EV(R) < 0$, so Column's best response is the smallest q choice of $q = 0$. Finally, if $p = \frac{2}{5}$, any q value suffices as the best response. These three line segments constitute Column's best response, which is in Fig. 1.3a dashed curve.

Two Fig. 1.3a best response curves cross at $(p, q) = (\frac{2}{5}, \frac{4}{7})$, creating an equilibrium. To determine whether this $(p, q) = (\frac{2}{5}, \frac{4}{7})$ mixed strategy is a *Nash* equilibrium requires showing whether, with these strategies, neither player can unilaterally do better. According to Eqs. 1.11, 1.13, $EV(T) = EV(B)$ if $q = \frac{4}{7}$ and $EV(L) = EV(R)$ if $p = \frac{2}{5}$. Thus, Eqs. 1.12 and 1.14 become

$$EV(Row) = 0p + EV(B), \quad EV(Column) = 0q + EV(R).$$

As both outcomes are constant, neither player has a unilateral improvement. Thus, the mixed strategy point is a Nash equilibrium.

1.4.2.1 Dynamics

A way to appreciate this equilibrium is to examine what happens with any other choice. The pure strategies, which are stripped of probabilistic effects, already have been examined. So, consider a mixed strategy such as bullet $\bullet_1$ in Fig. 1.3b.

- As indicated by Row's response curve, in this region $(EV(T) - EV(B)) > 0$. According to Eq. 1.14, Row's best response is $p = 1$, which leads to bullet $\bullet_2$.
- But in the $\bullet_2$ region, Column's best response curve indicates B, so $EV(L) < EV(R)$. As required by Eq. 1.12, Column's best response is $q = 0$, or Bottom, which leads to the third bullet $\bullet_3$.
- This connection reunites the analysis with the pure strategies, as illustrated with bullets 4, 5, and 6.

Not using this Nash mixed strategy leads to one of the four cells of Eq. 1.10 where one player does poorly. At the mixed equilibrium $(\frac{2}{5}, \frac{4}{7})$, Column's expected winnings are $2 - 2(\frac{2}{5}) = 1\frac{1}{5}$ (Eq. 1.13 and $p = \frac{2}{5}$) and Row can expect $6(\frac{4}{7}) - 2 = 1\frac{3}{7}$ (Eq. 1.11 and $q = \frac{4}{7}$). By being weighted averages, neither is as good as the best a player can attain, but then neither is as bad as the worst a player might suffer. Should a pure strategy be selected, one of the players does better with the expected Nash mixed strategy conclusion, which identifies the mixed outcome as a reasonable choice and even supports the notion that a player could honestly announce her or his probability value for a mixed equilibrium: the opposing player would be indifferent in making a selection.

1.4.3 Matching Pennies

A standard example of mixed strategies is the game of matching pennies—perhaps to determine who will pay for lunch. Here, each of the two players takes a penny and shows one side. Row wins if the sides agree (e.g., Heads–Heads or Tails–Tails), Column wins if they disagree. This game is captured by

$$
\mathcal{G}_{Penny} = \begin{array}{c} \\ \text{Heads} \\ \text{Tails} \end{array} \begin{array}{cc} \text{Heads} & \text{Tails} \\ \hline \begin{array}{cc} 1 & -1 \end{array} \begin{array}{cc} -1 & 1 \end{array} \\ \begin{array}{cc} -1 & 1 \end{array} \begin{array}{cc} 1 & -1 \end{array} \\ \end{array} \tag{1.15}
$$

where if the penny sides agree, Row wins, otherwise Column wins.

The same cyclic, twisting effect applies. If Column believes Row will play Heads (TL), then Column will play Tails (moving to TR). Row may suspect this is the case, so Row selects Tails (moving to BR). Maybe Column knew that Row would compute all of these mind boggling choices, so Column selects Heads (moving to BL). Now, Row can select Heads, which moves to the TL cell completing the cycle.

Trying to second-guess, with these "if she, then he, but then ..." mind games is too tiring; a mixed strategy is wiser. Following the lead of the previous subsection, the Nash mixed strategy is $(p, q) = (\frac{1}{2}, \frac{1}{2})$: Each player should find a random method of selecting Heads or Tails with the same 50–50 likelihood. This leads to the Nash mixed strategy approach where each player flips his or her penny.

A mixed strategy is a weighted sum of the choices, so, here, each player's expected outcome is zero reflecting an expected equal number of wins and loses. Well, at least the *expected* outcome is not negative! But the actual outcome will be negative for one player.

1.4.4 Another Source of Mixed Strategies

Figure 1.3 analysis strictly depends on the twisting form of Fig. 1.2a where arrows on opposite sides point in different directions (generated by the two hyperbolic-column and two hyperbolic-row cells). With this twisting property, if Column tends to prefer L, Row prefers one choice, if Column tends to prefer R, Row prefers the other choice.

Somewhere in the middle of Column's choices, Row's best response must change options; the best response curve is a horizontal line. Similarly, with a likelihood Row will play T, Column plays one choice, with a likelihood Row will play B, Column adopts the other choice. Somewhere in the middle is a transition represented by a vertical line. Because a horizontal and a vertical line in a square must meet at a point, this twisting on edges ensures that the best response curves intersect, which leads to Eqs. 1.11, 1.13.

It follows from Fig. 1.2 that the only similar twisting behavior occurs with Fig. 1.2d (with two Nash cells). Illustrating with $\mathcal{G}_{Sexes}$ (Eq. 1.7),

$$\mathcal{G}_{Sexes} = \begin{array}{c} \\ \text{Opera} \\ \text{Football} \end{array} \begin{array}{c} \text{Opera} \quad \text{Football} \\ \begin{array}{|cc|cc|} \hline 6 & 2 & 0 & 0 \\ 0 & 0 & 2 & 6 \\ \hline \end{array} \end{array},$$

the two Nash equilibria are TL and BR.

To describe the impact of the opposing arrows, suppose the wife, Row, and husband, Column, are wavering between the football game or the opera. In this decision process, let $p \geq 0$ and $q \geq 0$ represent, respectively, the likelihood Row and Column prefer opera. Thus $1 - p \geq 0$ and $1 - q \geq 0$ indicate, respectively, the likelihood that Row and Column will push for football. Large q values, for instance, mean that he is leaning toward opera, so her best response (Row) is to also select opera (T). A small q value (so a large $1 - q$ value) indicates his strong lean toward football, so her best response is to do the same. Somewhere between, there is a transition in her best response.

Following the lead of Eqs. 1.11, 1.13, with Column's choice of q, the wife's expected outcomes are

$$EV_{Wife}(Opera) = EV_{Row}(T) = 6q + 0(1 - q)$$
$$EV_{Wife}(Football) = EV_{Row}(B) = 0q + 2(1 - q).$$

Thus, the transition point is where $0 = EV_{Row}(T) - EV_{Row}(B) = 8q - 2$ or $q = \frac{1}{4}$. Similar to the computation of Eq. 1.12,

$$EV(Wife) = p(EV_{Row}(T) - EV_{Row}(B)) + EV_{Row}(B).$$

Consequently, her best response curve has $p = 0$ whenever $EV_{Row}(T) - EV_{Row}(B) < 0$, or $0 \leq q < \frac{1}{4}$. It has all p values for $q = \frac{1}{4}$. Finally, $p = 1$ when $EV_{Row}(T) - EV_{Row}(B) > 0$, or $\frac{1}{4} < q \leq 1$. This is depicted by the heavy solid line in Fig. 1.4a.

Similarly, for Row's given p value, Column's computations lead to

$$EV_{Col}(L) = 2p, \quad EV_{Col}(R) = 6(1 - p),$$

so

$$EV_{Col}(L) - EV_{Col}(R) = 2p - 6(1 - p) = 8p - 6$$

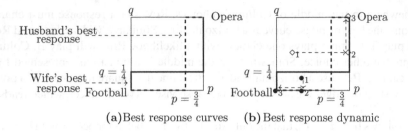

(a) Best response curves   (b) Best response dynamic

Fig. 1.4 A different mixed strategy

or $p = \frac{3}{4}$ is where $EV(L) - EV(R) = 0$. Again, similar to the above,

$$EV(Husband) = q(EV_{Col}(L) - EV_{Col}(R)) + EV_{Col}(R).$$

From this expression, the Husband's best response curve has $q = 0$ for $EV_{Col}(L) - EV_{Col}(R) < 0$, which is where $0 \le p < \frac{3}{4}$, all q values for $p = \frac{3}{4}$, and $q = 1$ for $EV_{Col}(L) - EV_{Col}(R) > 0$, which is where $\frac{3}{4} < p \le 1$. This is the dashed curve in Fig. 1.4a. This joint wavering about what to do, captured by the opposing arrows on opposite edges, forces the best response curves to cross. A computation similar to that of Eqs. 1.12, 1.14 proves that the crossing point, $(p, q) = (\frac{3}{4}, \frac{1}{4})$ is a Nash equilibrium.

As it must be expected, the $(EV(Wife), EV(Husband)) = (1\frac{1}{2}, 1\frac{1}{2})$ outcome is smaller than what either would receive with coordination. (This lower weighted average includes penalties from failed coordination.) Different from the Eq. 1.10 game of $\mathcal{G}_{Cyclic}$, at either pure Nash point, each player's outcome is better than the expected result from the mixed Nash point. But the mixed Nash expectations are better than what each player might achieve without coordination. Either pure Nash strategy (northeast or southwest from the Fig. 1.4a mixed strategy) has a higher outcome; the other two cells (northwest or southeast) are much lower. This creates a reward structure resembling a saddle with higher peaks in two directions and lower values in two other directions. Between any two pure Nash equilibria expect a mixed strategy.

The dynamics, as indicated in Fig. 1.4b, reflects this saddle structure. Any mixed starting point in one of the two rectangles leads to a reaction similar to that of $\bullet_1$. This point's p value is less than three-fourths, so Column's best response is $q = 0$ (Football) leading to the $\bullet_2$ point. With the $q = 0$ value, Row's best response is to select $p = 0$, which sends the two off to the football game! In fact, in either of the two rectangular regions, the first player to respond moves the strategy to a square edge; the other agent's best response leads to the region's vertex, which is a pure strategy of either football or opera.

A similar dynamic holds with a starting initial point in one of the squares, such as $\diamond_1$. The difference depends on which player is the first to react. The region with

$\diamond_1$, for instance, has $q > \frac{1}{4}$, so Row's best response is $p = 1$ leading to $\diamond_2$, where she strongly promotes the Opera. As $p = 1$, her husband's best choice is $q = 1$, and both set off for the opera. *But*, if the husband was the first to react, because $p < \frac{3}{4}$, his best response is $q = 0$, or football. Her best response of $p = 0$ is where both search for the football tickets.

This dynamic introduces an interesting concern. Each Nash cell is associated with a box that is defined by the best response curves. Do the box sizes contain information about the game? Is "bigger better" as suggested by the description of the dynamics? This issue is discussed in Sect. 2.6.

1.5   Nash Theorem and Defining Generic and Degenerate Settings

This discussion makes it clear that all 2×2 games have at least one Nash equilibrium. More precisely, in the *generic setting,* each 2×2 game has either one or three Nash equilibria. This statement requires showing that a game with one Nash cell cannot have a mixed equilibrium, which follows easily from Fig. 1.2b, c.[5]

This is an appropriate place to be more precise about what we mean by a *generic situation.* Remember, the goal is to emphasize robust settings; this is where very slight payoff changes do not affect the game's structure. With $\mathcal{G}_{Sexes}$ (Eq. 1.7), for instance, slightly changing any of the entries will not influence the analysis. This stress on robustness reflects our objective to consider classes of games; "classes" capture general behaviors. If changing payoffs with values so small that they resemble the width of a strand of hair can alter the outcomes, these conclusions represent anomalies rather than general behavior.

More explicitly, when analyzing the Nash structure for 2×2 games, a generic setting merely excludes games with a column where Row does not have a unique maximum and/or a row where Column does not have a unique maximum. An example of the forbidden is

$$\mathcal{G}_{Degenerate} = \begin{array}{|cc|cc|} \hline 4 & 7 & 3 & 6 \\ \hline 4 & 8 & 2 & 1 \\ \hline \end{array} \tag{1.16}$$

where in the first column, Row is ambivalent because selecting either T or B yields the same reward of 4. Without Row caring which way to move, an arrow on this left edge, indicating a preferred direction of change, disappears. What makes this game degenerate is that even the slightest change in either 4 value—even replacing one of the 4's with 4.00000000001—returns to the earlier discussion. This concentration on games where slight changes do not alter the analysis makes excellent sense. After

---

[5]A game with a double twist, on the top-and-bottom edges (reflecting changes of Column's best responses) and on the left-and-right edges (reflecting changes in Row's best responses), must have a mixed equilibrium; it serves as marker where to shift strategies. But if only one set has twists, the other player has a dominant strategy of preferring to remain with one choice, so there is no need for a shift.

all, rarely can a game's entries be absolutely precise. And so, with rare exceptions, we only consider *generic* (robust) settings.

These degenerate settings (e.g., Eq. 1.16) define a boundary, a shift, between what happens in neighboring generic cases. To see this tradeoff with $\mathcal{G}_{Degenerate}$, changing either 4 value ever so slightly creates an arrow on the left either moving upward or downward; these arrows identify, respectively, TL or TB as the Nash cell. Each of these new settings has *one Nash cell* and no mixed equilibrium. The degenerate case (where the values precisely equal 4) is an orderly transfer between the two regimes. This role of serving as a transition is why the Nash structure keeps *both* TL and BL as Nash cells and does not have a mixed strategy—it is a union of what happens with the two neighboring cases. This structure permits a slight change in one of the 4's to drop one of these two transitory (weak) Nash cells and return to a generic setting. Degenerate settings typically reflect changes between generic cases, so analyzing them is not difficult. As such, we tend to ignore them. (Readers wishing to learn more about this, see the material describing Eqs. 3.54 and 3.55.)

Returning to Nash equilibria, it is important to know whether such equilibria exist beyond 2×2 games. This comment underscores the importance of Nash's Theorem [18, 19] proving that

> every game with a finite number of players for which each player can choose from finitely many pure strategies has at least one Nash equilibrium.

This equilibrium might involve mixed strategies (as in Sect. 1.4), or even weak Nash points where a player's maximum value holds for several cells.

The advantage gained by including mixed strategies is that Nash enlarged the space of strategies from a finite collection of points to a geometrically smooth planar setting captured by probability simplexes. With this larger structure, Nash proved the theorem by means of fixed point theorems.

It is worth pointing out that Nash's result does *not* hold for all games. As it is well known, games that lie outside of Nash's specified structure, such as where there are an infinite number of strategies, or non-compact settings, need not have a Nash equilibrium. A commonly cited example is where two players simultaneously select an integer, and the largest one wins. Games bereft of Nash equilibria are not considered in what follows.

Chapter 2
Two-Player, Two-Strategy Games

2.1 Preview

While the analysis in this chapter is straightforward, a considerable amount of new information is developed. Thus, a quick preview will help to pull it together.

The material starts by introducing a decomposition of payoff entries (Sect. 2.3), which defines the promised coordinate system (Theorem 2.1). Similar to how a point in the plane is labeled with its x and y components, the coordinate system identifies how much of each of a game's payoffs is devoted toward individual actions ($\eta_{i,j}$ terms), conclusions requiring joint activity (β_j components), and redundancy terms (κ_j values) that can inflate payoffs but play no role in a typical analysis.

An effective use of a coordinate system requires developing intuition about what the components mean: This is the theme of Sect. 2.4.1. Which coordinates, for instance, cause Fig. 1.2 Nash dynamic description? Here, a simple addition technique is introduced to quickly identify the connections among hyperbolic, Nash, and repelling cells; a technique that will be particularly useful in Chap. 4 when confronted with the complexity introduced by admitting new strategies.

Coordinates are typically used to create examples. It can be difficult to design appropriate games for personal exploration, to illustrate a point in a paper or a lecture, or for lab experiments. But as emphasized in Sect. 2.4.4, the coordinate system makes doing so particularly easy. The idea is simple: Often the goal is to fix a particular feature of games while altering another attribute. Just as finding all locations that are one unit above the x-axis means holding $y = 1$ component fixed and varying the x values, a rich variety of games results by holding one interaction feature fixed and varying the others.

A typical ad hoc analysis of games can be difficult. An advantage of having an appropriate coordinate system is that it provides a systematic analytic approach: Analyze games in terms of their coordinates. Doing so, determines what aspect of a game is devoted to individual actions and what portion requires some form of group coordination, cooperation, and so forth. This is the theme of Sect. 2.5, where several games are analyzed in this manner.

© Springer Nature Switzerland AG 2019   21
D. T. Jessie and D. G. Saari, *Coordinate Systems for Games*, Static & Dynamic Game
Theory: Foundations & Applications, https://doi.org/10.1007/978-3-030-35847-1_2

Certain games require special techniques, such as the tit-for-tat strategy of forcing cooperation in a repeated Prisoner's Dilemma. As one might hope, the decomposition provides new insights into what happens, what causes these difficulties, and why they can be resolved (Sect. 2.5.1). Also, it is indicated (Sect. 2.6) how the decomposition clarifies concepts about games, such as the difference between risk and payoff dominance.

2.2 Introduction

Our development of games into me-we components is introduced with 2×2 games; this is where each of two players has two pure strategies. As described in subsequent chapters, the approach for more agents and multiple strategies remains essentially the same.

In order to appreciate what will happen, we invite the reader to analyze the game

$$\mathcal{G}_{Challenge} = \begin{array}{|c c|c c|} \hline -7 & 11 & 6 & 7 \\ \hline -3 & 0 & 0 & -2 \\ \hline \end{array} \tag{2.1}$$

and write down details for a later discussion. What role, for instance, does the -7 play in the game's structure? Similarly, how would you analyze the following three games, which appear to have nothing in common?

$$\mathcal{G}_1 = \begin{array}{|c c|c c|} \hline 0 & 0 & 4 & 2 \\ \hline 2 & 4 & 6 & 6 \\ \hline \end{array}, \quad \mathcal{G}_2 = \begin{array}{|c c|c c|} \hline 4 & 4 & -2 & 6 \\ \hline 6 & -2 & 0 & 0 \\ \hline \end{array}, \quad \mathcal{G}_3 = \begin{array}{|c c|c c|} \hline 8 & 0 & 0 & 2 \\ \hline 10 & 2 & 2 & 4 \\ \hline \end{array}, \tag{2.2}$$

It is easy to analyze $\mathcal{G}_1$ because each player's personal maximum resides in the BR cell. So $\mathcal{G}_1$'s simplicity is captured by the reality that even should each player totally ignore the other player, the BR Nash choice is the outcome.

The second game $\mathcal{G}_2$ is more complex; e.g., each player's optimal payoff is in a different cell (BL for Row and TR for Column). Moreover, the group's optimal cooperative choice of TL is a repelling cell, *not* a Nash cell. This repelling feature encourages each player to unilaterally renege to attain that personally better outcome, but only if the other player does not also default. This game, then, requires making assumptions about what the opponent will do. Contributing to the complexity is that this Prisoner's Dilemma (e.g., see Eq. 1.4) game's BR Nash cell offers each player the miserable outcome of zero, which again dashes expectations that Nash points constitute desired outcomes.

It is not clear what to say about $\mathcal{G}_3$, other than that BR is the sole Nash equilibrium. But should the rewards be transferrable, such as money, a side cooperative agreement of dividing the BL rewards of $10 + 2 = 12$ with six going to each player makes BL more attractive than the BR Nash point. Again, assumptions about what an opponent might do are important. After all, replacing BL with two 6's makes all cells hyperbolic, along with the complications that accompany games without a Nash cell.

The reader probably adopted a typical ad hoc approach of examining each of these games separately. One of our goals is to replace this standard method with a simple systematic procedure so that all games can be analyzed in the same manner. The method is the obvious one used with any coordinate system: Express what is being studied in terms of its coordinates, which here divides games into their "me" and "we" components. With these three games, for instance, the only common feature is the Nash structure where each game has a BR Nash cell and a repelling TL cell; as this Nash dynamic is based on unilateral actions, it constitutes the game's "me" portion. In turn, all commentary about cooperation, coordination, externalities, and so forth (which dominated the discussion of $\mathcal{G}_2$ and $\mathcal{G}_3$), belongs to the "we" part that is identified in Sect. 2.3.

For more examples, the two games

$$\mathcal{G}_4 = \begin{array}{|cc|cc|} \hline 6 & 6 & 0 & 4 \\ 4 & -4 & 2 & 2 \\ \hline \end{array}, \quad \mathcal{G}_5 = \begin{array}{|cc|cc|} \hline 0 & 18 & 20 & 16 \\ -2 & -6 & 22 & 0 \\ \hline \end{array}. \qquad (2.3)$$

have the same Nash "me" dynamic structure with Nash equilibria at TL and BR, so BL and TR are repelling cells (Theorem 1.1). Consequently, each game has a Nash mixed strategy equilibrium (Sect. 1.4.4) denoted as NME, which, while it is not clear from Eq. 2.3, agree.

Structures beyond a game's "me" dynamic are important. An attractive $\mathcal{G}_4$ outcome, for instance, is the TL Nash cell: Reaching this outcome requires the players to coordinate. In contrast, TR in $\mathcal{G}_5$ offers mutual benefits for both players. But by being a *repelling* cell, attaining this TR outcome requires cooperation rather than coordination. Clearly, it is of interest to be able to identify a game's coordination–cooperation structures.

How to do so is developed in Sect. 2.3, where it is shown how to divide a 2×2 game $\mathcal{G}$ into its Nash, Behavioral, and Kernel components

$$\mathcal{G} = \mathcal{G}^N + \mathcal{G}^B + \mathcal{G}^K. \qquad (2.4)$$

The Nash component, $\mathcal{G}^N$, contains the *minimal* amount of $\mathcal{G}$ information needed to find all Nash, hyperbolic, and repelling cells as well as all Nash equilibria. This structure captures the unilateral actions, so $\mathcal{G}^N$ is the game's "me" portion.

As indicated above and noted by Fudenberg and Tirole [7, p. 8],

...the analysis of some games ...is very sensitive to small uncertainties about the behavioral assumptions players make about each other.

The game's Behavioral component $\mathcal{G}^B$, more than any other term, affects these assumptions. It does so by identifying all relevant types of cooperative, coordinated behavior among the players. So, $\mathcal{G}^B$ is the game's "we" component; it is associated with any aspect of externalities, cooperation, coordination, or even why players need not play Nash terms (which is a theme of behavioral game theory).

The remaining $\mathcal{G}^K$, the "kernel" component, plays no substantive role in strategic or cooperative aspects; they are "leftovers." It does have an impact should a game's rewards be transferable, such as money or land or ..., or when designing games for lab experiments where negative entries are shunned because players tend to avoid them.

2.3   The Nash Decomposition

Using $\mathcal{G}_4$ to introduce the coordinates, represent Column's L-R strategy by probabilities $(q, 1 - q)$, $q \in [0, 1]$. With this choice, Row's preferences between T and B are determined by the expected rewards where more is better. Following the Sect. 1.4 description, with q, Row's expected outcome by playing T is $EV(T) = 6q + 0(1 - q)$ and from playing B is $EV(B) = 4q + 2(1 - q)$. The difference (in a best response analysis) is

$$EV(T) - EV(B) = \{6q + 0(1 - q)\} - \{4q + 2(1 - q)\} = [6 - 4]q + [0 - 2](1 - q). \tag{2.5}$$

If Eq. 2.5 has a positive value, Row should play T; with a negative value, Row should play B; with a zero value, it does not matter what Row plays.

Here is a key question: What roles do the 6 and 4 and the 0 and 2 play?

2.3.1   The Nash, or "me" Component

Only the final Eq. 2.5 *bracket values* are used in the Nash analysis. More precisely, the only role played by the 6 and 4 entries is to define the bracket value of 2, which is the q coefficient. Similarly, for the second bracket, the contribution made by the 0 and 2 values is their -2 difference that determines the $(1 - q)$ coefficient.

Consequently, the game's *Nash strategic analysis remains unchanged for any entries that share the same bracket values.* Replacing 6 and 4 with 0 and -2, and 0 and 2 with 20 and 22 (these are Row's $\mathcal{G}_5$ entries) makes no change in the Nash analysis!

Our decomposition is simple; retain the barest amount of information needed to determine a game's bracket values. In particular, because Row's terms in the L column are 6 and 4 with an average of $\frac{6+4}{2} = 5$, Row's bracket values in Eq. 2.5 remain unchanged by replacing each term by how it differs from the pair's average. Replacing 6 with $6 - \frac{6+4}{2} = 1$ and 4 with $4 - \frac{6+4}{2} = -1$ (Row's entries in the L column of Eq. 2.7) does not affect the q coefficient in Eq. 2.5. Similarly, replacing 0 with $0 - \frac{0+2}{2} = -1$ and 2 with $2 - \frac{0+2}{2} = 1$ keeps the same $(1 - q)$ coefficient.

The same argument holds when comparing Column's expected values for L and R in response to the row player's mixed strategies based on probabilities p and $(1 - p)$ for T and B. Namely, $EV(L) = 6p - 4(1 - p)$ and $EV(R) = 4p + 2(1 - p)$, so

$$EV(L) - EV(R) = [6 - 4]p + [-4 + 2](1 - p). \hspace{2cm} (2.6)$$

Again, the brackets' values, rather than the entries, are used in a strategic analysis. The average of the p terms is $\frac{6+4}{2} = 5$, so replace each term with how it differs from 5; i.e., replace 6 with $6 - 5 = 1$ and 4 with $4 - 5 = -1$. Similarly, the average of the $(1 - p)$ terms is -1, so replace -4 with $-4 - (-1) = -3$ and 2 with $2 - (-1) = 3$.

This construction defines the much simpler game G_4^N, which has precisely the same Nash structure as G_4.

$$G_4^N = \begin{array}{|c c|c c|} \hline 1 & 1 & -1 & -1 \\ \hline -1 & -3 & 1 & 3 \\ \hline \end{array} \hspace{2cm} (2.7)$$

As described later, this Nash component G_4^N *extracts all G_4 Nash strategic information*; it simplifies the analysis by eliminating all Eqs. 2.5, 2.6 redundancies. Notice: only two G_4^N cells, TL and BR, have positive entries, so they are Nash cells for both G_4^N and G_4. Similarly, BL and TR have negative entries, so they are repellers. Computing the common NME can use the simpler G_4^N entries rather than those from G_4.

Why replace the G_4 entries with differences from the average? After all, the same strategic structures result from

$$G_4^* = \begin{array}{|c c|c c|} \hline 1+x & 1+z & -1+y & -1+z \\ \hline -1+x & -3+u & 1+y & 3+u \\ \hline \end{array} \hspace{2cm} (2.8)$$

for any x, y, z, and u values. Using $x = y = z = 1$ and $u = 3$, for instance, leads to a particularly clean matrix

$$\begin{array}{|c c|c c|} \hline 2 & 2 & 0 & 0 \\ \hline 0 & 0 & 2 & 6 \\ \hline \end{array}$$

with no negative entries and the desired bracket values.

The problem with the x, y, z, u terms is that, whatever their choice, they are *redundant* for the Nash analysis. Remember, the intent of the decomposition is to retain *only entries that are relevant* for the analysis. When Row computes for Eq. 2.8, the best response

$$EV(T) - EV(B) = [(1 + x) - (-1 + x)]q + [(-1 + y) - (1 + y)](1 - q),$$

the x and y values cancel, which means that they clutter the equation without contributing to the Nash analysis. Similarly, for Column, in the Eq. 2.6 computation

$$EV(L) - EV(R) = [(1 + z) - (-1 + z)]p + [(-3 + u) - (3 + u)](1 - p),$$

the z and u terms are not needed; they subtract off to have no part in the Nash analysis.[1] The goal is *not* to find a comfortable representation for computations (as in Sect. 4.8), but to identify what informational aspects of a game explain individual versus group tensions.

Even more, while redundant terms do not distort the Nash computations, they distract from the Nash analysis. For instance, a useful result (derived later) is that

> with any finite number of players where each has any finite number of strategies, a Nash cell must have all positive entries in $\mathcal{G}^N$.

As true with Eq. 2.7, this fact significantly simplifies finding these equilibria. But with $x = y = -2$, $z = u = 1$ for Eq. 2.8, the two Nash cells

-1	2	-3	0
-3	-2	-1	4

are distorted with negative values. To eliminate complexities, remove all unnecessary terms.

This comparison leads to an important, subtle fact:

> $\mathcal{G}^N$ extracts all of a game's Nash information while dismissing redundancies; entries of $\mathcal{G}^N$ replace entries of $\mathcal{G}$ by how they deviate from the pair's average.

2.3.2 The Behavioral "we" Factor, and Kernel Terms

Each $\mathcal{G}_4^N$ entry is defined by how it differs from a specified $\mathcal{G}_4$ average, so these averages define the $\mathcal{G}_4 - \mathcal{G}_4^N$ entries. To be specific, as 5 is Row's average of the TL and BL cells, this 5 value must be Row's TL and BL entries in $\mathcal{G}_4 - \mathcal{G}_4^N$.

Indeed, Row's term in the TL cell of $\mathcal{G}_4 - \mathcal{G}_4^N$ is $6 - 1 = 6 - [6 - \frac{6+4}{2}] = \frac{6+4}{2} = 5$, which is the average. Computing these averaged entries leads to the decomposition

$$\mathcal{G}_4 = \begin{array}{|cc|cc|} \hline 6 & 6 & 0 & 4 \\ 4 & -4 & 2 & 2 \\ \hline \end{array} = \mathcal{G}_4^N + \mathcal{G}_4^{Averages} = \mathcal{G}_4^N + \begin{array}{|cc|cc|} \hline 5 & 5 & 1 & 5 \\ 5 & -1 & 1 & -1 \\ \hline \end{array}. \quad (2.9)$$

An interesting $\mathcal{G}_4^{Averages}$ feature (the second bimatrix at the end of Eq. 2.9) is that no player has a unilateral advantage. If Column selects L, for instance, then whatever Row does the payoff remains 5; Row enjoys a positive externality. If Column selected R, then whether Row selects T or B, the outcome is 1; here Row experiences a reduced externality. Similarly, whatever Row selects, it is immaterial what Column

---

[1] From the mathematical perspective developed Sect. 3.4 and in Chap. 7, $\mathcal{G}^N$ is the projection of $\mathcal{G}$ into a lower dimensional subspace that contains all of the Nash information. This projection eliminates redundancies for the Nash analysis. The Nash subspace is characterized by the sum of Row's entries in any column, and the sum of Column's entries in any row, must equal zero.

does because the outcome remains the same. This property holds for any $\mathcal{G}$ because $\mathcal{G}^{Averages}$ consists of appropriate "averages" of $\mathcal{G}$ entries from rows and columns.

Stated differently, because of its structure, $\mathcal{G}_4^{Averages}$ has absolutely no Nash information: All information about a game's Nash behavior is contained in $\mathcal{G}^N$. Conversely, all of a game's information involving cooperation, coordination, externalities, and so forth, must reside in $\mathcal{G}^{Averages}$.

To motivate the next reduction, if the same value, say 6, is added to each of a player's payoffs, it inflates the values but has no impact on the game's structure. So, the next step is to remove these "inflation" effects; e.g., remove $\mathcal{G}^{Averages}$ redundant values.[2]

Inflated values are reflected by a player's average outcome. So let κ_j be the jth player's average of $\mathcal{G}_4$ entries; i.e., for Row, $\kappa_1 = (6+4+0+2)/4 = 3$, and for Column, $\kappa_2 = 2$. The Kernel component, $\mathcal{G}^K$, which contains the leftovers that merely inflate or deflate each of a player's payoffs by the same amount, is the bimatrix where a player's entry in each cell is the player's κ value. That is,

$$\mathcal{G}_4^K = \begin{array}{|c c|c c|} \hline 3 & 2 & 3 & 2 \\ \hline 3 & 2 & 3 & 2 \\ \hline \end{array}, \text{ with the abbreviated notation } \mathcal{G}_4^K(3, 2). \qquad (2.10)$$

In general, $\mathcal{G}^K(\kappa_1, \kappa_2)$ lists each player's average payoff in the game $\mathcal{G}$.

What remains is the behavioral term $\mathcal{G}_4^B$: A player's $\mathcal{G}^B$ entry in a cell is the difference between the $\mathcal{G}^{Averages}$ entry and the κ_j average. This means that

$$\mathcal{G}_4^B = \mathcal{G}_4^{Averages} - \mathcal{G}_4^K = [\mathcal{G}_4 - \mathcal{G}_4^N] - \mathcal{G}_4^K = \mathcal{G}_4 - [\mathcal{G}_4^N + \mathcal{G}_4^K]. \qquad (2.11)$$

Stated in a simpler computational manner, replace each of a player's $\mathcal{G}_4^{Averages}$ entries by how it differs from the player's average payoff. For instance, Row's column entries in $\mathcal{G}_4^{Averages}$ are 5 and 1 and $\kappa_1 = 3$. So, to define $\mathcal{G}_4^B$, replace each of Row's 5's in $\mathcal{G}_4^{Averages}$ with $5 - 3 = 2$ and each of Row's 1's with $1 - 3 = -2$. Doing the same for Column with $\kappa_2 = 2$ leads to the following $\mathcal{G}_4$ decomposition:

$$\mathcal{G}_4 = \begin{array}{|c c|c c|} \hline 6 & 6 & 0 & 4 \\ \hline 4 & -4 & 2 & 2 \\ \hline \end{array} = \begin{array}{|c c|c c|} \hline 1 & 1 & -1 & -1 \\ \hline -1 & -3 & 1 & 3 \\ \hline \end{array} + \begin{array}{|c c|c c|} \hline 2 & 3 & -2 & 3 \\ \hline 2 & -3 & -2 & -3 \\ \hline \end{array} + \mathcal{G}_4^K(3, 2). \quad (2.12)$$

As true with $\mathcal{G}_4^{Averages}$ and $\mathcal{G}_4^K$, the structure of

$$\mathcal{G}_4^B = \begin{array}{|c c|c c|} \hline 2 & 3 & -2 & 3 \\ \hline 2 & -3 & -2 & -3 \\ \hline \end{array} \qquad (2.13)$$

ensures that it does not have any Nash information. *However,* certain $\mathcal{G}^B$ entries are appealing to the *coalition* of both players! The attractive TL entry of $\mathcal{G}_4^B$, where

---

[2]Similar to the previous footnote, this removal of redundancies is equivalent to a projection of $\mathcal{G}^{Averages}$, or $\mathcal{G}$, into an appropriate lower dimensional subspace. See Sect. 3.4 and Chap. 7.

both players enjoy a positive externality, cannot be attained with the players' uni-lateral actions. Instead, reaping the rewards of this TL cell of $\mathcal{G}_4^B$ requires coopera-tion/coordination between them; it captures what "we" can do rather than what "me" individual efforts can achieve.

To explain, by stressing competitive comparisons, $\mathcal{G}^N$ *ignores most information about the payoff magnitudes.* As the $\mathcal{G}^B$ entries retrieve this information (with payoff averages), the $\mathcal{G}^B$ cells identify where larger and smaller opportunities are located: This is precisely the kind of information that motivates coordination, cooperation, etc. Eq. 2.13, for instance, indicates that $\mathcal{G}_4^B$ offers advantages in the TL direction, but disadvantages in the BR positioning. These $\mathcal{G}_4^B$ entries, which distinguish between the rewards of the two Nash cells, indicate desired coordinating efforts.

2.3.3 More Examples

As this decomposition is central to what will be discussed, it is worth reviewing the construction by carrying out the decomposition for $\mathcal{G}_5$. In doing so, features shared by $\mathcal{G}_5$ and $\mathcal{G}_4$ (Eq. 2.3) are noted.

Start with Row's entries in $\mathcal{G}_5$'s L column. Remember, in computing $\mathcal{G}^N$, terms of importance are how the $\mathcal{G}$ entries differ from an average. Equivalently (to reduce the arithmetic), rather than computing the average, $\mathcal{G}^N$ values split the difference between the $\mathcal{G}$ entries. With both $\mathcal{G}_4$ and $\mathcal{G}_5$, the difference for Row between the T and B values is 2 in the L column and -2 in the R column. Thus, Row has identical Nash information for both games. Similarly, with both games, the difference for Column between Column's L and R entries in the T row is 2 and -6 in the B row. By splitting these differences, we have that

$$\mathcal{G}_4^N = \mathcal{G}_5^N = \begin{array}{|c c|c c|} \hline 1 & 1 & -1 & -1 \\ \hline -1 & -3 & 1 & 3 \\ \hline \end{array}. \tag{2.14}$$

Both games have identical Nash structures, so both share the same three Nash equi-libria.

Finding $\mathcal{G}_5^K$ is simple: The sum of Row's entries is 40, so $\kappa_1 = \frac{40}{4} = 10$. Similarly, the total of Column's entries is 28, so $\kappa_2 = \frac{28}{4} = 7$. Thus, $\mathcal{G}_5^K = \mathcal{G}_5^K(10, 7)$.

Armed with the κ_j values, $\mathcal{G}_5^B$ is $\mathcal{G}_5^B = \mathcal{G}_5 - [\mathcal{G}_5^N + \mathcal{G}_5^K(10, 7)]$. For readers who dislike matrix algebra, to compute $\mathcal{G}_5^B$, replace a player's $\mathcal{G}_5^{Averages} = \mathcal{G}_5 - \mathcal{G}_5^N$ entry with how it differs from the player's average payoff. As the average of Row's entries in the L column of $\mathbf{G}_5$ is -1, both of Row's entries in the L column of $\mathcal{G}_5^B$ must be $-1 - \kappa_1 = -1 - 10 = -11$. Similarly, Row's average in the R column is 21, so both of Row's entries in the R column of $\mathcal{G}_5^B$ are $21 - 10 = 11$. As for Column, the average of the entries in the T row is 17, so both of Column's T entries in $\mathcal{G}_5^B$ are $17 - \kappa_2 = 17 - 7 = 10$. For the B row, the two entries are $-3 - 7 = -10$. Thus, the decomposition of $\mathcal{G}_5$ is

$$\mathcal{G}_5 = \begin{array}{|cc|cc|} \hline 0 & 18 & 20 & 16 \\ -2 & -6 & 22 & 0 \\ \hline \end{array} = \begin{array}{|cc|cc|} \hline 1 & 1 & -1 & -1 \\ -1 & -3 & 1 & 3 \\ \hline \end{array} + \begin{array}{|cc|cc|} \hline -11 & 10 & 11 & 10 \\ -11 & -10 & 11 & -10 \\ \hline \end{array} + \mathcal{G}_4^K (10, 7).$$

(2.15)

Although $\mathcal{G}_4$ and $\mathcal{G}_5$ differ significantly, the decomposition proves that all differences reflect their clashing $\mathcal{G}^B$ components, which identify possible we-actions. According to Eq. 2.13, $\mathcal{G}_4$'s coordination effort is determined by the attractive $\mathcal{G}_4^B$ cell with positive entries; this directs the we-factor attention toward TL. In contract, $\mathcal{G}_5^B$'s potentially cooperative effort is pointed toward TR; it is the general direction of stronger $\mathcal{G}_5^B$ rewards. Thus, the complexity and difference between these games reflect their $\mathcal{G}^B$ terms.

Incidentally, there is an alternative computational approach that emphasizes averages: First, determine $\mathcal{G}^K$, then use averages of appropriate $[\mathcal{G} - \mathcal{G}^K]$ terms to define $\mathcal{G}^B$. Finally, compute $\mathcal{G}^N = [\mathcal{G} - \mathcal{G}^K] - \mathcal{G}^B$. To illustrate with $\mathcal{G}_1$ (Eq. 2.2), the sum of payoffs for each player is 12 with an average of 3, so $\mathcal{G}_1^K = \mathcal{G}_1^K (3, 3)$. Removing these averages from each $\mathcal{G}_1$ entry leads to

$$\mathcal{G}_1 - \mathcal{G}_1^K (3, 3) = \begin{array}{|cc|cc|} \hline -3 & -3 & 1 & -1 \\ -1 & 1 & 3 & 3 \\ \hline \end{array}$$

Row's $\mathcal{G}_1^B$ entries in the L and R columns are, respectively, the average of Row's entries in $\mathcal{G}_1 - \mathcal{G}_1^K (3, 3)$, or -2 and 2. Similarly, Column's entries in the T and D rows are, respectively, the average of Column's entries in $\mathcal{G}_1 - \mathcal{G}_1^K (3, 3)$, or -2 and 2. Therefore,

$$\mathcal{G}_1^B = \begin{array}{|cc|cc|} \hline -2 & -2 & 2 & -2 \\ -2 & 2 & 2 & 2 \\ \hline \end{array}.$$

The decomposition $\mathcal{G} = \mathcal{G}^N + \mathcal{G}^B + \mathcal{G}^K$ requires that $\mathcal{G}_1^N = \mathcal{G}_1 - [\mathcal{G}_1^B + \mathcal{G}_1^K]$ or

$$\mathcal{G}_1^N = \begin{array}{|cc|cc|} \hline -1 & -1 & -1 & 1 \\ 1 & -1 & 1 & 1 \\ \hline \end{array}$$

which leads to the decomposition

$$\mathcal{G}_1 = \begin{array}{|cc|cc|} \hline -1 & -1 & -1 & 1 \\ 1 & -1 & 1 & 1 \\ \hline \end{array} + \begin{array}{|cc|cc|} \hline -2 & -2 & 2 & -2 \\ -2 & 2 & 2 & 2 \\ \hline \end{array} + \mathcal{G}_1^K (3, 3).$$

(2.16)

Using a preferred approach, the decompositions of the two remaining games, $\mathcal{G}_2$ and $\mathcal{G}_3$, are given next in the $\mathcal{G} = \mathcal{G}^N + \mathcal{G}^B + \mathcal{G}^K$ order

$$\mathcal{G}_2 = \begin{array}{|cc|cc|} \hline -1 & -1 & -1 & 1 \\ 1 & -1 & 1 & 1 \\ \hline \end{array} + \begin{array}{|cc|cc|} \hline 3 & 3 & -3 & 3 \\ 3 & -3 & -3 & -3 \\ \hline \end{array} + \mathcal{G}_2^K (2, 2).$$

(2.17)

$$\mathcal{G}_3 = \begin{array}{|cc|cc|} \hline -1 & -1 & -1 & 1 \\ 1 & -1 & 1 & 1 \\ \hline \end{array} + \begin{array}{|cc|cc|} \hline 4 & -1 & -4 & -1 \\ 4 & 1 & -4 & 1 \\ \hline \end{array} + \mathcal{G}_3^K (5, 2).$$

(2.18)

The commonality where $\mathcal{G}_1$, $\mathcal{G}_2$, $\mathcal{G}_3$ have identical Nash, or "me" structures deserves recognition.

Definition 2.1 If two games $\mathcal{G}_i$ and $\mathcal{G}_j$ have the same Nash structure, $\mathcal{G}_i^N = \mathcal{G}_j^N$, they are said to be Nash equivalent. Denote this equivalence relationship by $\mathcal{G}_i \sim_{Nash} \mathcal{G}_j$.

Nash equivalent games have identical Nash equilibria; e.g., $\mathcal{G}_1 \sim_{Nash} \mathcal{G}_2 \sim_{Nash} \mathcal{G}_3$, so all three games have the same Nash dynamic structures and cells. Thus, *all* notable differences about these games (which are significant!) reflect their differing $\mathcal{G}^B$ components. Similarly, although $\mathcal{G}_4$ and $\mathcal{G}_5$ differ appreciably, they are Nash equivalent $\mathcal{G}_4 \sim_{Nash} \mathcal{G}_5$, so their remarkable differences, along with any contrasting assumptions about an opponent's tactics, totally reflect differences between the Behavioral $\mathcal{G}_4^B$ and $\mathcal{G}_5^B$.

More generally, and different from what normally is stated,

expect the complexity of games to reflect their Behavioral $\mathcal{G}^B$ components.

$$(2.19)$$

Intuition about the meaning of these central $\mathcal{G}^B$ terms starts next.

2.4   Gaining Intuition; Comparing Examples

Again, $\mathcal{G}^B$ components contain no Nash information because whatever column Column selects, all of Row's entries are the same; whatever row Row chooses, all of Column's entries remain the same. Consequently, $\mathcal{G}^B$ never offers advantage with *unilateral individual actions*. Instead, with $\mathcal{G}^B$, a player's action imposes either a positive or negative externality on the opponent. By emphasizing appropriate averages of a game's *actual rewards*, which are the differing externalities, $\mathcal{G}^B$ identifies *group opportunities and difficulties*.

In generic settings (the $\mathcal{G}^B$ entries are nonzero), $\mathcal{G}^B$ has a cell that *both players* find attractive. With $\mathcal{G}_2^B$, only TL offers both players a positive reward. But $\mathcal{G}_2^B$ is missing a Nash structure, so the players must coordinate to attain this $\mathcal{G}_2^B$ TL outcome. To do so, each player selects an action that creates a positive externality for the other. Similarly, the $\mathcal{G}_5^B$ TR cell is attractive to both players, but to enjoy this conclusion, they must cooperate. This feature, where the $\mathcal{G}^B$ behavioral component requires joint behavior to attain (or avoid) an outcome, identifies $\mathcal{G}^B$ as the game's we-component.

The $\mathcal{G}^N$ and $\mathcal{G}^B$ components incorporate significantly different information. Although $\mathcal{G}^N$ has all the information needed for Nash strategic action, it is becoming increasingly clear from observed behavior that Nash terms need not govern what players do, which is an issue in behavioral game theory. What does? As all non-Nash information resides in $\mathcal{G}^B$, it follows that this behavioral component is central for any such analysis. It should; larger $\mathcal{G}^B$ entries identify where stronger rewards are lurking. To reap these prizes, *joint action* is required. More specifically, comments about coordination, externalities, cooperation, collaboration, and so forth refer to aspects of $\mathcal{G}^B$.

Of interest is how the Nash and Behavioral components—the me and we features—interact. The $\mathcal{G}_1^N$ Nash structure of the simple $\mathcal{G}_1$ identifies BR as a Nash cell, which makes BR a unilateral objective for each individual. Similarly, the BR cell of $\mathcal{G}_1^B$ offers both players a positive reward, so BR also is a cooperative objective. The simplicity of $\mathcal{G}_1$ derives from the agreement between what the players want individually (from $\mathcal{G}_1^N$) and cooperatively (from $\mathcal{G}_1^B$). Without conflict, even myopic players can achieve success; the $\mathcal{G}_1^N$ and $\mathcal{G}_1^B$ agreement allows the players to ignore any cooperative needs.

Everything changes with $\mathcal{G}_2$ and $\mathcal{G}_3$; both share the $\mathcal{G}_1$ Nash structure (Eqs. 2.16, 2.17, 2.18), but their Behavioral components emphasize cells different from the BR Nash cell. This conflict between individual activity (from $\mathcal{G}^N$) and cooperative actions (from $\mathcal{G}^B$) generates the noted differences among these games. Indeed, many of the much studied concerns about games involve a conflict between the game's me and we features.

Even stronger, the decomposition demonstrates that most games experience some tension between individual and collective opportunities (the $\mathcal{G}^N$ and $\mathcal{G}^B$ components). To build upon this observation (which reflects Eq. 2.19 comment), a first step is to specify general $\mathcal{G}^N$ and $\mathcal{G}^B$ features as listed in Theorem 2.1.[3] This theorem defines a coordinate system for games.

Theorem 2.1 *In the following way, a 2×2 game $\mathcal{G}$ can be uniquely written as*

$$\mathcal{G} = \mathcal{G}^N + \mathcal{G}^B + \mathcal{G}^K. \tag{2.20}$$

All information about $\mathcal{G}$'s Nash structure is contained in the game's Nash $\mathcal{G}^N$ component; $\mathcal{G}^N$ is defined by two values for each player, $\eta_{i,j}$, where $j = 1, 2$ identifies the player and $i = 1, 2$ are the different strategies. In particular,

$$\mathcal{G}^N = \begin{array}{|c|c|} \hline \eta_{1,1} \quad \eta_{1,2} & \eta_{2,1} \quad -\eta_{1,2} \\ \hline -\eta_{1,1} \quad \eta_{2,2} & -\eta_{2,1} \quad -\eta_{2,2} \\ \hline \end{array}, \tag{2.21}$$

Conversely, any game with this Eq. 2.21 structure is the Nash component of some game.[4]

Each player has a unique β_j value for the game's behavior component $\mathcal{G}^B$. This component has the form

$$\mathcal{G}^B = \begin{array}{|c|c|} \hline \beta_1 \quad \beta_2 & -\beta_1 \quad \beta_2 \\ \hline \beta_1 \quad -\beta_2 & -\beta_1 \quad -\beta_2 \\ \hline \end{array}, \tag{2.22}$$

---

[3] The $\mathcal{G}^N$ and $\mathcal{G}^B$ components are well defined from the computational approach. For purposes of exposition, proofs and stronger mathematical descriptions are deferred to Sect. 3.4.2 and Chap. 7.
[4] For readers familiar with the term, notice how the set of $\mathcal{G}^N$ bimatrices forms an additive group where the identity is the weak Nash setting with zero terms. The same holds for the $\mathcal{G}^B$ and $\mathcal{G}^K$ components.

where, if the β_j values are nonzero, then one cell is Pareto superior (both entries are positive so it cannot be improved without hurting some player). Diametrically opposite is a Pareto inferior cell (both entries are negative). Component $\mathcal{G}^B$ contains no Nash information: for each column, Row's payoff is the same for T and B; for each row, Column's outcome is the same for L or R.

All of a player's entries in the game's kernel component, $\mathcal{G}^K$, equal the player's average payoff in $\mathcal{G}$. This component has the form

$$\mathcal{G}^K = \begin{array}{|cc|cc|} \hline \kappa_1 & \kappa_2 & \kappa_1 & \kappa_2 \\ \hline \kappa_1 & \kappa_2 & \kappa_1 & \kappa_2 \\ \hline \end{array}. \tag{2.23}$$

Not only does $\mathcal{G}^N$ identify all Nash cells (the ones with only positive entries), but it assigns values, different than payoffs, to these cells. It will become clear what added information comes from these $\eta_{i,j}$ values.

2.4.1 The Nash Component and Game Dynamic

The above decompositions have a peculiarity where Row's entries in any $\mathcal{G}^N$ column differed only by sign; the same was true for Column's entries in any $\mathcal{G}^N$ row. According to Eq. 2.21, this property holds for all $\mathcal{G}^N$. Consequently, $\mathcal{G}^N$ cells with positive entries (Nash cells) cannot be adjacent to each other; should there be two pure Nash equilibria, they are in diagonal cells. According to Eq. 2.21, these are general generic properties.

Even more, the $\mathcal{G}^N$ component captures all of $\mathcal{G}$'s Nash game dynamics including the Theorem 1.1 assertions. To make the dynamics explicit, let $\mathcal{G}^{N,signs}$ be where each $\eta_{i,j}$ is replaced by its sign. As an example,

$$\text{if } \mathcal{G}^N = \begin{array}{|cc|cc|} \hline 0.6 & 2 & -3 & -2 \\ \hline -0.6 & -5 & 3 & 5 \\ \hline \end{array}, \text{ then } \mathcal{G}^{N,signs} = \begin{array}{|cc|cc|} \hline 1 & 1 & -1 & -1 \\ \hline -1 & -1 & 1 & 1 \\ \hline \end{array}. \tag{2.24}$$

By replacing each $\eta_{i,j}$ with its sign, $\mathcal{G}^{N,signs}$ becomes a special case of Eq. 2.21. Figure 2.1 illustrates how to identify $\mathcal{G}^{N,signs}$ with Fig. 1.2 arrows.

Fig. 2.1 Equivalence of $\mathcal{G}^{N,signs}$ and Nash dynamics

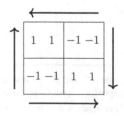

Arrows along the vertical edges represent Row's preferred choices. In the L column, Row's entries show a preference for T over B, which corresponds to the upward pointing arrow on the left. Similarly, the downward arrow on the right identifies Row's preferred action in the R column. The horizontal arrows capture Column's choices depending on the selected row. In this way, the $\mathcal{G}^{N,signs}$ structure identifies the Nash dynamics.

To further illustrate, a game with only one Nash cell at TL has the form

$$\mathcal{G}^{N,signs} = \begin{array}{|cc|cc|} \hline 1 & 1 & x & -1 \\ \hline -1 & y & -x & -y \\ \hline \end{array}$$

where $|x| = |y| = 1$ and $-x$ and/or $-y$ is negative. If $x = y = 1$, then BR is repelling and the other two cells are hyperbolic. If one variable is positive and the other negative, then either BL (if $y = -1$) or TR (if $x = -1$) is repelling and the other two cells are hyperbolic.

While the following definition and theorem are for 2×2 games, both extend to all games with any finite number of players and strategies.

Definition 2.2 The value of a cell in $\mathcal{G}^{N,signs}$ is the sum of its entries.

Both $\mathcal{G}^N$ entries in a Nash cell are positive, so a Nash cell's value is 2. Similarly, a repelling cell has all arrows moving away, so both $\mathcal{G}^{N,signs}$ entries are -1 giving the cell value of -2. Finally, a hyperbolic cell has value $1 + (-1) = 0$.

Theorem 2.2 In a 2×2 game, the sum of cell values in $\mathcal{G}^{N,signs}$ is zero.

Proof According to Eq. 2.21, the sum of $\mathcal{G}^N$ entries is zero, so the sum of $\mathcal{G}^{N,signs}$ entries also is zero. The sum of cell values is a reordering of the summation. □

This theorem permits a quick proof of Theorem 1.1. Of interest will be how this approach helps to analyze more complicated games with any number of players and strategies.

A proof of Theorem 1.1. Only a Nash cell has a positive value, so without any Nash cells, the sum of cells cannot be positive. According to Theorem 2.2, the sum value is zero, which means there can be no repelling cells; otherwise, the sum would be negative. Thus, all four cells must be hyperbolic.

One Nash cell contributes 2 to the cell sum, which requires a repelling cell with value -2. The other two are hyperbolic.

Two Nash cells contribute 4 to the cell sum. According to Theorem 2.2, there must be two repelling cells. There cannot be more than two Nash cells. □

The degenerate case with weak Nash cells has zeros in $\mathcal{G}^{N,signs}$. An example is $\mathcal{G}_{sofa}$ (Eq. 2.25) where two roommates must lift a sofa into a room to watch TV. Once done (TL), both are rewarded. But if even one person fails to help, neither gets anything.

$$\mathcal{G}_{sofa} = \begin{array}{|cc|cc|} \hline 1 & 1 & 0 & 0 \\ \hline 0 & 0 & 0 & 0 \\ \hline \end{array} \qquad (2.25)$$

Here,

$$\mathcal{G}^{N,signs}_{sofa} = \begin{array}{|cc|cc|} \hline 1 & 1 & 0 & -1 \\ -1 & 0 & 0 & 0 \\ \hline \end{array}$$

identifies TL and BR as Nash cells, but missing two of the game's dynamic arrows. Although the BR Nash cell has value zero rather than two, Theorem 2.2 still holds after slight, natural adjustments. Notice how the behavioral term

$$\mathcal{G}^{B}_{sofa} = \begin{array}{|cc|cc|} \hline 0.25 & 0.25 & -0.25 & 0.25 \\ 0.25 & -0.25 & -0.25 & -0.25 \\ \hline \end{array}$$

stresses the importance of cooperation.

Earlier (Sect. 1.5), weak Nash equilibria were described as transitions among generic settings. To demonstrate with $\mathcal{G}_{sofa}$, a slightly perturbed game replaces the BR entries with ϵ_1, ϵ_2 reflecting attitudes of avoiding the backbreaking lifting of that sofa. Those happy about not having to lift have a positive entry, otherwise, it is negative. The three possibilities are where both ϵ_1, ϵ_2 are positive, both ϵ_1, ϵ_2 are negative, or they differ in sign where, say, $\epsilon_1 > 0$. This leads to three different $\mathcal{G}^{N,signs}_{sofa}$ given, respectively, as

$$\begin{array}{|cc|cc|} \hline 1 & 1 & -1 & -1 \\ -1 & -1 & 1 & 1 \\ \hline \end{array} \quad \begin{array}{|cc|cc|} \hline 1 & 1 & 1 & -1 \\ -1 & 1 & -1 & -1 \\ \hline \end{array} \quad \begin{array}{|cc|cc|} \hline 1 & 1 & -1 & -1 \\ -1 & 1 & 1 & -1 \\ \hline \end{array}.$$

The first has Nash cells at TL and BR, while the other two have one Nash cell at TL. Again, the weak Nash properties of $\mathcal{G}^{N,signs}_{sofa}$ reflect the transition among these choices.

2.4.2 Behavior Component and Cooperation

Row's entry in a column for each of the $\mathcal{G}^{B}_j$, $j = 1, \ldots, 5$, games differs only by sign from Row's entry in the other column, with a similar assertion for Column's entries in $\mathcal{G}^B$ rows. Rather than a coincidence, Eq. 2.22 ensures that this is a standard property.

Of particular interest is that $\mathcal{G}^B$ *always* has a Pareto superior cell; it offers the best $\mathcal{G}^B$ outcome (both payoffs are positive) for both players. This Pareto superior cell identifies cooperative opportunities and behavior for the two players. According to Eq. 2.22, $\mathcal{G}^B$'s Pareto inferior cell has negative entries, so it carries negative consequences. As it will become clear in Sect. 2.5.3, this "reward versus punishment" feature (of Pareto superior and inferior cells) plays a surprising role in encouraging cooperation in repeated games.

The $\mathcal{G}^B$ Pareto superior cell's location depends on the β values. With $\beta_1 > 0$, $\beta_2 > 0$, TL is the Pareto superior cell and BR is Pareto inferior; with $\beta_1 > 0$, $\beta_2 < 0$, the

Fig. 2.2 Different
contributions of $\mathcal{G}^N$ and $\mathcal{G}^B$

(a) $\mathcal{G}^N$ and $\mathcal{G}^B$ (b)Prisoner's Dilemma

Pareto superior cell drops to BL (with the inferior cell at TR), with $\beta_1 < 0$, $\beta_2 > 0$, the Pareto superior cell is TR, and with $\beta_1 < 0$, $\beta_2 < 0$, the Pareto superior cell is at BR.

Differences in how the $\mathcal{G}^N$ and $\mathcal{G}^B$ components affect the game are reflected by Fig. 2.2. The "me" Nash action from a cell allows Row to explore only options in the column defined by the cell (the solid vertical arrows in Fig. 2.2a), while Column examines options only in the defined row (Fig. 2.2a horizontal arrow). In contrast, the collective "we" factor of $\mathcal{G}^B$ introduces added opportunities because its Pareto superior point can be in *any* cell. This added flexibility is captured by the Fig. 2.2a dashed arrow that points to a Pareto superior cell, where coordinated action is required to attain any $\mathcal{G}^B$ benefit.

Figure 2.2b captures these differences with the $\mathcal{G}_2$ Prisoner's Dilemma game. The solid arrows indicate the Nash dynamic, which is restricted to a specific row and column that identifies BR as a Nash cell. The dashed arrow is the $\mathcal{G}_2^B$ direction from its Pareto inferior cell (BR) to its Pareto superior choice (TL). The $\mathcal{G}^B$ entries, then, can enhance payoffs to create interest in cells that are hyperbolic or even repelling! Of importance, $\mathcal{G}^B$ components have *no effect* on the Nash dynamic.

2.4.3 What Does All of This Mean?

This is an appropriate place to take a breather to assimilate what has been done. To do so, review how you analyzed the game $\mathcal{G}_{Challenge}$ (Eq. 2.1). With a little exploration, it becomes obvious that each payoff entry, by itself, is not meaningful, so the goal becomes to understand how the parts, the various payoffs, interact to constitute a whole, the game's structure. What whole? What should be the objective?

Themes identified in Chap. 1 suggest finding whether Nash equilibria or cooperative opportunities exist. This requires parsing each payoff in terms of how it interacts with other payoffs. Row's -7 in the TL cell of $\mathcal{G}_{Challenge}$, for instance, must be dissected into its separate portions where each contributes to one of these structures. How? Relying on a usual ad hoc search to discover answers explains why game theory can be so difficult.

While the ad hoc and coordinates approaches have related objectives, coordinates resolve the issue by explicitly partitioning each payoff into portions that contribute to a game's different defining features. With $\mathcal{G}_{Challenge}$, we have in the $\mathcal{G}^N + \mathcal{G}^B + \mathcal{G}^K$ order

$$\mathcal{G}_{Challenge} = \begin{array}{|cc|cc|} \hline -7=-2-4-1 & 11=2+5+4 & 6=3+4-1 & 7=-2+5+4 \\ \hline -3=2-4-1 & 0=1-5+4 & 0=-3+4-1 & -2=-1-5+4 \\ \hline \end{array} \tag{2.26}$$

So the -7 payoff in TL is divided into a -2 that contributes to the Nash structure, a -4 that is a negative externality imposed by Column, and a -1 deflation (kernel) value. Coordinates replace ad hoc approaches by offering a systematic parsing process.

Coordinates even assist in modeling where variables capture different attributes. As later described when supporting Eq. 3.38, the $\eta_{i,j}$ terms can be subdivided to reflect a driver's preference of which route to drive to work, perhaps based on expected travel time, coupled with a driver's attitude about sharing the road with another driver. These traits govern individual decisions, so they appropriately combine to create $\eta_{i,j}$ values. The β_j values could be externalities generated by what the *other* driver brings to the driving experience; a $\beta_1 > 0$ might mean that Column is driving one of those fancy sports cars that Row would enjoy watching on the free-flowing first route, but Row would find to be a danger on the curvy second route. And so, the six non-kernel variables can model six different, independent attributes (three for each driver).

2.4.4   Creating Exotic Examples

Examples in this book not from the literature were created by using Theorem 2.1 (and natural extensions for games with more strategies and players). Of importance, the three bimatrix components are independent of each other, *so any combination of choices is admissible.*

To illustrate, return to the Sect. 1.3.1 concerned about creating games representing a free-riding coauthor, say Column. What we want is a game with a hyperbolic-column cell that is diametrically opposite to a Nash cell, so the Nash form is

$$\begin{array}{|cc|cc|} \hline 1 & -1 & -1 & 1 \\ \hline -1 & -1 & 1 & 1 \\ \hline \end{array} \quad \text{or, more generally} \quad \begin{array}{|cc|cc|} \hline \eta_{1,1} & \eta_{1,2} & \eta_{2,1} & -\eta_{1,2} \\ \hline -\eta_{1,1} & \eta_{2,2} & -\eta_{2,1} & -\eta_{2,2} \\ \hline \end{array}$$

(from Eq. 2.21). Here, $\eta_{2,1}$ and $\eta_{2,2}$ are negative (to make BR a Nash cell), and $\eta_{1,1} > 0$ and $\eta_{1,2} < 0$ (so that TL is a hyperbolic-column cell). These choices determine all possible ways to construct a $\mathcal{G}^N$ with the desired Nash properties.

The $\mathcal{G}^B$ component (Eq. 2.22) is what can make the hyperbolic cell TL attractive. For this to happen, each agent must find the TL cell of $\mathcal{G}^N + \mathcal{G}^B$ to be more appealing than the Nash BR cell. This requires $\beta_j + \eta_{1,j} > -\beta_j - \eta_{2,j}$, or $2\beta_j > -[\eta_{1,j} + \eta_{2,j}]$, $j = 1, 2$. Here, $\beta_1 = \beta_2 = 2$ suffices leading to

$$\begin{array}{|cc|cc|} \hline 1 & -1 & -1 & 1 \\ \hline -1 & -1 & 1 & 1 \\ \hline \end{array} + \begin{array}{|cc|cc|} \hline 2 & 2 & -2 & 2 \\ \hline 2 & -2 & -2 & -2 \\ \hline \end{array} = \begin{array}{|cc|cc|} \hline 3 & 1 & -3 & 3 \\ \hline 1 & -3 & -1 & -1 \\ \hline \end{array}$$

or, more generally with TL being a hyperbolic-column cell,

$$
\begin{array}{|cc|cc|}
\hline
|\eta_{1,1}| + |\beta_1| & -|\eta_{1,2}| + |\beta_2| & -|\eta_{2,1}| - |\beta_1| & |\eta_{1,2}| + |\beta_2| \\
-|\eta_{1,1}| + |\beta_1| & -|\eta_{2,2}| - |\beta_2| & |\eta_{2,1}| - |\beta_1| & |\eta_{2,2}| - |\beta_2| \\
\hline
\end{array}
\tag{2.27}
$$

where $2|\beta_1| > |\eta_{2,1}| - |\eta_{1,1}|$ and $2|\beta_2| > |\eta_{2,2}| + |\eta_{1,2}|$.

For a different example, suppose we want a game with Nash equilibria at BL and TR where the BL cell dominates. This condition requires the BL and TR cells of $\mathcal{G}^N$ to have positive values where BL's entries are larger. From Eq. 2.21, one such example is

$$
\mathcal{G}^N = \begin{array}{|cc|cc|}
\hline
-10 & -1 & 1 & 1 \\
10 & 11 & -1 & -11 \\
\hline
\end{array}
$$

Next, adjust this game so that the TL cell can attract *collective attention* away from the Nash equilibria. Here, $\mathcal{G}^B$ must have large entries in a TL Pareto superior cell; e.g.,

$$
\mathcal{G}^B = \begin{array}{|cc|cc|}
\hline
20 & 30 & -20 & 30 \\
20 & -30 & -20 & -30 \\
\hline
\end{array}
$$

If the negative terms of $\mathcal{G}^N + \mathcal{G}^B$, such as the BR entry of $(-21, -41)$, are distracting, then to reduce the number of cells with this property, add $\mathcal{G}^K (20, 40)$ to create

$$
\mathcal{G} = \mathcal{G}^N + \mathcal{G}^B + \mathcal{G}^K (20, 40) = \begin{array}{|cc|cc|}
\hline
30 & 69 & 1 & 71 \\
50 & 11 & -1 & -1 \\
\hline
\end{array}
$$

While the above $\mathcal{G}^N$ and $\mathcal{G}$ differ radically in appearance (and possible reactions by players and the assumptions players have about each other), they are Nash equivalent!

Let's expand on this. Suppose games are to be designed for the classroom or lab experiments to capture whether a particular theory accurately explains why players do, or do not, select certain Nash points. Doing so, requires designing a large number of games with *an identical Nash structure* but accompanied by other features that may distract the players. Normally, finding such examples can be a difficult challenge; with the decomposition, it is almost trivial!

All that is needed is to select a common desired $\mathcal{G}^N$, maybe

$$
\mathcal{G}^N_{experiment} = \begin{array}{|cc|cc|}
\hline
-1 & -3 & 3 & 3 \\
1 & 1 & -3 & -1 \\
\hline
\end{array}
$$

with Nash cells at TR and BL, where TR is more attractive. Next, select an accompanying $\mathcal{G}^B$ to focus attention on whichever cell you wish. That is, choose β_1, β_2 values to ensure that an appropriate $\mathcal{G}^B$ cell is Pareto Superior. To focus on BL, select $\beta_1 > 0$ and $\beta_2 < 0$; perhaps $\beta_1 = 3$, $\beta_2 = -2$. To add appeal to TL, select $\beta_1, \beta_2 > 0$, such as $\beta_1 = \beta_2 = 4$. How about BR? Here, both β values should be negative, such as $\beta_1 = \beta_2 = -6$.

To illustrate these selected choices, the three resulting games are, respectively,

2	−5	0	1		3	1	−1	7		−7	−9	9	−3
4	3	−6	1		5	−3	−7	−5		−5	7	3	5

which, even with identical Nash structures, most likely would elicit different responses from players. Jessie and Kendall (e.g., [11]) used this approach to generate a rich trove of Nash equivalent games for which players, in experimental economics labs, display radically different reactions. The observed behavior, where students won or lost actual money, supports Eq. 2.19 comment that the $\mathcal{G}^B$ component can cause complexity in a game.

2.4.5 Does Size Matter?

A gift of the coordinates is how they partition a game's payoffs into components with different roles of individual opportunities (Nash), group possibilities (Behavioral), and inflation (kernel). Do the magnitudes of these terms matter? Of course! Size matters!

This answer clearly applies to the β_j terms. Depending on how they are interpreted (e.g., average payoff, a positive or negative externality imposed by the other player, and providing coordination or anti-coordination pressures), the sign and size of each β_j can transform the game's meaning and analysis. In the previous section, for instance, different signs and sizes of β_j values changed the focus about which cell of a game is the most, or least attractive. Starting in Sect. 2.5, it will become apparent how these behavioral features play a significant role in defining the characteristics of widely used games.

A similar comment applies to the $\eta_{i,j}$ values. Their signs already determine a game's Nash dynamic as manifested by the $\mathcal{G}^{N,signs}$ properties. Also, a cell of 2×2 game $\mathcal{G}$ is a pure Nash cell if and only if both of its $\mathcal{G}^N$ entries are positive. Even more, a weak Nash equilibrium is where a $\eta_{i,j}$ entry equals zero so it is a transition between a cell being Nash or not. Think about this, if a zero η value of a Nash equilibrium represents a weak Nash, then larger η values constitute "stronger" Nash cells. Stated differently, much of the literature identifies Nash cells by comparing payoffs; the decomposition appends the analysis by assigning values to the $\mathcal{G}^N$ Nash cell's entries. This suggests that the η sizes offer a new tool to develop various refinements of Nash equilibria.

Indeed, the size of $\eta_{i,j}$ entries will be important in comparing Nash cells (e.g., Sect. 2.6) and when determining which Nash cell is more apparent to players (e.g., Sect. 3.5). Although not developed here (the theme differs from that of this book), the size of $\eta_{i,j}$ values are critical when motion and issues of "speed" are introduced, such as the rapidity of approach to an equilibrium when a game $\mathcal{G}$ is converted into a replicator dynamic setting, or when using a learning algorithm to find Nash information.

Consequences of the size and signs of the κ_j terms are obvious; when creating games for experiments, they can ensure that a game avoids negative payoffs. In games with transferrable assets, the κ_j values are, literally, valuable.

Next, standard games are reexamined with the decomposition. In doing so, the signs and sizes of these variables in creating each game's flavor become apparent.

2.5 Structure of Standard Games

A way to demonstrate the value added by the coordinates is to apply them to standard games. One contribution is that rather than requiring a typical ad hoc analysis, *all* of these games can be analyzed in an identical manner: Compute the $\mathcal{G}^N$ and $\mathcal{G}^B$ components and then determine how they interact. That is, check whether a game has me-we tensions triggered by preferred but conflicting $\mathcal{G}^N$ and $\mathcal{G}^B$ cells.

2.5.1 Prisoner's Dilemma

The Eq. 1.4 form of the Prisoner's Dilemma does not immediately capture the discord between individual and group interests: It takes analysis and expertise to tease out these features. But the decomposition immediately separates the factors as

$$
\mathcal{G}_{PD} = \begin{array}{|cc|cc|} \hline 4 & 4 & -4 & 6 \\ \hline 6 & -4 & -2 & -2 \\ \hline \end{array} = \begin{array}{|cc|cc|} \hline -1 & -1 & -1 & 1 \\ \hline 1 & -1 & 1 & 1 \\ \hline \end{array} + \begin{array}{|cc|cc|} \hline 4 & 4 & -4 & 4 \\ \hline 4 & -4 & -4 & -4 \\ \hline \end{array} + \mathcal{G}^K(1,1),
$$
(2.28)

which instantly demonstrates a distinct conflict between the Nash and Behavioral terms. After all, the most attractive cell for one component (i.e., the BR Nash cell for $\mathcal{G}^N_{PD}$ and the TL Pareto superior cell for $\mathcal{G}^B_{PD}$) is the least attractive cell for the other component. Thus, the "me" perspective governed by the BR Nash cell of $\mathcal{G}^N_{PD}$ creates a tension with the "we" attitude determined by the diametrically opposite TL cell of $\mathcal{G}^B_{PD}$.

This conflict between individual opportunities and mutual benefits generates the game's complexity. Large enough β_1, β_2 values (e.g., "size matters," Sect. 2.4.5) for $\mathcal{G}^B$

$$
\begin{array}{|cc|cc|} \hline -1+\beta_1 & -1+\beta_2 & -1-\beta_1 & 1+\beta_2 \\ \hline 1+\beta_1 & -1-\beta_2 & 1-\beta_1 & 1-\beta_2 \\ \hline \end{array} = \begin{array}{|cc|cc|} \hline -1 & -1 & -1 & 1 \\ \hline 1 & -1 & 1 & 1 \\ \hline \end{array} + \begin{array}{|cc|cc|} \hline \beta_1 & \beta_2 & -\beta_1 & \beta_2 \\ \hline \beta_1 & -\beta_2 & -\beta_1 & -\beta_2 \\ \hline \end{array}
$$
(2.29)

converts $\mathcal{G}^N$'s lousy TL cell, a repeller, into an attractive opportunity. The resulting appeal of this TL term is a $\mathcal{G}^B$ feature—not a $\mathcal{G}^N$ property. Indeed, TL is a $\mathcal{G}$ repelling cell whereby either agent can renege from cooperative efforts to be *rewarded* with a personally better conclusion—at the expense of the former partner. Assumptions that either player may have about the opponent are influenced by the $\mathcal{G}^B$ component.

2.5.2 Collective Action

Closely related to the Prisoner's Dilemma are collective action games where some players, but not necessarily all (so, in two-person games, one person), can defect. A standard choice is a public good situation where players can, but need not, contribute to a joint project such as a community park. These settings could be modeled where some players cooperate in a game that has a Prisoner's Dilemma flavor, while another possibility is to position a hyperbolic cell diametrically opposite the Nash cell.

This distinction makes it easy to create any number of examples by following the lead of Eq. 2.27—just select the Nash component to have a desired structure. As examples, let the two Nash structures be

$$\mathcal{G}_6^N = \begin{array}{|cc|cc|} \hline -1 & -1 & -1 & 1 \\ \hline 1 & -1 & 1 & 1 \\ \hline \end{array} \quad \text{and} \quad \mathcal{G}_7^N = \begin{array}{|cc|cc|} \hline 1 & -1 & -1 & 1 \\ \hline -1 & -1 & 1 & 1 \\ \hline \end{array} \tag{2.30}$$

where $\mathcal{G}_6^N$ has a TL repelling cell opposite the BR Nash cell, while $\mathcal{G}_7^N$ has a TL hyperbolic-column cell. With this difference, if $\mathcal{G}^B$ makes TL attractive, both players have incentives to defect in $\mathcal{G}_6$, but only Column is so motivated with $\mathcal{G}_7$.

To encourage comparisons, the same $\mathcal{G}^B$ is used with both Nash components. The choice of $\beta_1 = \beta_2 = 5$, for instance, leads to the two games

$$\mathcal{G}_6 = \begin{array}{|cc|cc|} \hline 4 & 4 & -6 & 6 \\ \hline 6 & -6 & -4 & -4 \\ \hline \end{array} \quad \text{and} \quad \mathcal{G}_7 = \begin{array}{|cc|cc|} \hline 6 & 4 & -6 & 6 \\ \hline 4 & -6 & -4 & -4 \\ \hline \end{array} \tag{2.31}$$

where $\mathcal{G}_6$ is a Prisoner's Dilemma game supplying an incentive for each player to defect. With $\mathcal{G}_7$, only Column experiences such a stimulant.

This structure suggests exploring other possibilities. After all, the $\mathcal{G}^B$ component is what can make non-Nash cells interesting, so consider $\mathcal{G}^B$ where $\beta_1 = 5, \beta_2 = -5$ to focus attention on cell BL. Adding this $\mathcal{G}^B$ to the two Eq. 2.30 $\mathcal{G}^N$ components creates

$$\mathcal{G}_6^* = \begin{array}{|cc|cc|} \hline 4 & -6 & -6 & 4 \\ \hline 6 & 4 & -4 & 6 \\ \hline \end{array} \quad \text{and} \quad \mathcal{G}_7^* = \begin{array}{|cc|cc|} \hline 6 & -6 & -6 & -4 \\ \hline 4 & 4 & -4 & 6 \\ \hline \end{array} \tag{2.32}$$

By design, in both games both players find BL to be attractive. But BL is hyperbolic for $\mathcal{G}_6^N$, so only Column has an incentive to defect in $\mathcal{G}_6^*$. On the other hand, BL is the $\mathcal{G}_7^N$ repelling cell, so both players have a reason to defect in $\mathcal{G}_7^*$.

Only Row would find the BR Nash cell in the Eq. 2.32 games to be repulsive; both players could find BR to be disgusting in Eq. 2.31 games. An interesting new class of games would result by making the Nash BR loathsome and TR attractive, which suggests experimenting with $\mathcal{G}^B$ choices with a Pareto superior cell at TR. But the second paragraph of Theorem 2.3, which identifies restrictions that are imposed by $\mathcal{G}^B$, outlaws this possibility.

2.5.2.1 Prisoner's Dilemma Flavor

This discussion underscores a standard concern where a game's attractive group outcome experiences a perilous status by providing an incentive for some player not to cooperate. This feature is captured by the following definition.

Definition 2.3 Game $\mathcal{G}$ has a "Prisoner's Dilemma flavor" if $\mathcal{G}$ has a group preferred cell where each player has a stronger payoff than in any of $\mathcal{G}$'s Nash cells.

A group's preferred cell is not Nash, so cooperation is required to attain these benefits. A troubling consequence of failing to be Nash is the reality that some player has an incentive to not cooperate. The roles of $\mathcal{G}^N$ and $\mathcal{G}^B$ in such settings, which are described next, are illustrated with Eqs. 2.31, 2.32.

Theorem 2.3 *If a 2×2 game $\mathcal{G}$ has a Prisoner's Dilemma flavor, then $\mathcal{G}$ has a single Nash cell and the $\mathcal{G}^B$ Pareto inferior cell has the same location as the $\mathcal{G}^N$ Nash cell.*[5]

In $\mathcal{G}$, the group preferred cell must be diametrically opposite the Nash cell. This cell can be a repeller, where each player has an incentive to defect, or hyperbolic, where only one specific player has the incentive to defect.

If the group preferred cell is a repeller, the two β_j parameters defining $\mathcal{G}^B$ satisfy

$$\frac{1}{2}\{|\eta_{1,j}| + |\eta_{2,j}|\} < |\beta_j|, \ j = 1, 2. \tag{2.33}$$

If the group preferred cell is hyperbolic, then Eq. 2.33, in terms of a player's average $|\eta_{i,j}|$ entries for $\mathcal{G}^N$, holds for the player with an incentive to defect. The β_k value for player, k, who has no incentive to defect, satisfies

$$\frac{1}{2}\{|\eta_{i,k}| - |\eta_{j,k}|\} < \beta_k \tag{2.34}$$

where $|\eta_{i,k}|$ and $|\eta_{j,k}|$ come, respectively, from $\mathcal{G}^N$'s Nash and group preferred cells.

Thus, to be a group preferred cell, what is offered to each player through cooperation (the size of the β value) must exceed the average of the strategic terms (the η values). A surprise is how this setting limits attention to games with a single Nash cell.

---

[5]This statement supports footnote #3 in Chap. 1.

Proof The Nash dynamic for $\mathcal{G} = \mathcal{G}^N + \mathcal{G}^B + \mathcal{G}^K$ is strictly determined by the $\mathcal{G}^N$ component. The group preferred cell cannot be adjacent to a Nash cell because, by definition, both players would prefer it, which would negate the Nash cell's Nash structure. (Suppose, for instance, that TL is a Nash cell. If the group preferred cell is at TR, then Column would prefer TR to TL, which contradicts TL's assumed Nash cell status.) Thus, the group preferred cell is diametrically opposite the Nash cell.

This structure is impossible to attain with two Nash cells because it would require the group preferred cell to be adjacent to both Nash cells. Thus, there can be only one Nash cell, where the group preferred cell is diametrically opposite.

The $\mathcal{G}^N + \mathcal{G}^B$ structure is given by

$$
\mathcal{G}^N + \mathcal{G}^B =
\begin{array}{|cc|cc|}
\hline
\eta_{1,1} & \eta_{1,2} & \eta_{2,1} & -\eta_{1,2} \\
\hline
-\eta_{1,1} & \eta_{2,2} & -\eta_{2,1} & -\eta_{2,2} \\
\hline
\end{array}
+
\begin{array}{|cc|cc|}
\hline
\beta_1 & \beta_2 & -\beta_1 & \beta_2 \\
\hline
\beta_1 & -\beta_2 & -\beta_1 & -\beta_2 \\
\hline
\end{array}
$$

Select any $\mathcal{G}^N$ cell to be its sole Nash point. By symmetry of the $\mathcal{G}^N$ and $\mathcal{G}^B$ structures, the choice does not matter, so let it be TL, which means that $\eta_{1,1}$ and $\eta_{1,2}$ are positive. If BR (the diametrically opposite cell) is repelling, then $\eta_{2,1}$ and $\eta_{2,2}$ also are positive.

For BR to be a group preferred cell, each player's BR payoffs must be better than what the player gets from TL. That is

$$
\eta_{1,1} + \beta_1 < -\eta_{2,1} - \beta_1, \quad \eta_{1,2} + \beta_2 < -\eta_{2,2} - \beta_2. \tag{2.35}
$$

So, β_1, β_2 are negative, and Eqs. 2.35 lead to Eq. 2.33.

What remains is if BR is a hyperbolic cell; assume it is attracting for Row (so $-\eta_{2,1} > 0$, or $\eta_{2,1} < 0$) and repelling for Column (so, $-\eta_{2,2} < 0$, or $\eta_{2,2} > 0$). The condition for Column to have a better BR payoff than in TL is as in Eq. 2.35. For Row, the equations are

$$
\eta_{1,1} + \beta_1 < |\eta_{2,1}| - \beta_1,
$$

which leads to Eq. 2.34.   □

2.5.2.2   Examples

To have a group preferred cell, a player's β value from $\mathcal{G}^B$ must be larger than the average magnitude of the player's two η terms from $\mathcal{G}^N$ (Eq. 2.33). A first example has the Nash cell at BL and the repelling group preferred cell at TR. These conditions dictate the signs (but not the magnitudes) of the $\mathcal{G}^N$ entries; an example is

$$
\mathcal{G}^N =
\begin{array}{|cc|cc|}
\hline
-1 & 2 & -1 & -2 \\
\hline
1 & 4 & 1 & -4 \\
\hline
\end{array}.
$$

For TR to be a group preferred cell, it must be (Eqs. 2.33) that $|\beta_1| > \frac{1+1}{2} = 1$ and $|\beta_2| > \frac{2+4}{2} = 3$, so let $|\beta_1| = 2$, $|\beta_2| = 4$. The signs for these terms are selected to position the $\mathcal{G}^B$ Pareto Superior cell at TR, which leads to the game

$$\mathcal{G} = \begin{array}{|cc|cc|} \hline -3 & 6 & 1 & 2 \\ \hline -1 & 0 & 3 & -8 \\ \hline \end{array} = \begin{array}{|cc|cc|} \hline -1 & 2 & -1 & -2 \\ \hline 1 & 4 & 1 & -4 \\ \hline \end{array} + \begin{array}{|cc|cc|} \hline -2 & 4 & 2 & 4 \\ \hline -2 & -4 & 2 & -4 \\ \hline \end{array}. \qquad (2.36)$$

If negative entries are bothersome, drop them by adding $\mathcal{G}^K(3, 8)$.

To alter the example so that TR is hyperbolic, where only one agent can defect, change the signs for Column's top row $\mathcal{G}^N$ entries. The restrictions on the β_j values are the same, leading to

$$\mathcal{G} = \begin{array}{|cc|cc|} \hline -3 & 2 & 1 & 6 \\ \hline -1 & 0 & 3 & -8 \\ \hline \end{array} = \begin{array}{|cc|cc|} \hline -1 & -2 & -1 & 2 \\ \hline 1 & 4 & 1 & -4 \\ \hline \end{array} + \begin{array}{|cc|cc|} \hline -2 & 4 & 2 & 4 \\ \hline -2 & -4 & 2 & -4 \\ \hline \end{array}.$$

2.5.2.3   Without a Nash Cell

Definition 2.3 compares each player's outcome in a group preferred cell to that in a Nash cell. The comparison, however, could be with expected outcomes from a Nash equilibrium, which extends this Prisoner's Dilemma flavor to games with hyperbolic cells and a mixed strategy equilibrium. The analysis is the same as shown by the example

$$\mathcal{G}_{Mixed} = \begin{array}{|cc|cc|} \hline 2 & 4 & -2 & 2 \\ \hline 4 & -4 & 4 & -2 \\ \hline \end{array} = \begin{array}{|cc|cc|} \hline -1 & 1 & 1 & -1 \\ \hline 1 & -1 & -1 & 1 \\ \hline \end{array} + \begin{array}{|cc|cc|} \hline 3 & 3 & -3 & 3 \\ \hline 3 & -3 & -3 & -3 \\ \hline \end{array}. \qquad (2.37)$$

where TL is the group preferred cell. This is because the mixed strategy (from $\mathcal{G}_{Mixed}^N$ of $p = q = \frac{1}{2}$) has the expected values of 0 and 0. Because TL is a hyperbolic-row cell of $\mathcal{G}_{Mixed}^N$, only Row has an incentive to defect (to BL).

2.5.3   Cooperation

This ability to default reflects the reality that $\mathcal{G}^B$ is missing a self-enforcing Nash structure. More precisely, although $\mathcal{G}^B$ attributes can convert a non-Nash cell into an attractive $\mathcal{G}$ option, the $\mathcal{G}^N$ unilateral game structure can corrupt opportunities by allowing a cooperative attempt to collapse. Ways to ensure cooperation become important.

One approach is to impose a new structure on the game whereby the Behavioral terms inherit a self-enforcing Nash flavor. This comment sounds contradictory, but it is, in fact, commonly used. A preferred community action while driving is to wear seat belts, or not to talk on a cell phone, or to stay out of carpool lanes when not eligible. Experience proves that cooperation can be spotty at best. So a Nash quality is inserted into these Behavioral terms by fining offenders: Doing so, converts adherence into a best response. Thus, a way to view certain regulations is that they are societal mechanisms imposed to support $\mathcal{G}^B$ cooperative benefits. Even more, introducing some sort of punishment is a natural way to convert $\mathcal{G}^B$'s free-flowing aspects into a choice that now has a Nash behavior.

Similarly, treaties between countries with noncompliance consequences, regulations on companies for pollution control or creating safe products are supported by penalties, are intended to convert cooperation into the best response. The same is true for the area of mechanism design. Social norms, such as with a fishing community, are enforced by punishment, which, if strong enough, makes cooperation desirable. Notice how these comments provide theoretical support for the works of T. Schelling and E. Ostrom.

A natural approach to impose this change on a game's behavioral aspects comes from an earlier (Sect. 1.3.1) Prisoner's Dilemma example of a free-riding coauthor on a joint project. The response is obvious: Never work with that *person again!* This *Grim Trigger* strategy is where should a partner defect, then defect on the partner forever after. A less extreme "tit-for-tat" approach also applies if a game is played repeatedly; e.g., this may be daily interactions with a neighbor concerning parking spots. The idea is that if you hurt me, expect me to retaliate: Tomorrow I will do to you what you did to me today. If the cost is high enough, the behavioral terms acquire a Nash flavor discouraging unilateral actions and encouraging cooperation.

The need for cooperation occurs only if the $\mathcal{G}^B$ Pareto superior cell makes a non-Nash cell attractive. Here is a surprise; the $\mathcal{G}^B$ *Pareto inferior cell* introduces a way to handle the difficulty! It does so by ensuring that certain outcomes are unappealing! To be precise, return to $\mathcal{G}_{Mixed}$ (Eq. 2.37), which requires cooperation to enjoy the TL rewards. Row has the incentive to defect by playing B, which collapses the outcome to the BL cell. Column most surely wants to punish Row, and Column can do so by playing R. Here, the best Row can obtain (-2 by playing T) is smaller than what Row would receive (2) by cooperating.

This ability for Column to punish *is not* a $\mathcal{G}^N_{Mixed}$ feature. Row's dire situation of being stuck in the R column of $\mathcal{G}_{Mixed}$ is strictly due to Row's R entries in $\mathcal{G}^B_{Mixed}$; these $\mathcal{G}^B$ properties (Theorem 2.1) ensure that Row wants to avoid this column. The following helps us understand how $\mathcal{G}^B$ terms allow the use of the grim trigger.

To punish, two items are involved:

1. There must be an option where the non-cooperating player truly is punished. That is, there is an option where the best non-cooperating player can obtain is smaller than what could be attained by cooperation. This feature is what converts cooperation into the best response Nash-like behavior.
2. The punishment works; the non-cooperating player is affected by the punishment. In repeated games, this means that the non-cooperating player cares enough about future losses with tit-for-tat or grim trigger.

To avoid a morass of subscripts, let player 1 (Row) be the non-cooperating player where cooperation involves a cell in the jth column. To be attractive, the β_1 value for a non-Nash cooperating cell must be positive ($\mathcal{G}^B$'s Pareto superior term). In this jth column, Row is evaluating (Theorem 2.1) between $-|\eta_{j,1}| + |\beta_1|$ and $|\eta_{j,1}| + |\beta_1|$, where cooperation requires selecting the *smaller* value. (With Eq. 2.37, this comparison is $-1 + 3$ with $1 + 3$.) The only punishment Column can impose is to select another strategy, the ith column, where Row selects between $-|\eta_{i,1}| - |\beta_1|$ and $|\eta_{i,1}| - |\beta_1|$. (With Eq. 2.37, this comparison is $-1 - 3$ with $1 - 3$.) Here, Row prefers the larger $|\eta_{i,1}| - |\beta_1|$ value. To be a punishment, this $|\eta_{i,1}| - |\beta_1|$ value must be smaller than what Row would get from cooperation, or

$$- |\eta_{j,1}| + |\beta_1| > |\eta_{i,1}| - |\beta_1| \text{ which is } |\beta_1| > \frac{1}{2}\{|\eta_{1,1}| + |\eta_{2,1}|\}. \qquad (2.38)$$

Look familiar? It should; this inequality agrees with Eq. 2.33, which determines when a non-Nash cell is attractive. Size matters. The surprising fact is that

for 2 × 2 games, whenever Pareto superior properties of $\mathcal{G}^B$ make a non-Nash cell attractive, the $\mathcal{G}^B$ Pareto inferior cell makes punishment possible!

Thus, $\mathcal{G}^B$ generates a positive externality to make a non-Nash cell appealing *and* creates a negative externality that ensures a non-cooperating player can be punished. Stated more simply, $\mathcal{G}^B$ both gives and takes.

The second concern involves embracing the future; this is modeled with a discounting $\delta \in (0, 1)$ term. Treat δ as reflecting the present value of money; e.g., having a \$100 tomorrow is worth $\delta \times$ \$100 today. It is well known that sufficiently large δ values, which reflect an appreciation for future payoffs, encourage cooperation. What has not been explored is how these δ values reflect a tension between strategic and cooperative terms.

The following result ties together these variables to determine for repeated games when tit-for-tat after a first defection or grim converts cooperation into a player's best response.

Theorem 2.4 *In a* 2 × 2 *repeated game, suppose there is a targeted non-Nash cell in the* j*th column, where cooperation requires Player 1 to select the* $|\eta_{j,1}| + |\beta_1|$ *value over the* $|\eta_{j,1}| + |\beta_1|$ *choice. Suppose Player 1's entries in the two columns satisfy Eq. 2.38. If Player 2 uses the grim trigger strategy, it is in Player 1's best interest to cooperate if*

$$\delta > \frac{|\eta_{j,1}|}{|\beta_1| + \frac{1}{2}\{|\eta_{j,1}| - |\eta_{i,1}|\}} \qquad (2.39)$$

If Player 2 uses tit-for-tat, it is in Player 1's best interest (in the special case) to cooperate after defecting once if

$$\delta > \frac{|\eta_{j,1}|}{|\beta_1| + \frac{1}{2}\{-|\eta_{j,1}| + |\eta_{i,1}|\}} \qquad (2.40)$$

One of our concerns is whether conclusions reflect general principles of a class of games over a particular game's specific entries. Here, the inequalities reflect general lessons: According to Eqs. 2.39, 2.40, cooperation requires an interest in the future (a larger δ value) *and* a concern for cooperation as reflected by the β value. Notice:

the larger the $\mathcal{G}^B$ behavioral incentives (the larger the $|\beta_j|$ values), the more likely players will cooperate.

This makes sense; size matters. Larger $|\beta_j|$ values enhance the payoffs, which makes cooperation more lucrative. To see this with $\mathcal{G}_{Mixed}$, as $|\eta_{1,1}| = |\eta_{2,1}| = 1$ and $|\beta_1| = 3$, it follows that with either strategy, cooperation is in Player 1's best interest if Row's δ value is greater than one-third. With the Eq. 2.36 game, Eq. 2.38 inequality is satisfied, so, with Column playing tit-for-tat or grim trigger, cooperation of keeping TR is in Row's interest with $\delta > \frac{1}{2}$. It is in Column's interest to cooperate for $\delta > \frac{2}{4+\frac{1}{2}\{2-4\}} = \frac{2}{3}$ if Row plays grim trigger, and for $\delta > \frac{2}{4+\frac{1}{2}\{-2+4\}} = \frac{2}{5}$ if Row plays tit-for-tat. The larger the $|\beta|$ value, the smaller the lower bound for δ.

The smaller δ value for this game is associated with the special case of tit-for-tat used here, which may be surprising because it might seem that the drastic grim trigger punishment would force a player to become cooperative with a smaller δ. While this is not true here, with other games, it can be the case. To illustrate with Eq. 2.36, interchange Column's 2 and 4 values in $\mathcal{G}^N$. Keeping everything else the same leads to

$$\mathcal{G}^* = \begin{array}{|cc|cc|} \hline -3 & 8 & 1 & 0 \\ -1 & -2 & 3 & -6 \\ \hline \end{array} = \begin{array}{|cc|cc|} \hline -1 & 4 & -1 & -4 \\ 1 & 2 & 1 & -2 \\ \hline \end{array} + \begin{array}{|cc|cc|} \hline -2 & 4 & 2 & 4 \\ -2 & -4 & 2 & -4 \\ \hline \end{array},$$

which, as with $\mathcal{G}$ in Eq. 2.36 has a BL Nash cell and TR repelling cell. But here, the $\eta_{j,2}$ and $\eta_{i,2}$ values are reversed, which reverses the δ values for grim trigger and tit-for-tat. Namely, it is in Column's interest to cooperate for $\delta > \frac{2}{4+\frac{1}{2}\{-4+2\}} = \frac{2}{3}$ if Row plays tit-for-tat, and for $\delta > \frac{2}{4+\frac{1}{2}\{4-2\}} = \frac{2}{5}$, if Row plays grim trigger.

Proof of Theorem 2.4 The standard proof compares what a player would obtain by cooperating with what would be gained by defecting and then being punished. The argument here is the same except that the role of the Nash and behavioral terms are identified.

Starting with grim trigger, to ensure cooperation, the payoff obtained from cooperating must be better than what would be received by defecting and then always getting the non-cooperative outcome. The gains for the first player from cooperation are $\sum_{k=0}^{\infty} \delta^k(-|\eta_{j,1}| + |\beta_1|)$; it is the sum of the $(-|\eta_{j,1}| + |\beta_1|)$ payoff as adjusted with the discount term. If this player defects, the first payoff is $|\eta_{j,1}| + |\beta_1|$, and the sum of the subsequent punished payoffs is $\sum_{k=1}^{\infty} \delta^k(|\eta_{i,1}| - |\beta_1|)$. The goal is to determine when $\sum_{k=0}^{\infty} \delta^k(-|\eta_{j,1}| + |\beta_1|) > |\eta_{j,1}| + |\beta_1| + \sum_{k=1}^{\infty} \delta^k(|\eta_{i,1}| - |\beta_1|)$.

By using the standard $\sum_{k=0}^{\infty} \delta^k = \frac{1}{1-\delta}$, the cooperation value is $\frac{1}{1-\delta}(-|\eta_{j,1}| + |\beta_1|)$. Similarly, the outcome for defecting and then being punished is $|\eta_{j,1}| + |\beta_1| + \frac{\delta}{1-\delta}(|\eta_{i,1}| - |\beta_1|)$. Solving for δ leads to Eq. 2.39.

The difference in the computations for our special case of tit-for-tat is that the outcome for cooperation, $\frac{1}{1-\delta}(-|\eta_{j,1}| + |\beta_1|)$, must exceed the outcome from defection plus the cost of punishment plus the cost of continued cooperation. This second term is $|\eta_{j,1}| + |\beta_1| + \delta(-|\eta_{i,1}| - |\beta_1|) + \sum_{k=2}^{\infty} \delta^k(-|\eta_{j,1}| + |\beta_1|)$. The summation equals $\frac{\delta^2}{1-\delta}(-|\eta_{j,1}| + |\beta_1|)$. Standard algebra leads to Eq. 2.40.   □

2.5.4 Hawk–Dove

Possible conflict or coordinated actions are not immediately obvious from Eq. 1.8 form of the Hawk–Dove game. The decomposition clarifies all of this with the $\mathcal{G}^B$ positioning of the Pareto superior cell (BR), which differs from both Nash cells (BL and TR).

$$\mathcal{G}_{Hawk,Dove} = \begin{array}{|cc|cc|} 0 & 0 & 6 & 2 \\ 2 & 6 & 4 & 4 \end{array} = \begin{array}{|cc|cc|} -1 & -1 & 1 & 1 \\ 1 & 1 & -1 & -1 \end{array} + \begin{array}{|cc|cc|} -2 & -2 & 2 & -2 \\ -2 & 2 & 2 & 2 \end{array} + \mathcal{G}^K(3,3)$$

$$(2.41)$$

Notice how the $\mathcal{G}^B_{Hawk,Dove}$ Pareto inferior cell (TL) discourages aggression, while Pareto superior cell embraces a cooperative effort—both players should act like doves. Should a Nash cell be selected, treat the coordination as reflecting which player defects from the game's cooperative $\mathcal{G}^B_{Hawk,Dove}$. This is a standard $\mathcal{G}^B$ feature where answers about appropriate coordinating efforts follow from the $\mathcal{G}^B$ structure.

The original version of this game [33] had the form

$$\mathcal{G}_{Hawk,Dove} = \begin{array}{c} \\ \text{Hawk} \\ \text{Dove} \end{array} \begin{array}{|cc|cc|} \multicolumn{2}{c}{\text{Hawk}} & \multicolumn{2}{c}{\text{Dove}} \\ \hline \frac{V-C}{2} & \frac{V-C}{2} & V & 0 \\ 0 & V & \frac{V}{2} & \frac{V}{2} \end{array}$$

where V is the reward, or value of the resource, and C is the cost of conflict. If both players are doves (BR), the rewards are split, otherwise the hawk takes all. Well, unless both players are hawks where, after the battle, they split what is left, if anything.

The decomposition of this game is

$$\begin{array}{|cc|cc|} \frac{V-C}{4} & \frac{V-C}{4} & \frac{V}{4} & -\frac{V-C}{4} \\ -\frac{V-C}{4} & \frac{V}{4} & \frac{V}{4} & -\frac{V}{4} \end{array} + \begin{array}{|cc|cc|} -\frac{4V+C}{8} & -\frac{4V+C}{8} & \frac{4V+C}{8} & \frac{4V+C}{8} \\ -\frac{4V+C}{8} & \frac{4V+C}{8} & \frac{4V+C}{8} & \frac{4V+C}{8} \end{array} \quad (2.42)$$

with the Kernel component $\mathcal{G}^K(\frac{V}{2} - \frac{C}{8}, \frac{V}{2} - \frac{C}{8})$. As a strict Nash cell has positive entries in $\mathcal{G}^N$, BL and TR are Nash cells only should $V - C < 0$, which requires $V < C$ where the cost of conflict exceeds rewards.

This comment along with Eq. 2.42 captures a nice duality between the combative Prisoner's Dilemma and Hawk–Dove games. If the rewards of victory exceed the cost of conflict ($V - C > 0$), then Eq. 2.42 becomes a Prisoner's Dilemma game with the TL Nash cell. A costly combat ($V - C < 0$) converts the TR and BL hyperbolic cells into Nash cells, and TL into a repelling cell. In both the Prisoner's Dilemma and Hawk–Dove characterizations, the $\mathcal{G}^B$ Pareto superior location identifies Dove–Dove as the appropriate cooperative action, while the $\mathcal{G}^B$ Pareto inferior position (with negative entries) discourages both players from being aggressive. (The transition, where $V = C$, has weak Nash cells at TL, BL, and TR, which, again, reflects combined Nash properties of the two general cases.)

A common description of this game is of anti-coordination where one player acts dovish and the other hawkish. Who should do what? According to $\mathcal{G}^B_{Hawk,Dove}$, the choice of one Nash cell over the other does *not* depend on $\mathcal{G}^N$; it depends on which agent reneges from the $\mathcal{G}^B_{Hawk,Dove}$ Pareto superior choice of Dove–Dove. And so, anti-coordination resembles the defecting player in the Prisoner's Dilemma. It should; the $\mathcal{G}^B$ components are similar. While this description might explain birds, or even teenagers speeding toward each other, it remains far from being a desired setting should this game model international brinkmanship settings where annihilation clearly designates $V - C << 0$.

Although it is impossible (Theorem 2.3) to design a $\mathcal{G}^B$ component to convert this desired BR outcome into a group preferred cell, $\mathcal{G}^B$ terms can make BR sufficiently attractive so that participants would consider BR rather than the available Nash cells that have personally better outcomes. Doing so, requires cooperation, or, in a repeated setting, pressure to induce a Nash-like structure. With brinkmanship, peace of some sort is desired, so Theorem 2.4 strategies become relevant. The general form of this game is

$$
\mathcal{G}^N + \mathcal{G}^B = \begin{array}{|cc|cc|}\hline -|\eta_{1,1}| & -|\eta_{1,2}| & |\eta_{2,1}| & |\eta_{1,2}| \\ \hline |\eta_{1,1}| & |\eta_{2,2}| & -|\eta_{2,1}| & -|\eta_{2,2}| \\ \hline \end{array} + \begin{array}{|cc|cc|}\hline -|\beta_1| & -|\beta_2| & |\beta_1| & -|\beta_2| \\ \hline -|\beta_1| & |\beta_2| & |\beta_1| & |\beta_2| \\ \hline \end{array}
$$

where $|\beta_j| > \frac{1}{2}\{|\eta_{1,j}| + |\eta_{2,j}|\}$ (Eq. 2.38). Thus, (Theorem 2.4) it is in player k's best interest to cooperate if $\delta > \frac{|\eta_{2,k}|}{|\beta_k| + \frac{1}{2}\{|\eta_{2,k}| - |\eta_{1,k}|\}}$ should the other player use grim trigger, and $\delta > \frac{|\eta_{2,k}|}{|\beta_k| + \frac{1}{2}\{-|\eta_{2,k}| + |\eta_{1,k}|\}}$ should the other player use tit-for-tat. Illustrating with Eq. 2.41, cooperation is in each player's best interest with either strategy should both have $\delta > \frac{1}{2 + \frac{1}{2}\{1-1\}} = \frac{1}{2}$. With either strategy, the punishing approach is to become hawkish, which corresponds to the commonly heard saber-rattling that can accompany international negotiations between antagonists.

With Eq. 2.42 and tit-for-tat, cooperation becomes a choice with $\delta > \frac{V}{V+C}$. As it probably should be anticipated, higher costs of conflict encourage cooperation.

2.5.5 Battle of the Sexes

Similar to Hawk–Dove, the Battle of the Sexes has two Nash cells. But the contrasting interpretations of these games makes it reasonable to explore why they differ.

A standard comparison is that the Hawk–Dove requires an anti-coordination while the Battle of the Sexes involves coordination. Maybe, but this is more of a semantic than substantive difference because, by interchanging the $\mathcal{G}_{Sexes}$ columns so that each player's first option is the player's favorite, we have

$$\mathcal{G}_{Sexes} = \begin{array}{c} \\ \text{Opera} \\ \text{Football} \end{array} \begin{array}{cc} \text{Opera} & \text{Football} \\ \begin{array}{c|c|cc} 6 & 2 & 0 & 0 \\ 0 & 0 & 2 & 6 \end{array} \end{array} \rightarrow \begin{array}{c} \\ \text{Opera} \\ \text{Football} \end{array} \begin{array}{cc} \text{Football} & \text{Opera} \\ \begin{array}{cc|c|c} 0 & 0 & 6 & 2 \\ 2 & 6 & 0 & 0 \end{array} \end{array}, \quad (2.43)$$

which now assumes an anti-coordination flavor. Indeed, in both Eqs. 2.41 and 2.43, the top row and left column indicate each player's preferred choice.

The behavioral component $\mathcal{G}^B_{Sexes}$ from

$$\mathcal{G}_{Sexes} = \begin{array}{|cc|cc|} \hline 6 & 2 & 0 & 0 \\ 0 & 0 & 2 & 6 \\ \hline \end{array} = \begin{array}{|cc|cc|} \hline 3 & 1 & -1 & -1 \\ -3 & -3 & 1 & 3 \\ \hline \end{array} + \begin{array}{|cc|cc|} \hline 1 & -1 & -1 & -1 \\ 1 & 1 & -1 & 1 \\ \hline \end{array} + \mathcal{G}^K(2,2) \quad (2.44)$$

identifies the game's cooperative structure (given by $\mathcal{G}^B_{Sexes}$'s BL Pareto superior cell) where each player is solicitous about the partner's wishes. Similarly, the Pareto inferior TR cell carries the message that the game's "we" structure discourages the players from pushing for personal favorite outcomes.

One way this game differs from Hawk–Dove is that $\mathcal{G}^B_{Sexes}$ is not sufficiently strong enough to produce an attractive repelling cell. This is because in Eq. 2.44,

$$|\beta_j| = 1 < \frac{1}{2}\{|\eta_{1,j}| + |\eta_{2,j}|\} = 2,$$

rather than the reversed inequality required to generate an attractive non-Nash cell. This lack of cooperative impact of $\mathcal{G}^B_{Sexes}$ focusses attention on the selection of a Nash cell.

To see what would emerge should $\mathcal{G}^N_{Sexes}$ could be coupled with a strong enough $\mathcal{G}^B$ to generate an attractive BL, we have

$$\begin{array}{|cc|cc|} \hline 3+|\beta_1| & 1-|\beta_2| & -1-|\beta_1| & -1-|\beta_2| \\ -3+|\beta_1| & -3+|\beta_2| & 1-|\beta_1| & 3+|\beta_2| \\ \hline \end{array} = \begin{array}{|cc|cc|} \hline 3 & 1 & -1 & -1 \\ -3 & -3 & 1 & 3 \\ \hline \end{array} + \begin{array}{|cc|cc|} \hline |\beta_1| & -|\beta_2| & -|\beta_1| & -|\beta_2| \\ |\beta_1| & |\beta_2| & -|\beta_1| & |\beta_2| \\ \hline \end{array}$$

While the BL cell becomes inviting with sufficiently large $|\beta_j|$ values, say $|\beta_j| = 4$, notice what it means: each person gains pleasure by promoting the partner's choice. There are even retaliatory grim trigger or tit-for-tat actions where the associated warnings assume a quirky form where the husband might threaten his wife,

> If you don't agree that we should do what *you* really prefer to do, go to the opera, then I am going to insist on football!!

Using different words, a strong $\mathcal{G}^B$ (Sect. 2.4.5) can convert the loving Battle of the Sexes into a weird Hawk–Dove conflict demonstrating how an overly strong worry about a partner (here, $\mathcal{G}^B$'s BL entries) can induce difficulties. This brings to mind O. Henry's "Gift of the Magi" story where the wife cuts and sells her hair to buy a chain for her husband's pocket watch, which he had sold to buy a comb for his wife's hair.

As a final experiment, it is reasonable to wonder how $\mathcal{G}_{Sexes}$ would change should the $\mathcal{G}^B$ structure change from where each player is solicitous of the other, to where each player has more of a self-interest. Here, the $\mathcal{G}^B_{Sexes}$ Pareto superior cell is in

TR rather than BL. Using $\beta_1 = \beta_2 = -1$ and $\kappa_1 = \kappa_2 = 4$, this "flipped battle of the sexes" becomes

$$\mathcal{G}_{Flipped} = \begin{array}{|cc|cc|} \hline 6 & 4 & 2 & 2 \\ \hline 0 & 0 & 4 & 6 \\ \hline \end{array}$$

indicating a certain pleasure (the TR cell) in disagreeing. Aha! By interchanging rows,

$$\mathcal{G}_{Flipped} = \begin{array}{|cc|cc|} \hline 0 & 0 & 4 & 6 \\ \hline 6 & 4 & 2 & 2 \\ \hline \end{array}$$

the result resembles a Hawk–Dove game! Not quite, but close enough to notice that an important difference between Hawk–Dove and Battle of the Sexes is whether the $\mathcal{G}^B$ Pareto superior cell expresses self-interest or support for the other player.

Finally, if the cooperative interest leans toward Opera (let $\beta_1 = \beta_2 = 1$ so that TL is the $\mathcal{G}^B$ Pareto superior cell, and let $\kappa_1 = 2$, $\kappa_2 = 4$ to eliminate negative entries), TL becomes the clearly preferred Nash cell for the game where the husband prospers as long as there is agreement

$$\mathcal{G}_{Opera} = \begin{array}{|cc|cc|} \hline 6 & 6 & 0 & 4 \\ \hline 0 & 0 & 2 & 6 \\ \hline \end{array} \tag{2.45}$$

2.5.6 Stag Hunt

A third type of 2×2 games with two Nash cells is the Stag Hunt. To review, the $\mathcal{G}^B$ Pareto superior cell for Hawk–Dove and for the Battle of the Sexes is positioned over a $\mathcal{G}^N$ repelling cell, where the difference is whether the $\mathcal{G}^B$ Pareto superior cell conflicts, or supports, the other player. It remains to place the $\mathcal{G}^B$ Pareto superior cell over a Nash cell, which is precisely what happens with the Stag Hunt (and Eq. 2.45).

The $\mathcal{G}^B_{Stag}$ cooperative force accentuates the appeal of a particular Nash cell (here TL). The BR Nash cell, however, has an attractive feature; it eliminates risk. No matter what the other player does, if a player in Eq. 2.46 selects B or R, the player is guaranteed the payoff of 3. Going for the big TL reward carries the danger of ending up with -1.

$$\mathcal{G}_{Stag} = \begin{array}{|cc|cc|} \hline 7 & 7 & -1 & 3 \\ \hline 3 & -1 & 3 & 3 \\ \hline \end{array} = \begin{array}{|cc|cc|} \hline 2 & 2 & -2 & -2 \\ \hline -2 & -2 & 2 & 2 \\ \hline \end{array} + \begin{array}{|cc|cc|} \hline 2 & 2 & -2 & 2 \\ \hline 2 & -2 & -2 & -2 \\ \hline \end{array} + \mathcal{G}^K(3,3) \tag{2.46}$$

A generalized version of Eq. 2.46 using $\mathcal{G}^N_{Stag}$

$$\begin{array}{|cc|cc|} \hline 2 + |\beta_1| & 2 + |\beta_2| & -2 - |\beta_1| & -2 + |\beta_2| \\ \hline -2 + |\beta_1| & -2 - |\beta_2| & 2 - |\beta_1| & 2 - |\beta_2| \\ \hline \end{array}$$

demonstrates how the $\mathcal{G}^B$ component not only adds interest to one Nash cell (here, TL) at the expense of the other (BR), but it enhances the rewards in the L column for Row and the T row for Column. So, central Stag Hunt features are derived from $\mathcal{G}^B$ properties.

2.5.6.1 Summary

There are two general types of 2×2 games with two Nash cells. (By having two Nash cells, the other two must be repelling cells.)

1. The game's $\mathcal{G}^B$ component places the cooperative effort on a repelling cell.

   (a) If $\mathcal{G}^B$ is strong enough to make the repelling cell attractive to both players, then, in a repeated version, a retaliatory action can encourage cooperation.
   (b) If the $\mathcal{G}^B$ term does not convert a repelling cell into an interesting cooperative option, then the emphasis is to coordinate on a particular Nash cell.

2. The game's $\mathcal{G}^B$ component places cooperative attention on one of the two Nash cells. Because the Pareto superior and inferior terms are over the two Nash cells, the role of $\mathcal{G}^B$ is to accentuate interest in one Nash cell at the expense of the other.

2.6 Risk Versus Payoff Dominance

The stag hunt is often cited as illustrating two refinements of Nash equilibria that were introduced in 1988 by Harsanyi and Selten [8]. To describe them, suppose TL and BR are Nash cells with the game

$$
\begin{array}{|cc|cc|}
\hline
A_1 & A_2 & B_1 & B_2 \\
C_1 & C_2 & D_1 & D_2 \\
\hline
\end{array}
=
\begin{array}{|cc|cc|}
\hline
|\eta_{1,1}| & |\eta_{1,2}| & -|\eta_{2,1}| & -|\eta_{1,2}| \\
-|\eta_{1,1}| & -|\eta_{2,2}| & |\eta_{2,1}| & |\eta_{2,2}| \\
\hline
\end{array}
+
\begin{array}{|cc|cc|}
\hline
\beta_1 & \beta_2 & -\beta_1 & \beta_2 \\
\beta_1 & -\beta_2 & -\beta_1 & -\beta_2 \\
\hline
\end{array}
+ \mathcal{G}^K,
$$

$$(2.47)$$

where absolute values, $|\eta_{i,j}|$, are used to underscore which cells are Nash. To resolve the issue about which Nash cell should be selected, Harsanyi and Selten introduce two criteria.

Payoff dominance is the Nash cell where each player receives at least as much as from any other Nash cell, and at least one player gets more. In terms of Eq. 2.47, TL is payoff dominant if and only if

$$A_1 \geq D_1, \quad A_2 \geq D_2 \quad \text{and at least one inequality is strict.} \qquad (2.48)$$

Social welfare. A game could have two Nash cells where neither is payoff dominant. With Eq. 2.47, TL and BR are Nash cells with entries $A_1 = 5$, $A_2 = 4$, and $D_1 = 4$, $D_2 = 7$. Row gets a better outcome in TL while Column's better choice is

BR, so neither cell is payoff dominant. Yet, BR has a flavor of being a better choice with the *social welfare* offering of $D_1 + D_2 = 4 + 7 = 11 > A_1 + A_2 = 5 + 4 = 9$. So, where payoff dominance fails to apply, a social welfare comparison (summing the cell's payoffs) is a possible substitute. Of course, if a cell is payoff dominant, it also is social welfare dominant.

Risk dominance is intended to identify which Nash cell minimizes risk. According to Harsanyi and Selten, BR satisfies this condition if the inequality

$$(B_1 - D_1)(C_2 - D_2) > (C_1 - A_1)(B_2 - A_2) \qquad (2.49)$$

holds. This means that the joint deviation from BR (given by the product $(B_1 - D_1)(C_2 - D_2)$) exceeds the joint deviation from TL (given by $(C_1 - A_1)(B_2 - A_2)$).

Game theoretic concepts are commonly expressed in terms of payoff inequalities. But, without an accompanying analysis, the inequalities need not reflect much about a game's underlying structure. Be honest; it is not immediately clear from Eqs. 2.48 and 2.49 how these dominance conditions differ, nor how to create as many examples as desired illustrating where different Nash cells have a competing status.

This concern is similar to the one raised about Eq. 1.6 inequalities characterizing Prisoner's Dilemma games. Fortunately, the coordinate system for games (Sect. 2.5.1) permits us to reach beyond the inequalities to extract new interpretations; e.g., the Prisoner's Dilemma reflects an explicit tension between individual (the η values) and group benefits (the location of the $\mathcal{G}^B$ Pareto superior cell).

2.6.1   Payoff Dominance

A similar approach applies to most (if not all) definitions and constraints that are defined with payoff inequalities. To demonstrate with Eqs. 2.48 and 2.49, start with Row and Eq. 2.48. The coordinates convert the $A_1 \geq D_1$ inequality into $|\eta_{1,1}| + \beta_1 \geq |\eta_{2,1}| - \beta_1$, or the familiar $2\beta_1 \geq |\eta_{2,1}| - |\eta_{1,1}|$ where Row's behavioral term must be larger than the difference between Row's two Nash entries. If, for instance, $|\eta_{1,1}| > |\eta_{2,1}|$, then β_1 could be zero, or even negative, and still focus attention on TL rather than BR. But if $|\eta_{2,1}| > |\eta_{1,1}|$, then for TL to be payoff dominant, $\mathcal{G}^B$'s TL cell must have sufficiently large β values; that is, the payoff dominance concept is dominated by β.

Similarly, the $A_2 \geq D_2$ inequality leads to the condition that TL is payoff dominant over BR if and only if

$$2\beta_1 \geq |\eta_{2,1}| - |\eta_{1,1}| \text{ and } 2\beta_2 \geq |\eta_{2,2}| - |\eta_{1,2}| \text{ where at least one inequality is strict.} \qquad (2.50)$$

In particular, payoff dominance relates the size of $\mathcal{G}^N$ and $\mathcal{G}^B$ entries (Sect. 2.4.5).

If payoff dominance is replaced with social welfare where TL is the better choice, then $A_1 + A_2 > D_1 + D_2$, or $(|\eta_{1,1}| + \beta_1) + (|\eta_{1,2}| + \beta_2) > (|\eta_{2,1}| - \beta_1) + (|\eta_{2,2}| - \beta_2)$. Thus,

$$2\beta_1 + 2\beta_2 > (|\eta_{2,1}| - |\eta_{1,1}|) + (|\eta_{2,2}| - |\eta_{1,2}|). \tag{2.51}$$

If payoff dominance fails (one of Eq. 2.50 inequalities is in the opposite direction), and the difference in the satisfied inequality larger than the negative difference in the other, then Eq. 2.51 is satisfied and the social welfare function successfully replaces payoff dominance.

2.6.2 Risk Dominance

Turn now to Eq. 2.49 where $B_1 - D_1 = [-|\eta_{2,1}| - \beta_1] - [|\eta_{2,1}| - \beta_1] = -2|\eta_{2,1}|$. Similarly, $(C_2 - D_2) = -2|\eta_{2,2}|, (C_1 - A_1) = -2|\eta_{1,1}|$, and $(B_2 - A_2) = -2|\eta_{1,2}|$. These expressions convert Eq. 2.49 into the more understandable condition that BR is risk dominant (over TL) if and only if

$$|\eta_{2,1}|\,|\eta_{2,2}| > |\eta_{1,1}|\,|\eta_{1,2}|. \tag{2.52}$$

This refinement of a Nash equilibrium compares the sizes of the $\eta_{i,j}$ values of the Nash cells where bigger is better (Sect. 2.4.5).

This result, which requires only $\mathcal{G}^N$ information, is stated formally and more generally in Theorem 2.5. Again, the we-$\mathcal{G}^B$ terms play no role.

Theorem 2.5 *In a* 2×2 *game* $\mathcal{G}$ *with two Nash cells, the first is risk dominant over the second if and only if in* $\mathcal{G}^N$ *the product of the first Nash cell's entries is greater than the product of the second Nash cell's entries.*

An intent of risk dominance is that when faced with uncertainty about an opponent's actions, a player should select the less risky option. To explore and expand upon this comment, it follows from $\mathcal{G}^N$ that the NME of Eq. 2.47 game is

$$p^* = \frac{|\eta_{1,2}|}{|\eta_{1,2}| + |\eta_{2,2}|}, \quad q^* = \frac{|\eta_{1,1}|}{|\eta_{1,1}| + |\eta_{2,1}|}. \tag{2.53}$$

The associated best response curves are in Fig. 2.3. They are computed in the manner of the derivation of Fig. 1.4a.

The best response dynamics of the Battle of the Sexes is illustrated in Fig. 1.4b. The same argument applies to Fig. 2.3 where any starting point in box $\mathcal{A}$ ends at the top-left

Fig. 2.3 Best response and risk dominance

Nash corner, while any dynamic starting in box $\mathcal{B}$ terminates at the bottom-right Nash corner. Consequently, in an appropriate random sense of using only $\mathcal{G}^N$ information, *the size, or area, of these two boxes—these basins of attraction—determines which Nash point is more likely to occur.* The area of box $\mathcal{A}$ is p^*q^*, while that of box $\mathcal{B}$ is $(1 - p^*)(1 - q^*)$. Using Eq. 2.53, this leads to

$$\text{Area}(\mathcal{A}) = \frac{|\eta_{1,2}||\eta_{1,1}|}{(|\eta_{1,2}| + |\eta_{2,2}|)(|\eta_{1,1}| + |\eta_{2,1}|)}, \ \text{Area}(\mathcal{B}) = \frac{|\eta_{2,2}||\eta_{2,1}|}{(|\eta_{1,2}| + |\eta_{2,2}|)(|\eta_{1,1}| + |\eta_{2,1}|)}.$$
$$(2.54)$$

Equation 2.54 denominators agree, so the size difference of the two boxes, the two basins of attraction, is uniquely determined by the numerators. But each numerator is the product of the specified $\mathcal{G}^N$'s entries, which determines which Nash cell is risk dominant (Theorem 2.5). Thus, in a random sense of what a player may do, expect the risk dominant outcome to be more likely to occur.

Corollary 2.1 *In a best response curve Nash analysis of a game with two Nash cells, if one of the rectangular regions with a Nash cell is larger than the other, then that is the risk dominant strategy.*

For an even simpler argument, all that is needed to determine which of two Nash cells is risk dominant are the cell locations and the NME (p^*, q^*). This is because the box sizes are uniquely determined by the (p^*, q^*) values. If, for instance, BL and TR are Nash cells, then BL is the risk dominant choice if $(1 - p^*)q^* > p^*(1 - q^*)$. With this observation, this "box size" way to identify a risk dominance cell extends to any number of players and options. But there remain subtle features. For instance, a 3×3 game with three Nash cells need not have a mixed strategy relating them, so this risk dominance approach would not apply. Even if it did, a cell that is risk dominant over the triplet need not be risk dominant when compared with a particular other Nash cell.

A related issue that is not difficult to analyze (so it is left for the reader) is to interpret the meaning of the box areas. (Lowering Fig. 2.3 horizontal line creates a smaller $\mathcal{B}$.) If a larger box identifies risk dominance, what does it mean if, for instance, both boxes have small areas? Reflecting the theme of Sect. 2.4.5, do their sizes matter, or is it how they are used? What does it mean if the area of $\mathcal{B}$ is essentially that of $\mathcal{A}$, or if the areas differ significantly? (This changes NME values.) In a meaningful manner, one Nash cell is "strongly risk dominant" if, for instance, the area of its associated box is at least twice that of the other. The two ignored boxes belong to repellers with defecting behaviors, which suggests there exist expressions similar to those of risk dominance that compare and reflect these tendencies.

2.6.3 Examples

The comment that the Stag Hunt illustrates differences between payoff and risk dominance should be taken with care. For instance, BR is *not* risk dominant in Eq. 2.46 Stag Hunt because $|\eta_{2,1}|\,|\eta_{2,2}| = |\eta_{1,1}|\,|\eta_{1,2}| = 2 \times 2 = 4$. This equality identifies these η terms as belonging to the boundary separating whether TL or BR is the risk dominant cell. Slightly changing the $\eta_{1,1}$ and $\eta_{1,2}$ values from 2 to 2.1 creates the Stag Hunt

$$
\mathcal{G}^*_{stag} = \begin{array}{|cc|cc|}
\hline
7.1 & 7.1 & -1 & 2.9 \\
\hline
2.9 & -1 & 3 & 3 \\
\hline
\end{array}
$$

where TL is both payoff and risk dominant.

Although BR is not the risk dominant cell in $\mathcal{G}^*_{stag}$, a risk adverse player might prefer to hunt hare and accept the lower payoff of 2.9 if it is accompanied with a guarantee of avoiding the danger of being saddled with -1. Namely, should a player be leery of the partner, those hares begin to look more attractive. But rather than Eq. 2.49, an appropriate defining condition for this type of risk avoidance must involve payoff values such as

$$A_1 > D_1 \text{ while } C_1, D_1 > \max(0, B_1) \text{ and } A_2 > D_2 \text{ while } B_2, D_2 > \max(0, C_2).$$

For Row (with Eq. 2.47), this becomes $|\eta_{1,1}| + \beta_1 > |\eta_{2,1}| - \beta_1$ and $-|\eta_{1,1}| + \beta_1, |\eta_{2,1}| - \beta_1 > -|\eta_{2,1}| - \beta_1$, which reduce to

$$2\beta_1 > \left|(|\eta_{2,1}| - |\eta_{1,1}|)\right| \text{ and } \kappa_1 > \max(|\eta_{1,1}| - \beta_1, -|\eta_{2,1}| + \beta_1), \qquad (2.55)$$

with a similar expression for Column. Because payoff values are involved (as with payoff dominance), Eq. 2.55 combines $\mathcal{G}^N$ and $\mathcal{G}^B$ terms to identify sufficiently strong β values; the sole role of $\mathcal{G}^K$ is to guarantee that certain payoffs are positive.

The difference between the two dominance concepts reduces to a simple size observation:

A Nash cell is payoff dominant if it has sufficiently large β values; a Nash cell is risk dominant if it has sufficiently large η values.

As payoff dominance depends on $\mathcal{G}^B$, but risk dominance does not, it becomes doable to characterize *all possible* examples demonstrating their differences. Rather than writing down the obvious inequalities, to illustrate with an example, start with

$$
\mathcal{G}^N = \begin{array}{|cc|cc|}
\hline
2 & 2 & -3 & -2 \\
\hline
-2 & -3 & 3 & 3 \\
\hline
\end{array}
$$

that anoints BR as the risk dominant Nash cell. According to Eq. 2.52, BR retains that risk dominant status *independent* of the choice of $\mathcal{G}^B$. As payoff dominance involves $\mathcal{G}^B$, it now is easy to create examples displaying differences between the two concepts. Namely, to make the TL Nash cell more appealing and payoff dominant, select a $\mathcal{G}^B$ where its Pareto superior cell hovers over TL. Doing so, adds to each player's TL payoffs while *subtracting* from each player's BR offerings (because of $\mathcal{G}^B$'s Pareto inferior cell). According to Eq. 2.50, it suffices to select $\beta_1, \beta_2 > \frac{1}{2}[3 - 2] = \frac{1}{2}$. Avoiding fractions, an example would be $\beta_1 = \beta_2 = 3$ leading to

$$\mathcal{G} = \begin{array}{|cc|cc|} \hline 2 & 2 & -3 & -2 \\ -2 & -3 & 3 & 3 \\ \hline \end{array} + \begin{array}{|cc|cc|} \hline 3 & 3 & -3 & 3 \\ 3 & -3 & -3 & -3 \\ \hline \end{array} = \begin{array}{|cc|cc|} \hline 5 & 5 & -6 & 1 \\ 1 & -6 & 0 & 0 \\ \hline \end{array}$$

Notice how the risk dominant BR Nash cell loses its allure by offering *nothing* to either player. The situation further deteriorates with larger β values such as $\beta_1 = \beta_2 = 4$: Now, each player receives an enhanced TL payoff of 6, while the risk dominant BR *punishes* each player with a negative payoff of -1. This, of course, is an argument asserting that risk dominance by itself is lacking; it should be replaced with risk avoidance approaches perhaps of Eq. 2.55 type, or, maybe of more interest, where risk dominance is accompanied with payoff considerations. Writing such a definition is immediate.

The message: To involve β, or not to involve β, the legitimacy of the intent is the difference.

2.6.4 A 50–50 Choice

Occasionally a comment can be found suggesting that expected values should be used to select between Nash cells. The idea is that, absent any information about what an opponent might do, it seems reasonable to assume there is a 50–50 chance the opponent will prefer one or the other Nash cell. So, using this assumption, the player should select the strategy that yields the optimal expected value.

On first glance, this 50–50 approach appears to involve the behavioral terms. It does not. To see why this is so with Eq. 2.47, suppose Row assumes that Column will play L and R each with probability $\frac{1}{2}$. For Row, the expected return of playing T is

$$EV(T) = \frac{1}{2}[|\eta_{1,1}| - |\eta_{2,1}|] + \frac{1}{2}[\beta_1 - \beta_1] + \frac{1}{2}[\kappa_1 + \kappa_1],$$

or $EV(T) = \frac{1}{2}[|\eta_{1,1}| - |\eta_{2,1}|] + \kappa_1$. Similarly, $EV(B) = \frac{1}{2}[-|\eta_{1,1}| + |\eta_{2,1}|] + \kappa_1$. Notice, the β values cancel, so they play no role in this analysis. Instead, Row's 50–50 choice is strictly determined by which of the two Nash cells has Row's larger η value. A similar assertion holds for Column's choice.

This information permits creating examples. With

$$\begin{array}{|cc|cc|} \hline 10 & 11 & 2 & 7 \\ 8 & 4 & 8 & 6 \\ \hline \end{array} = \begin{array}{|cc|cc|} \hline 1 & 2 & -3 & -2 \\ -1 & -1 & 3 & 1 \\ \hline \end{array} + \begin{array}{|cc|cc|} \hline 2 & 2 & -2 & 2 \\ 2 & -2 & -2 & -2 \\ \hline \end{array} + \mathcal{K}(7, 7)$$

cells TL and BR are Nash.

In the full game, each player has a better payoff in TL than in BR, so TL is payoff dominant. The product of the $\mathcal{G}^N$ entries of TL and BR are, respectively, $(1)(2) = 2$ and $(3)(1) = 3$, so BR, with the larger product value, is the risk dominant cell. As for the 50–50 approach, Row prefers B (because it has Row's larger $\mathcal{G}^N$ Nash cell entry of 3) while Column prefers L. These choices lead to the BL outcome, which is a repeller, *not* a Nash cell.

2.7   Recap

Decomposing a game into its three components is computationally elementary. Of interest is how all information needed for a game's Nash analysis is assembled in a simple, clean form by clipping away all values redundant for the analysis. The $\mathcal{G}^N$ structure identifies the game's Nash dynamic with repellers, hyperbolic cells, and Nash equilibria. By specifying a cell's $\eta_{i,j}$ values, refinements of Nash cells, comparisons, and other options become possible.

It has long been recognized that the complexity exhibited by many games is how a group tension can overshadow the game's Nash structure. But a typical ad hoc analysis emphasizes features of specific classes of games. The decomposition replaces this local study with a general analysis, where an emphasis now is placed on how changes in the $\mathcal{G}^N$ and $\mathcal{G}^B$ coordinate directions create differences.

A surprise is how this easily computed $\mathcal{G}^B$ component introduces so much of the complexity experienced in game theory. Fortunately for 2×2 games, this term's simple structure involves only the parameters β_1 and β_2. As a result, it becomes possible to discover the source of the various complications; this chapter provides a sample of what can be done. The next chapter delves deeper into consequences.

Chapter 3
Consequences

3.1 Introduction

As indicted in the Preface, a measure of whether a coordinate system is appropriately
designed is if it simplifies discovering and proving new results. A reason this feature
holds with our decomposition is that the associated coordinates explicitly separate
and display the inherent tension between individual preferences and group advantages or disadvantages. Indeed, as shown in Chap. 2, the $\mathcal{G}^B$ component can launch
unexpected complexity into a game. But what about other choices of coordinates?
Several places in this chapter, modified coordinates are introduced to reflect new
needs.

The coordinate system introduces new tools to uncover a trove of consequences,
which is the theme of the current chapter. As an example, "best response" is a
me-action, so expect its impact to be restricted to $\mathcal{G}^N$ components. This suggests
the surprising fact, as developed in Scct. 3.5, that theories based on best response
(and there are *many* of them) must miss all of a game's complexities caused by $\mathcal{G}^B$
components! According to the previous chapter, this includes all of those $\mathcal{G}^B$ features
that distinguish the Prisoner's Dilemma, Hawk–Dove, Battle of the Sexes, and the
Stag Hunt.

Many different games could have been used to illustrate what follows, but our
choices, such as symmetric, zero-sum, identical play, matching pennies, CoCo, congestion, and potential games, were selected because their familiarity can make the
accompanying technical analysis more transparent. It is reasonable to wonder what
new can be said about these commonly discussed choices: Surprisingly, a lot! (The
reader is encouraged to apply these techniques to games of personal interest.)

An appropriately designed coordinate system facilitates "seeing" and comparing
points. To do so here, we create a roadmap of *all* 2×2 games, which displays (e.g.,
Fig. 3.3) how games are connected and how to identify new ones. Of interest, this
roadmap reduces the analysis from eight to three dimensions! (The eight 2×2 game
entries define a point in the eight-dimensional space $\mathbb{R}^8$. Conversely, *any* $\mathbb{R}^8$ point
defines a 2×2 game.) Armed with this map, questions, such as finding the portion

© Springer Nature Switzerland AG 2019
D. T. Jessie and D. G. Saari, *Coordinate Systems for Games*, Static & Dynamic Game
Theory: Foundations & Applications, https://doi.org/10.1007/978-3-030-35847-1_3

of all games with, say, one pure Nash cell, now are easy to answer. The chapter starts with symmetric games. Unless otherwise specified, treat the results as being for generic settings. An analysis for degenerate cases, such as with weak Nash cells, etc., usually are immediate.

3.2 Symmetric Games

An easy starting place is with symmetric games, where the game is the same for Row and Column. This feature is satisfied by earlier examples of the Prisoner's Dilemma, Hawk–Dove, and the Stag Hunt. To be precise, a 2×2 game is symmetric if and only if it can be represented as

$$\mathcal{G}_{symmetric} = \begin{array}{|c c|c c|} \hline A & A & B & C \\ \hline C & B & D & D \\ \hline \end{array} \tag{3.1}$$

Both players have the same outcome for TL and BR. If either player moves in an allowed direction from TL, the player's outcome is C; if either player moves in an allowed direction from BR, the player's outcome is B.

Certain properties of symmetric games follow directly from the coordinate system. While our emphasis is on 2×2 games (the general decomposition is in the next chapter), everything extends to games with more strategies and players.

Theorem 3.1 *For a nonzero 2×2 symmetric game $\mathcal{G}$, the components $\mathcal{G}^N$, $\mathcal{G}^B$, and $\mathcal{G}^K$ are symmetric.*

Generically, a 2×2 symmetric game has at least one Nash cell. If it has one Nash cell, it is either at TL or BR; the other cell is a repeller. If a symmetric 2×2 game has two Nash cells (at TL and BR, or BL and TR), the NME is symmetric in that $p = q$.

The $\mathcal{G}^B$ Pareto superior cell always is positioned at TL or at BR. If $\mathcal{G}$ has a single Nash cell, then $\mathcal{G}^B$ either enhances the $\mathcal{G}$ value of the Nash cell, or it subtracts from the appeal of the Nash cell while adding attraction to the only repelling cell. (A strong enough $\mathcal{G}^B$ term creates a Prisoner's Dilemma game.)

If the game has Nash cells at TL and BR, then $\mathcal{G}^B$ enhances one Nash cell while hurting the other (as in the Stag Hunt). If the game has Nash cells at BL and TR, then, with a strong $\mathcal{G}^B$ term, a game of the Hawk–Dove type occurs.

Proof The first assertion follows directly from the coordinates. For instance, $\eta_{1,1} = \eta_{1,2} = A - \frac{A+C}{2}$, $\eta_{2,1} = \eta_{2,2} = B - \frac{B+D}{2}$, and $\kappa_i = \frac{1}{4}[A + B + C + D]$. Finally, $\beta_1 = \beta_2 = \frac{1}{2}[A + C] - \kappa_i$. An alternative proof is given in Theorem 3.5.

If Row's $\mathcal{G}^N$ TL entry is A^*, the symmetric nature and properties of $\mathcal{G}^N$ (Theorem 2.1) define four of the eight payoffs as

$$\mathcal{G}^N = \begin{array}{|c c|c c|} \hline A^* & A^* & -A^* & \\ \hline -A^* & & & \\ \hline \end{array}.$$

If Row's BR entry of $\mathcal{G}^N$ is B^*, then $\mathcal{G}^N$ must be

$$\mathcal{G}^N = \begin{array}{|cc|cc|} \hline A^* & A^* & -B^* & -A^* \\ -A^* & -B^* & B^* & B^* \\ \hline \end{array}, \tag{3.2}$$

which means that all $\mathcal{G}^N$ properties are determined by the two variables A^* and B^*.

The A^* and B^* values either have the same sign, or they disagree. If A^*, $B^* > 0$, then TL and BR are Nash cells where the cell with the larger value is risk dominant (Theorem 2.5). If A^*, $B^* < 0$, then BL and TR are Nash cells; neither is risk dominant (the product of cell entries agree). If A^* and B^* differ in sign, there can be only one Nash cell. If $A^* > 0$, this Nash cell is TL, so BR, with two negative entries, is repelling; if $B^* > 0$, the Nash cell is BR and TL is repelling. Equation 3.2 symmetry guarantees a symmetric NME; $p = q$.

With a symmetric game, the $\mathcal{G}^B$ structure requires that it is determined by a single value C^*; this is because the $\mathcal{G}^B$ TL entries define all $\mathcal{G}^B$ cells. So for a symmetric game

$$\mathcal{G}^B = \begin{array}{|cc|cc|} \hline C^* & C^* & -C^* & C^* \\ C^* & -C^* & -C^* & -C^* \\ \hline \end{array}, \tag{3.3}$$

where if $C^* > 0$, the Pareto superior cell is TL; if $C^* < 0$, it is at BR. □

An appealing feature is how, by using symmetry arguments, Theorem 3.1 conclusions extend to describe properties of other classes of games. The reason is that games, such as the Battle of the Sexes, can be converted into a symmetric format by interchanging columns and/or rows as in Eq. 2.3 (and below in Eq. 3.4). Thus Theorem 3.1 conclusions hold by applying the same symmetry exchange to the assertions. In Eq. 2.43 Battle of the Sexes, interchanging the columns place the game in a symmetric format: If such a game has one Nash cell, then, rather than being at TL or BR, it must be at TR or BL. With two Nash cells, the NME has $q = (1 - p)$. (See Fig. 1.3a.)

More generally, the Eq. 3.4 chain of exchanges starts with a symmetric game, interchanges the rows, then the columns, then the rows. (A last interchange of columns returns to the initial symmetric matrix.)

$$\begin{array}{|cc|cc|} \hline A & A & B & C \\ C & B & D & D \\ \hline \end{array} \rightarrow \begin{array}{|cc|cc|} \hline C & B & D & D \\ A & A & B & C \\ \hline \end{array} \rightarrow \begin{array}{|cc|cc|} \hline D & D & C & B \\ B & C & A & A \\ \hline \end{array} \rightarrow \begin{array}{|cc|cc|} \hline B & C & A & A \\ D & D & C & B \\ \hline \end{array} \tag{3.4}$$

This symmetry structure is represented by $S_2 \times S_2$. For readers unfamiliar with this notation, the first S_2 merely represents interchanging rows while the second captures the columns. Symmetric $k_1 \times k_2 \times \ldots \times k_n$ games, where each of the n players has $k_1 = k_2 = \ldots = k_n = k \geq 2$ strategies, admit related symmetry groups $S_k \times S_k \times \ldots \times S_k$ representing the wide variety of exchange of rows, columns, etc. (This symmetry plays a central role in Chap. 7.) Equation 3.4 example motivates the following easily proved theorem that holds with more strategies and players.

Theorem 3.2 *If a game can be represented by interchanging rows of a symmetric game $\mathcal{G}$, Theorem 3.1 assertions hold by interchanging T and B, where, for a mixed strategy, $p = (1 - q)$. By interchanging columns of $\mathcal{G}$, the assertions of Theorem 3.1 hold by interchanging R and L, where with a mixed strategy $p = (1 - q)$. By interchanging columns and then the rows of $\mathcal{G}$, the assertions of Theorem 3.1 hold by interchanging T and B, and R and L, where, for a mixed strategy, $p = q$.*

Thus, if a game has the structure of the second choice in Eq. 3.4, it has a single Nash cell that must be either BL or TR. Should the game have two Nash cells, the NME has $p = (1 - q)$. A game, with the third equation, Eq. 3.4 form is symmetric, so Theorem 3.2 holds. Indeed, as shown in Eq. 3.4, there are only two types.

3.3 Zero-Sum and Identical Play

Two widely used, closely related games are zero-sum and identical play. The first identifies a competitive environment where the winner wins what the loser loses. The second ensures cooperation because, for each cell, both players receive the same payoff.

Definition 3.1 For any finite number of players and strategies, a zero-sum game is where the sum of each cell's entries equals zero. An identical play game is where all of a cell's entries are the same.

These two classes of games nicely complement each other. This is because any game can be uniquely expressed as a sum of an identical play and a zero-sum game. The way this is done shares some of the flavor of our decomposition. (See Chaps. 2 and 7.) Namely, the average of a cell's entries becomes the identical play cell's common entry; how the entries differ from the average become the zero-sum cell's entries. As a two-player example

$$
\begin{array}{|cc|cc|}
\hline
10 & 2 & 8 & -4 \\
\hline
-6 & 0 & 6 & 4 \\
\hline
-6 & 2 & 4 & 6 \\
\hline
\end{array}
=
\begin{array}{|cc|cc|}
\hline
6 & 6 & 2 & 2 \\
\hline
-3 & -3 & 5 & 5 \\
\hline
-2 & -2 & 5 & 5 \\
\hline
\end{array}
+
\begin{array}{|cc|cc|}
\hline
4 & -4 & 6 & -6 \\
\hline
-3 & 3 & 1 & -1 \\
\hline
-4 & 4 & -1 & 1 \\
\hline
\end{array}
\tag{3.5}
$$

3.3.1 CoCo Solutions

Here is a thought: The introductory chapter started with a game (Eq. 1.1) that was presented to decision-makers. Recall, after being told they were row players for

$$
\mathcal{G}_{Lecture} = \begin{array}{c|cc|cc}
 & L & & R & \\
\hline
T & \$10,000 & -\$20 & -\$100 & \$250 \\
B & \$100 & \$50 & \$150 & \$100 \\
\end{array},
$$

all participants selected T in a quest to enjoy that $10,000. Surprisingly, after they learned the folly of their choice, *nobody* suggested the obvious: Offer $1,000 to Column to play L. After all, with this side deal, Column leaves with $980, which exceeds anything she could win. Row would end up with $9,000, which is much better than Row's Nash payoff of $150.

But suppose Column recognizes her bargaining power and demands more. How much should each be paid? Kalai and Kalai [15] address this issue by decomposing a game into its identical play and zero-sum components. With this example, it is

$$
\begin{array}{|cc|cc|}
\hline
10,000 & -20 & -100 & 250 \\
\hline
100 & 50 & 150 & 100 \\
\hline
\end{array}
=
\begin{array}{|cc|cc|}
\hline
4990 & 4990 & 75 & 75 \\
\hline
75 & 75 & 125 & 125 \\
\hline
\end{array}
+
\begin{array}{|cc|cc|}
\hline
5010 & -5010 & -175 & 175 \\
\hline
25 & -25 & 25 & -25 \\
\hline
\end{array}
$$

They suggest that each player's base value is the maximum *cooperative* outcome (identical play), which endows each player with $4990. For this to happen, the players play TR and equally split the $10,000 - 20 = 9,980$ payoff.

Row, who would get $10,000 from the original game by playing TR, might object to the equality. To handle this problem, these values are adjusted with the minimax values from the *competitive* zero-sum component. As the minimax solution for a zero-sum game agrees with the Nash equilibrium, the BR minimax values are $25, -$25. By adding these competitive values to the cooperative term, Row receives $4990 + $25 = $5015 while Column leaves with $4990 - $25 = $4965. For details, see [15].

While this method is attractive, a troubling feature is that had the BR cell of $\mathcal{G}_{Lecture}$ been (110, 140), the zero-sum BR entry would be $(-15, 15)$ favoring *Column!* (Row would now receive $4990 - 15 = 4985$ while Column would waltz away with $4990 + 15 = 5005$.) What causes this feature is that CoCo solutions emphasize individual cells rather than the game's full structure.

Incidentally, the $\mathcal{G}_{Lecture}^{B}$ Pareto superior cell (TL) of our decomposition

$$
\mathcal{G}_{Lecture} =
\begin{array}{|cc|cc|}
\hline
4950 & -135 & -125 & 135 \\
\hline
-4950 & -25 & 125 & 25 \\
\hline
\end{array}
+
\begin{array}{|cc|cc|}
\hline
2512.5 & 20 & 2512.5 & 20 \\
\hline
2512.5 & -20 & -2512.5 & -20 \\
\hline
\end{array}
+ \mathcal{G}^{K}(2537.50, 95)
$$

indicates a cooperative leaning for this TL outcome. In contrast, the $\mathcal{G}_{Lecture}^{B}$ Pareto inferior cell (BR) positioned over the Nash cell ensures a me-we tension.

3.3.2 Zero-Sum Games

Zero-sum games are so often used that it is reasonable to determine their decomposition.

Theorem 3.3 *If $\mathcal{G}$ is a 2×2 zero-sum game, then $\kappa_1 = -\kappa_2$. The sum of its $\mathcal{G}^N$ entries in TL is the negative of the sum of its $\mathcal{G}^N$ entries in BR, or*

$$
\eta_{1,1} + \eta_{1,2} = \eta_{2,1} + \eta_{2,2}, \text{ which in terms of each agent is } \eta_{1,1} - \eta_{2,1} = \eta_{2,2} - \eta_{1,2}.
\tag{3.6}
$$

Conversely, if Eq. 3.6 is satisfied, the zero-sum games in its Nash equivalence class are uniquely defined by $\mathcal{G}^K(\kappa, -\kappa)$ for any κ value, and

$$\beta_1 = -\frac{1}{2}[\eta_{1,2} + \eta_{2,2}], \quad \beta_2 = -\frac{1}{2}[\eta_{1,1} + \eta_{2,1}]. \tag{3.7}$$

While a proof based on vector analysis is in Sect. 3.4.3, a different proof (see Theorem 4.2) is to solve the four equations defined by the four cells. For instance, by being zero-sum, the TL entries satisfy $[\eta_{1,1} + \beta_1 + \kappa_1] = -[\eta_{1,2} + \beta_2 + \kappa_2]$.

Notice the peculiarity where each player's β value is uniquely determined by the *other* player's η values (Eq. 3.7). A way to appreciate this equality is with a story where both players want to have a zero-sum game while retaining their personal Nash $\eta_{i,j}$ values. To realize this scenario, the players must exhibit cooperation and coordination, which is a $\mathcal{G}^B$ trait. More precisely, each player selects an appropriate β value that *depends* on the opponent's η values: This is what Eq. 3.7 specifies.

Zero-sum games differ from symmetric games in that the $\mathcal{G}^N$ component of a zero-sum game typically is *not* zero-sum. (An example is constructed below in Eq. 3.8.) Indeed, a necessary and sufficient condition for $\mathcal{G}^N$ to be zero-sum is that $\eta_{1,1} = \eta_{2,2} = -\eta_{2,1} = -\eta_{1,2}$, which, according to Eq. 3.7, requires $\mathcal{G}^B$ to be identically zero. Consequently,

Corollary 3.1 *Generically, the $\mathcal{G}^N$ component of a 2×2 zero-sum game $\mathcal{G}$ is zero-sum if and only if $\mathcal{G}$ is a "matching pennies" game Eq. 1.15.*

The structure of a zero-sum game, where one player is the winner and the other the loser, might incorrectly suggest that the game does not have a Nash cell. What changes the answer is that $\mathcal{G}^N$ need not be zero-sum, which permits a Nash cell.

Corollary 3.2 *Generically, a 2×2 zero-sum game has zero or one Nash cell.*

Proof A zero-sum example without Nash cells is the matching pennies game. To create an example with one Nash cell located at, say TL, select positive values for $\eta_{1,1}$ and $\eta_{1,2}$. Whatever the choice, the remaining values, $\eta_{2,1}$ and $\eta_{2,2}$, are free to be selected as long as they satisfy Eq. 3.6, which always is possible. (So, any cell can be the Nash cell.)

To prove there cannot be two Nash cells, suppose the contrary: If TL and BR are Nash cells, then the sum of entries in each of these two $\mathcal{G}^N$ cells would be positive, which contradicts Eq. 3.6. Similarly, if BL and TR were Nash cells, then $\eta_{1,1}, \eta_{1,2}$ would be negative and $\eta_{2,1}$ and $\eta_{2,,2}$ would be positive, which would force the two sides of Eq. 3.6 to have opposite signs. This, again, leads to a contradiction. $\qquad\square$

To create examples, arbitrarily choose values for three of the $\eta_{i,j}$ values, say $\eta_{1,1} = 10$, $\eta_{2,1} = 8$, and $\eta_{2,2} = 3$, and use Eq. 3.6 to find the remaining value; here $\eta_{1,2} = 1$. The β_j values (which depend on the other player's $\eta_{i,j}$ values) come from Eq. 3.7; here $\beta_1 = -\frac{1}{2}(\eta_{1,2} + \eta_{2,2}) = -2$ and $\beta_2 = -\frac{1}{2}[10 + 8] = -9$. Thus

$$\begin{array}{|c c|c c|}\hline 10 & 1 & 8 & -1 \\ \hline -10 & 3 & -8 & -3 \\ \hline \end{array} + \begin{array}{|c c|c c|}\hline -2 & -9 & 2 & -9 \\ \hline -2 & 9 & 2 & 9 \\ \hline \end{array} = \begin{array}{|c c|c c|}\hline 8 & -8 & 10 & -10 \\ \hline -12 & 12 & -6 & 6 \\ \hline \end{array}, \tag{3.8}$$

which has a Nash cell at TL. Adding any κ value to Row's entries and $-\kappa$ to Column's entries defines a class of zero-sum games that differ only by the kernel component.

Another use of Theorem 3.3 is to examine whether zero-sum games exhibit me-we tensions. As the first line of the following suggests, this is true.

Corollary 3.3 *A Nash cell in a zero-sum game never is the $\mathcal{G}^B$ Pareto superior cell.*

The decomposition of a zero-sum game is determined by selecting values for any three of the six $\eta_{i,j}$, β_j variables. Should the choices be $\beta_1, \beta_2, \eta_{1,1}$, the game is

$$
\begin{array}{|c c|c c|}
\hline
\eta_{1,1} & -[\eta_{1,1} + \beta_1 + \beta_2] & -2\beta_2 - \eta_{1,1} & \eta_{1,1} + \beta_1 + \beta_2 \\
-\eta_{1,1} & -\beta_1 + \eta_{1,1} + \beta_2 & 2\beta_2 + \eta_{1,1} & \beta_1 - \eta_{1,1} - \beta_2] \\
\hline
\end{array}
+
\begin{array}{|c c|c c|}
\hline
\beta_1 & \beta_2 & -\beta_1 & \beta_2 \\
\beta_1 & -\beta_2 & -\beta_1 & -\beta_2 \\
\hline
\end{array}
$$

$$(3.9)$$

Proof To prove the first statement, select a cell and suppose it is both a Nash and $\mathcal{G}^B$ Pareto superior cell. To be Nash, or Pareto superior, both entries of the cell of the respective $\mathcal{G}^N$ and $\mathcal{G}^B$ components must be positive. The kernel values of this cell cancel, $\kappa_1 + \kappa_2 = 0$, so the sum of this cell's $\mathcal{G}^N$ and $\mathcal{G}^B$ components are positive, which disqualifies $\mathcal{G}$ as zero-sum.

The space of zero-sum games is four dimensional; one variable is $\kappa_1 = -\kappa_2$. Selecting any three of the six remaining $\eta_{i,j}$, β_j variables allows the last three to be expressed (with Eqs. 3.6, 3.7) in terms of the chosen three. Equation 3.9 was selected to explore how group interests (β_j values) influence the game. □

A way to illustrate Eq. 3.9 is to create a zero-sum game with a slight Prisoner's Dilemma flavor in that a repelling TL is the $\mathcal{G}^B$ Pareto superior cell (so $\beta_1, \beta_2 > 0$), and BR is a Nash cell. To be a repeller, TL's $\mathcal{G}^N$ entries are negative, so $\eta_{1,1} < 0$ and $\beta_1 + \beta_2 > |\eta_{1,1}|$. For BR to be Nash, both $\mathcal{G}^N$ entries are positive, so $2\beta_2 > |\eta_{1,1}|$ and $\beta_1 + |\eta_{1,1}| > \beta_2$. Thus the TL cell is $(\beta_1 - |\eta_{1,1}|, |\eta_{1,1}| - \beta_1)$ and BR is $(2\beta_2 - [\beta_1 + |\eta_{1,1}|], -[2\beta_2 - [\beta_1 + |\eta_{1,1}|]])$. An example with $\eta_{1,1} = -1$, $\beta_1 = \beta_2 = 2$ is (Eq. 3.9)

$$
\begin{array}{|c c|c c|}
\hline
1 & -1 & -5 & 5 \\
3 & -3 & 1 & -1 \\
\hline
\end{array}
$$

As the TL and BR entries agree, one might wonder whether these values can be slightly nudged to create a true Prisoner's Dilemma where both players prefer TL over BR. Yes (add 0.1 to both TL entries), but not with preserving the zero-sum status. This statement follows from the derived relationships. (For a trivial argument, let the TL and BR entries be, respectively, $(A, -A)$ and $(D, -D)$ where $A > 0$. For Column to prefer TL, it must be that $-A > -D$, or $D > A$, so Row prefers BR over TL.)

3.3.3 Identical Play Games

An identical play's cell entries agree while those of a zero-sum differ by sign, which makes it reasonable to anticipate that certain zero-sum results transfer to identical play games with only a sign change. This is the case.

Theorem 3.4 *If $\mathcal{G}$ is a 2×2 identical play game, then differences of entries in the TL and BR cells agree with*

$$\eta_{1,1} - \eta_{1,2} = \eta_{2,1} - \eta_{2,2}, \text{ which in terms of each agent is } \eta_{1,1} - \eta_{2,1} = \eta_{1,2} - \eta_{2,2}. \tag{3.10}$$

If Eq. 3.10 is satisfied, the identical play games in its Nash equivalence class are uniquely defined by $\mathcal{G}^K(\kappa, \kappa)$ for any κ value, and

$$\beta_1 = \frac{1}{2}[\eta_{1,2} + \eta_{2,2}], \quad \beta_2 = \frac{1}{2}[\eta_{1,1} + \eta_{2,1}]. \tag{3.11}$$

Again, each player's β value is uniquely determined by the other player's η values. This makes sense; suppose for some reason (e.g., to coordinate on a joint project) the players want to "be on the same page" by having an identical play game. To do so without sacrificing individuality represented by the player's η values, the coordination must come from β values that reflect the other player's characteristics.

Theorem 3.4 is proved in Sect. 3.4.3. As with zero-sum games, an identical play's $\mathcal{G}^N$ does not, in general, have an identical play format. Indeed, if $\mathcal{G}^N$ is identical play, then the $\mathcal{G}^N$ structure[1] requires $\eta_{1,1} = \eta_{1,2} = -\eta_{2,1} = -\eta_{2,2}$ with (Eq. 3.11) $\beta_1 = \beta_2 = 0$. Thus, an identical play game has an identical play $\mathcal{G}^N$ component if and only if for some η value

$$\mathcal{G} = \begin{array}{|cc|cc|} \hline \eta & \eta & -\eta & -\eta \\ \hline -\eta & -\eta & \eta & \eta \\ \hline \end{array},$$

which is a coordination game.

The following result parallels Corollary 3.2.

Corollary 3.4 *Generically, a 2×2 identical play game has one or two Nash cells.*

This assertion is easy to prove because the $\mathcal{G}$ cell with the largest common entry is a Nash cell. If the diametrically opposite cell has the second largest (or equal) entry, it is the second Nash cell. Otherwise (according to the decomposition), there is only one Nash cell.

As true with zero-sum games, any number of identical play games can be constructed (Theorem 3.4) with a desired Nash cell structure. An interesting choice is to create one that can be compared with Exercise 3.8 zero-sum game, so use the same

---

[1] If $\mathcal{G}^N$ is identical play, then from the TL cell, $\eta_{1,1} = \eta_{1,2}$. The BL cell requires $\eta_{2,2} = -\eta_{1,1}$, BR requires $\eta_{2,1} = \eta_{2,2}$, and TR requires $\eta_{2,1} = -\eta_{1,2}$ to complete the proof.

$\eta_{1,1} = 10$, $\eta_{2,1} = 8$, and $\eta_{2,2} = 3$ values. According to Eq. 3.10, the missing values are $\eta_{1,2} = 5$ with $\beta_1 = 4$ and $\beta_2 = 9$. The resulting identical play game is

$$\begin{array}{|cc|cc|}\hline 14 & 14 & 4 & 4 \\ \hline -6 & -6 & -12 & -12 \\ \hline \end{array},$$

which, by sharing the same ancestral roots (the $\eta_{1,1}$, $\eta_{2,1}$, $\eta_{2,2}$ values), is an unrecognizable, but not overly distant cousin of Eq. 3.8 zero-sum game.

Another example is to create an identical play game where TL is the only Nash cell. Select, for instance, $\mathcal{G}^N$ entries of, say, $\eta_{1,1} = 1$, $\eta_{1,2} = 2$; the positive values ensure that TL is a Nash cell. Select $\eta_{2,1} = 3 > 0$ to ensure that BR has a negative entry to disqualify BR as a Nash cell. The missing $\eta_{2,2}$ value comes from Eq. 3.10, which is $1 - 2 = 3 - \eta_{2,2}$, or $\eta_{2,2} = 4$. According to Eq. 3.11, $\beta_1 = \frac{1}{2}[\eta_{1,2} + \eta_{2,2}] = 3$, and $\beta_2 = \frac{1}{2}[\eta_{1,1} + \eta_{2,1}] = 2$, which defines the identical play game

$$\begin{array}{|cc|cc|}\hline 1 & 2 & 3 & -2 \\ \hline -1 & 4 & -3 & -4 \\ \hline \end{array} + \begin{array}{|cc|cc|}\hline 3 & 2 & -3 & 2 \\ \hline 3 & -2 & -3 & -2 \\ \hline \end{array} = \begin{array}{|cc|cc|}\hline 4 & 4 & 0 & 0 \\ \hline 2 & 2 & -6 & -6 \\ \hline \end{array}$$

This construction motivates an interesting thought. All games can be expressed as a sum of an identical and a zero-sum game (e.g., Eq. 3.5). Could it be that *any* selected game has the Nash structure of *either* a particular zero-sum or an identical play game?

In a limited sense, this conjecture has been established for NME in 2×2 games. Monderer and Shapley [16] commented that

> ...every nondegenerate 2-person 2×2 game is best response equivalent in mixed strategies either to [an identical play] or to a zero-sum game.

The "best response equivalent in mixed strategies" statement means that the mixed strategies for the two games have identical p and q values.

This quote raises several questions. For instance:

- *Which* games are best response equivalent to zero-sum games, and *which* games are best response equivalent to identical play games?
- Can the result be extended beyond NME to reflect more general structures such as the repelling, hyperbolic-column, and hyperbolic-row cells?

Positive answers are developed in Sect. 3.6 after developing a roadmap of all 2×2 games.

3.4 Using Vectors to Analyze and Design Games

Pleasant advantages arise by identifying games with vectors: New issues and simpler proofs follow from elementary vector analysis. To do so, identify game

$$\mathbf{g} = \begin{array}{|cc|cc|} x_1 & y_1 & x_2 & y_2 \\ x_3 & y_3 & x_4 & y_4 \end{array} \text{ with the vector } \mathbf{g} = (x_1, x_2, x_3, x_4; y_1, y_2, y_3, y_4) \in \mathbb{R}^8.$$

$$(3.12)$$

3.4.1 Review of Vector Analysis

Representing games with vectors simplifies the analysis while introducing novel questions. It now makes sense, for instance, to determine the *angle* between two games, or how much of one game is going in another game's direction.

To review, for vectors $\mathbf{a} = (a_1, a_2, \ldots, a_n)$, $\mathbf{b} = (b_1, b_2, \ldots, b_n) \in \mathbb{R}^n$, the inner product (or dot product) is

$$[\mathbf{a}, \mathbf{b}] = \sum_{j=1}^{n} a_j b_j.$$

An example is $[(3, -1, 2), (-2, 5, 1)] = (3)(-2) + (-1)(5) + (2)(1) = -9$.

As an immediate application, recall from the right-triangle $a^2 + b^2 = c^2$ equation that the length of vector $\mathbf{c} = (a, b)$ is $||\mathbf{c}|| = \sqrt{a^2 + b^2}$. The same expression holds for higher dimensions; as $[\mathbf{d}, \mathbf{d}] = \sum_{j=1}^{n} d_j^2$, the length of vector $\mathbf{d}$ is $||\mathbf{d}|| = \sqrt{[\mathbf{d}, \mathbf{d}]}$.

Vector applications include finding the angle between two vectors $\mathbf{c}$ and $\mathbf{d}$ (Fig. 3.1a) and the amount of $\mathbf{c}$ in the $\mathbf{d}$ direction (Fig. 3.1b). (As a direction has no length, the direction defined by $\mathbf{d}$ is represented by the unit vector $\frac{1}{||\mathbf{d}||}\mathbf{d}$.) The cosine of the angle θ between vectors $\mathbf{c}$ and $\mathbf{d}$ is

$$\cos(\theta) = \frac{[\mathbf{c}, \mathbf{d}]}{||\mathbf{c}|| \, ||\mathbf{d}||}, \tag{3.13}$$

which means that two vectors are orthogonal if $\theta = 90°$, or if $[\mathbf{c}, \mathbf{d}] = 0$.

Similarly, if vector $\mathbf{c}$ is projected onto the line defined by $\mathbf{d}$, it can be shown from Eq. 3.13 and some trigonometry that the amount of $\mathbf{c}$ in the $\mathbf{d}$ direction is

$$\text{Amount of } \mathbf{c} \text{ in direction } \mathbf{d} = [\mathbf{c}, \frac{\mathbf{d}}{||\mathbf{d}||}]\frac{\mathbf{d}}{||\mathbf{d}||} = \frac{[\mathbf{c}, \mathbf{d}]}{||\mathbf{d}||^2}\mathbf{d}. \tag{3.14}$$

Fig. 3.1 Vector analysis

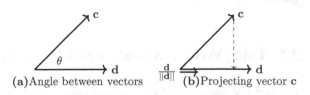

(a) Angle between vectors   (b) Projecting vector $\mathbf{c}$

To illustrate with the games

$$\mathbf{g}_1 = \begin{array}{|cc|cc|} \hline 4 & -8 & 12 & 4 \\ \hline 12 & 4 & -4 & 8 \\ \hline \end{array}, \quad \mathbf{g}_2 = \begin{array}{|cc|cc|} \hline 10 & 0 & 7 & -1 \\ \hline 4 & 8 & -8 & -1 \\ \hline \end{array}, \tag{3.15}$$

the amount of game $\mathbf{g}_2$ in game $\mathbf{g}_1$'s direction is $\frac{[\mathbf{g}_1,\mathbf{g}_2]}{[\mathbf{g}_1,\mathbf{g}_1]}\mathbf{g}_1$ where $[\mathbf{g}_1, \mathbf{g}_2] = (4)(10) + (-8)(0) + (12)(7) + (4)(-1) + (12)(4) + (4)(8) + (-4)(-8) + (8)(-1) = 320$, while $\|\mathbf{g}_1\|^2 = (4)^2 + (-8)^2 + (12)^2 + (4)^2 + (12)^2 + (4)^2 + (-4)^2 + (-8)^2 = 640$. Therefore, the amount of game $\mathbf{g}_2$ in $\mathbf{g}_1$'s direction is $\frac{320}{640}\mathbf{g}_1$, or $\frac{1}{2}\mathbf{g}_1$.

This expression introduces a novel way to comparing $\mathbf{g}_2$ with $\mathbf{g}_1$: Write $\mathbf{g}_2 = \mathbf{g}_3 + \frac{1}{2}\mathbf{g}_1$, where $\mathbf{g}_1$ and the unknown $\mathbf{g}_3$ are orthogonal (so they have no common game theoretic interactions). Solving for the unknown $\mathbf{g}_3$ yields

$$\mathbf{g}_3 = \mathbf{g}_2 - \frac{1}{2}\mathbf{g}_1 = \begin{array}{|cc|cc|} \hline 8 & 4 & 1 & -3 \\ \hline -2 & 6 & -6 & -5 \\ \hline \end{array}.$$

All ways game $\mathbf{g}_2$ can differ from game $\mathbf{g}_1$ are strictly caused by $\mathbf{g}_3$'s features. For instance, game $\mathbf{g}_1$ has a Nash cell at TR while $\mathbf{g}_2$ has it at TL; it is the TL position of $\mathbf{g}_3$'s Nash cell that forces the difference.

3.4.2 Intuition and a Vector Basis

Vector analysis can extract a variety of game theory topics while sharpening intuition about what else can be done. To illustrate, consider

$$\mathcal{G}_{intuition} = \begin{array}{|cc|cc|} \hline A_1 & A_2 & B_1 & B_2 \\ \hline C_1 & C_2 & D_1 & D_2 \\ \hline \end{array}. \tag{3.16}$$

If Column selects L and R with the probabilities q_1 and q_2 (so $q_1 + q_2 = 1$), then

$$EV(T) - EV(B) = (A_1 - C_1)q_1 + (B_1 - D_1)q_2.$$

It is clear how to represent the $A_1 - C_1$ coefficient for q_1 with a dot product; choose a vector that, for Row, has a 1 in the TL cell and -1 in the BL, which is

$$\mathbf{n}_{1,1} = \begin{array}{|cc|cc|} \hline 1 & 0 & 0 & 0 \\ \hline -1 & 0 & 0 & 0 \\ \hline \end{array}.$$

Here, $[\mathcal{G}_{intuition}, \mathbf{n}_{1,1}] = A_1 - C_1$. Similarly, to represent the $B_1 - D_1$ coefficient, use

$$\mathbf{n}_{2,1} = \begin{array}{|cc|cc|} \hline 0 & 0 & 1 & 0 \\ \hline 0 & 0 & -1 & 0 \\ \hline \end{array}.$$

Row has four payoff entries, so Row's portion of a game resides in a four-dimensional space: Two directions are given by $\mathbf{n}_{1,1}$ and $\mathbf{n}_{2,1}$, so two more basis vectors must be found. To be orthogonal to $\mathbf{n}_{1,1}$ and $\mathbf{n}_{2,1}$, two of Row's entries (for the new vectors) for either column must have the same value. This leads to the choices

$$\mathbf{b}_1 = \begin{array}{|c c|c c|} \hline 1 & 0 & -1 & 0 \\ \hline 1 & 0 & -1 & 0 \\ \hline \end{array}, \quad \mathbf{k}_1 = \begin{array}{|c c|c c|} \hline 1 & 0 & 1 & 0 \\ \hline 1 & 0 & 1 & 0 \\ \hline \end{array}$$

By being orthogonal to one another, these four vectors span Row's portion of the space of games. Consequently, Row's payoffs can be uniquely represented by these vectors as

$$\begin{array}{|c c|c c|} \hline A_1 & 0 & B_1 & 0 \\ \hline C_1 & 0 & D_1 & 0 \\ \hline \end{array} = c_1\mathbf{n}_{1,1} + c_2\mathbf{n}_{2,1} + c_3\mathbf{b}_1 + c_4\mathbf{k}_1.$$

The c_j coefficients describe the portion of $\mathcal{G}_{intuition}$ that lies in the respective direction. These c_j values can be computed with Eq. 3.14 where $\|\mathbf{n}_{1,1}\| = \|\mathbf{n}_{2,1}\| = 2$ and $\|\mathbf{b}_1\| = \|\mathbf{k}_1\| = 4$. Thus $c_1 = \frac{A_1-C_1}{2}$, $c_2 = \frac{B_1-D_1}{2}$, $c_3 = \frac{[A_1+C_1]-[B_1+D_1]}{4}$, and $c_4 = \frac{[A_1+C_1]+[B_1+D_1]}{4}$.

Items of importance:

1. Notice that $c_1 = \eta_{1,1}$, $c_2 = \eta_{2,1}$, $c_3 = \beta_1$,[2] and $c_4 = k_1$.
2. The bimatrix form of vector $\eta_{1,1}\mathbf{n}_1 + \eta_{2,1}\mathbf{n}_2$ is precisely Row's portion of $\mathcal{G}^N$. Row's portion of $\mathcal{G}^B$ and $\mathcal{G}^K$ is, respectively, $\beta_1\mathbf{b}_1$ and $\kappa_1\mathbf{k}_1$. As such, *a standard vector analysis captures the decomposition for Row!* As the same approach holds for Column, Chap. 2 decomposition can be obtained by using standard vector analysis. This methodology plays a major role in Sect. 5.4 when analyzing extensive form games.
3. The $A_1 - C_1$ and the $B_1 - D_1$ algebraic expressions are zero for $\beta_1\mathbf{b}_1$ and $k_1\mathbf{k}_1$. Thus information about the game in these directions is redundant for the Nash analysis.
4. All of Row's non-Nash information about a game is given by $\beta_1\mathbf{b}_1$ and $k_1\mathbf{k}_1$.

As a similar description holds for Column, the Nash basis vectors are

$$\mathbf{n}_{1,1} = \begin{array}{|c c|c c|} \hline 1 & 0 & 0 & 0 \\ \hline -1 & 0 & 0 & 0 \\ \hline \end{array}, \quad \mathbf{n}_{2,1} = \begin{array}{|c c|c c|} \hline 0 & 0 & 1 & 0 \\ \hline 0 & 0 & -1 & 0 \\ \hline \end{array},$$

$$\mathbf{n}_{1,2} = \begin{array}{|c c|c c|} \hline 0 & 1 & 0 & -1 \\ \hline 0 & 0 & 0 & 0 \\ \hline \end{array}, \quad \mathbf{n}_{2,2} = \begin{array}{|c c|c c|} \hline 0 & 0 & 0 & 0 \\ \hline 0 & 1 & 0 & -1 \\ \hline \end{array},$$

$$(3.17)$$

which span the four-dimensional Nash vector space $\mathbb{N}$. Again, Eq. 3.14 describes how to find the amount of game $\mathbf{g}_1$ that is in the $\mathbf{n}_{1,1}$ direction. Because $\|\mathbf{n}_{1,1}\|^2 = [\mathbf{n}_{1,1}, \mathbf{n}_{1,1}] = 2$, it follows (Eq. 3.14) that the amount of game $\mathbf{g}$ in the $\mathbf{n}_{i,j}$ direction is the $\mathcal{G}^N$ term

---

[2]To see that $c_3 = \beta_1$, the average of Row's payoffs in the first column is $\frac{A_1+C_1}{2}$. Once the average of all of Row's payoffs are subtracted from this value, we have $\frac{A_1+C_1}{2} - c_4 = c_3$.

$$\eta_{i,j} = \frac{1}{||\mathbf{n}_{i,j}||^2}[\mathbf{g}, \mathbf{n}_{i,j}] = \frac{1}{2}[\mathbf{g}, \mathbf{n}_{i,j}]; \quad i, j = 1, 2. \tag{3.18}$$

Similarly, vectors

$$\mathbf{b}_1 = \boxed{\begin{array}{cc|cc} 1 & 0 & -1 & 0 \\ 1 & 0 & -1 & 0 \end{array}}, \quad \mathbf{b}_2 = \boxed{\begin{array}{cc|cc} 0 & 1 & 0 & 1 \\ 0 & -1 & 0 & -1 \end{array}} \tag{3.19}$$

span the behavioral vector space $\mathbb{B}$. The amount of a game $\mathbf{g}$ in a $\mathbf{b}_j$ direction is $\mathcal{G}^B$'s β_j value

$$\beta_j = \frac{1}{||\mathbf{b}_j||^2}[\mathbf{g}, \mathbf{b}_j] = \frac{1}{4}[\mathbf{g}, \mathbf{b}_j], \quad j = 1, 2. \tag{3.20}$$

Illustrating with game $\mathbf{g}_1$, it follows that $\beta_1 = \frac{(4)(1)+(-1)(12)+(12)(1)+(-4)(-1)}{4} = 2.$ Finally, the kernel space $\mathbb{K}$ is spanned by the two vectors

$$\mathbf{k}_1 = \boxed{\begin{array}{cc|cc} 1 & 0 & 1 & 0 \\ 1 & 0 & 1 & 0 \end{array}}, \quad \mathbf{k}_2 = \boxed{\begin{array}{cc|cc} 0 & 1 & 0 & 1 \\ 0 & 1 & 0 & 1 \end{array}}. \tag{3.21}$$

With $[\mathbf{k}_j, \mathbf{k}_j] = ||\mathbf{k}_j||^2 = 4$, the amount $\mathbf{g}$ in the $\mathbf{k}_j$ directions (Eq. 3.14) are the $\mathcal{G}^K$ values

$$\kappa_j = \frac{1}{4}[\mathbf{g}, \mathbf{k}_j], \quad j = 1, 2. \tag{3.22}$$

For $\mathbf{g}_1$, $\kappa_1 = \frac{1}{4}[\mathbf{g}_1, \mathbf{k}_1] = \frac{1}{4}((4)(1) + (12)(1) + (12)(1) + (-4)(1)) = 6$, which is Row's average payoff.

Adding the amount of $\mathbf{g}_1$ that is in each of the eight directions leads to

$$\mathbf{g}_1 = [\ 4\mathbf{n}_{1,1} + 8\mathbf{n}_{2,1} - 6\mathbf{n}_{1,2} - 2\mathbf{n}_{2,2}] + [2\mathbf{b}_1 - 2\mathbf{b}_2] + [6\mathbf{k}_1 + 2\mathbf{k}_2],$$

which is the vector version of $\mathcal{G} = \mathcal{G}^N + \mathcal{G}^B + \mathcal{G}^K$. The basis for $\mathbb{N} + \mathbb{B}$ is given by the four $\mathbf{n}_{i,j}$ and two $\mathbf{b}_j$ vectors, with similar comments for other subspaces.

3.4.3 Designing Games

A convenient way to handle a specified class of games is to modify the basis to make it more appropriate for analysis. This is illustrated with symmetric games.

According to Eq. 3.1, a basis for this space is given by

$$\mathbf{s}_A = \boxed{\begin{array}{cc|cc} 1 & 1 & 0 & 0 \\ 0 & 0 & 0 & 0 \end{array}}, \quad \mathbf{s}_B = \boxed{\begin{array}{cc|cc} 0 & 0 & 1 & 0 \\ 0 & 1 & 0 & 0 \end{array}}, \quad \mathbf{s}_C = \boxed{\begin{array}{cc|cc} 0 & 0 & 0 & 1 \\ 1 & 0 & 0 & 0 \end{array}}, \quad \mathbf{s}_D = \boxed{\begin{array}{cc|cc} 0 & 0 & 0 & 0 \\ 0 & 0 & 1 & 1 \end{array}}, \tag{3.23}$$

which spans a four-dimensional subspace of $\mathbb{G}$ (the eight-dimensional space of 2×2 games). Indeed, Eq. 3.1 becomes $A\mathbf{s}_A + B\mathbf{s}_B + C\mathbf{s}_C + D\mathbf{s}_D$.

A four-dimensional subspace in an eight-dimensional space has four linearly independent vectors that are orthogonal to this subspace. An easy way to find these normal vectors is, with each $\mathbf{s}_x$, replace Column's entry of 1 with -1 to define $\mathbf{os}_x$. The next theorem, which completes the proof of Theorem 3.1, demonstrates how orthogonal vectors can be used to characterize properties of studied classes of games.

Theorem 3.5 *Game* **g** *is symmetric if and only if it is orthogonal to the four* $\mathbf{os}_x$ *vectors; that is* $[\mathbf{g}, \mathbf{os}_x] = 0$, $x = A, B, C, D$. *These conditions reduce to* $\eta_{1,1} = \eta_{1,2}$, $\eta_{2,1} = \eta_{2,2}$, $\beta_1 = \beta_2$, $\kappa_1 = \kappa_2$. *Thus, if* $\mathcal{G}$ *is symmetric, then* $\mathcal{G}^N$, $\mathcal{G}^B$, $\mathcal{G}^K$ *are symmetric games.*[3]

Another class of games, the *antisymmetric* games, emerges from the orthogonal vectors

$$
\mathbf{os}_A = \begin{array}{|cc|cc|} 1 & -1 & 0 & 0 \\ 0 & 0 & 0 & 0 \end{array}, \ \mathbf{os}_B = \begin{array}{|cc|cc|} 0 & 0 & 1 & 0 \\ 0 & -1 & 0 & 0 \end{array}, \ \mathbf{os}_C = \begin{array}{|cc|cc|} 0 & 0 & 0 & -1 \\ 1 & 0 & 0 & 0 \end{array}, \ \mathbf{os}_D = \begin{array}{|cc|cc|} 0 & 0 & 0 & 0 \\ 0 & 0 & 1 & -1 \end{array}.
$$
$$(3.24)$$

In a symmetric game, it does not matter if a player is Row or Column. In an antisymmetric game, a player changing from Row to Column faces the opposite payoffs as in

$$
\mathcal{G}_{antisymmetric} = \begin{array}{|cc|cc|} A & -A & B & -C \\ C & -B & D & -D \end{array}.
$$
$$(3.25)$$

Vectors $\{\mathbf{s}_x, \mathbf{os}_x\}_{x=A,B,C,D}$ are orthogonal and span $\mathbb{G}$, so we have the following.

Theorem 3.6 *A game is antisymmetric if* $[\mathbf{g}, \mathbf{s}_x] = 0$, $x = A, B, C, D$. *This reduces to* $(\eta_{1,1}, \eta_{2,1}) = -(\eta_{1,2}, \eta_{2,2})$, $\beta_1 = -\beta_2$, $\kappa_1 = -\kappa_2$.

Any 2×2 *game can be uniquely expressed as the sum of a symmetric and an antisymmetric game.*

Proof An orthogonal basis for $\mathbb{G}$, the space of 2×2 games, is given by the four Eq. 3.23 and four Eq. 3.24 games (vectors). Thus, any game can be uniquely represented as a linear sum of these basis vectors. Components using Eq. 3.23 vectors define a symmetric game, those using Eq. 3.24 define an antisymmetric game. This proves the theorem's last statement.

For the first statement, an antisymmetric game is uniquely expressed as a sum of Eq. 3.24 vectors. Each of these vectors is orthogonal to each of Eq. 3.23 vectors, so the first assertion follows. And so properties of a class of games can be derived by using the orthogonal space of games.   □

---

[3]The conclusion also follows because the symmetric $\mathbf{k}_1 + \mathbf{k}_2$, $\mathbf{b}_1 + \mathbf{b}_2$, $\mathbf{n}_{1,1} + \mathbf{n}_{1,2}$, and $\mathbf{n}_{2,1} + \mathbf{n}_{2,2}$ form a basis for the space of symmetric games.

Expressing a game as the sum of a symmetric and antisymmetric game resembles dividing a game $\mathcal{G}$ into a sum of zero-sum and identical play games. For the TL and BR cells, the averages of the $\mathcal{G}$ cell's entry are the respective cell's entries for the symmetric game; differences from averages are the antisymmetric cell's entries. The average of Row's TR and Column's BL entries define the players' common value for these symmetric game cells' entries while differences from averages give the antisymmetric game's entries. The same holds for Row's BL and Column's TR entries. All of this extends to any $k \times k$ game, or any game with n players where each has k strategies.

3.4.4 Utility Functions

As a shift in emphasis, certain games (described below) are simplified by expressing a player's $\mathcal{G}$ payoffs with a utility function. To identify these games (following this chapter's theme), create an appropriate coordinate system.

Toward this end, notice that each of

$$
\mathbf{a}_1 = \begin{array}{|cc|cc|} \hline 1 & 0 & 1 & 0 \\ \hline -1 & 0 & 1 & 0 \\ \hline \end{array}, \quad
\mathbf{a}_2 = \begin{array}{|cc|cc|} \hline 0 & 1 & 0 & -1 \\ \hline 0 & 1 & 0 & -1 \\ \hline \end{array} \tag{3.26}
$$

is in $\mathbb{N}$; e.g., $\mathbf{a}_1 = \mathbf{n}_{1,1} + \mathbf{n}_{2,1}$ and $\mathbf{a}_2 = \mathbf{n}_{1,2} + \mathbf{n}_{2,2}$. Here, player i's difference in $\mathbf{a}_i$ payoffs is the same no matter what the other agent selects. For instance, independent of whether Column selects R or L, the $\mathbf{a}_1$ difference of Row's payoff in T versus B is the same.

After selecting $\mathbf{a}_1$ and $\mathbf{a}_2$, two more $\mathbb{N}$ vectors (linearly independent to each other and to $\mathbf{a}_1$ and $\mathbf{a}_2$) create a different $\mathbb{N}$ basis. One choice inverts the R column entries of $\mathbf{a}_1$ and the B row entries of $\mathbf{a}_2$ to obtain

$$
\mathbf{d}_1 = \begin{array}{|cc|cc|} \hline 1 & 0 & -1 & 0 \\ \hline -1 & 0 & 1 & 0 \\ \hline \end{array}, \quad
\mathbf{d}_2 = \begin{array}{|cc|cc|} \hline 0 & 1 & 0 & -1 \\ \hline 0 & -1 & 0 & 1 \\ \hline \end{array} \tag{3.27}
$$

where $\mathbf{d}_1 = \mathbf{n}_{1,1} - \mathbf{n}_{2,1}$ and $\mathbf{d}_2 = \mathbf{n}_{1,2} - \mathbf{n}_{2,2}$ generating an alternative $\mathbb{N}$ basis of

$$
\{\mathbf{a}_1, \mathbf{a}_2, \mathbf{d}_1, \mathbf{d}_2.\} \tag{3.28}
$$

According to Eq. 3.28, rather than expressing a vector in $\mathbb{N}$ with the $\mathcal{G}^N$ form

$$
\eta_{1,1}\mathbf{n}_{1,1} + \eta_{2,1}\mathbf{n}_{2,1} + \eta_{1,2}\mathbf{n}_{1,2} + \eta_{2,2}\mathbf{n}_{2,1} = \begin{array}{|cc|cc|} \hline \eta_{1,1} & \eta_{1,2} & \eta_{2,1} & -\eta_{1,2} \\ \hline -\eta_{1,1} & \eta_{2,2} & -\eta_{2,1} & -\eta_{2,2} \\ \hline \end{array}, \tag{3.29}
$$

an alternative $\mathcal{G}^N$ representation is

$$a_1\mathbf{a}_1 + a_2\mathbf{a}_2 + d_1\mathbf{d}_1 + d_2\mathbf{d}_2 = \begin{array}{|cc|cc|} a_1 + d_1 & a_2 + d_2 & a_1 - d_1 & -a_2 - d_2 \\ -a_1 - d_1 & a_2 - d_2 & -a_1 + d_1 & -a_2 + d_2 \end{array}. \quad (3.30)$$

Interpreting Eq. 3.30 is immediate: The a_j values are the jth player's average $\mathcal{G}^N$ payoff for appropriate rows or columns, while the d_j capture deviates from these averages. Stated differently, the a_i terms identify what a player really wants to do; e.g., if $a_1 > 0$, Row prefers T independent of what Column does. Similarly, if $a_2 < 0$, Column prefers R independent of Row's choices. In contrast, the d_j variables identify costs or rewards associated with these a_j choices. Illustrating with $a_2 < 0$, Column prefers R. Here a $d_2 > 0$ value identifies a reward if Row selects B, but a cost if Row selects T. A concern is whether the d_2 costs, influenced by Row's actions, could cause Column to consider L rather than R. Thus a d_j value reflects how a player can influence the other player's individual Nash actions. Namely, these coordinates refine the Nash structure by identifying what a player really wants as modified by what the other player does.

A delight of Eq. 3.30 is how it allows $\mathcal{G}^N$ to be characterized by utility functions. First, let $0 \le s_j \le 1$, $j = 1, 2$, be the jth player's mixed strategy where $s_1 = 1$, $s_2 = 1$ represent, respectively, the pure strategies T and L. A convenient change of variables is

$$t_j = 2s_j - 1, \quad (3.31)$$

where $t_1 = 1$, $t_1 = -1$ represent, respectively, T and B, while $t_2 = 1$, $t_2 = -1$ are L and R. (The usual s_j and $1 - s_j$ values become t_j and $-t_j$.)

The two utility functions for the $\mathcal{G}^N$ portion of the game are

$$u_1^N(t_1, t_2) = a_1 t_1 + d_1 t_1 t_2, \quad u_2^N(t_1, t_2) = a_2 t_2 + d_2 t_1 t_2. \quad (3.32)$$

The full game has the utility functions

$$u_1(t_1, t_2) = \kappa_1 + t_2\beta_1 + u_1^N(t_1, t_2), \quad u_2(t_1, t_2) = \kappa_2 + t_1\beta_2 + u_2^N(t_1, t_2). \quad (3.33)$$

Theorem 3.7 *The $\mathcal{G}^N$ component of a 2×2 game can be described as in Eq. 3.30 along with Eq. 3.32 utility functions. The full game has Eq. 3.33 utility functions.*

Equation 3.32 coefficients follow from vector analysis. The a_j values, for instance, represent the amount of $\mathbf{g}$ in the $\mathbf{a}_j$ directions (Eq. 3.14). Because $[\mathbf{a}_j, \mathbf{a}_j] = ||\mathbf{a}_j||^2 = 4$, for game $\mathbf{g}_1$ (Eq. 3.15)

$$a_1 = \frac{[\mathbf{g}_1, \mathbf{a}_1]}{4} = \frac{24}{4} = 6, \quad a_2 = \frac{[\mathbf{g}_1, \mathbf{a}_2]}{4} = 2.$$

Similarly, $d_1 = -6$, $d_2 = -4$, $\kappa_1 = 6$, $\kappa_2 = 2$, $\beta_1 = 2$, $\beta_2 = -4$, leading to the utility functions $u_1(t_1, t_2) = 6 + 2t_2 + 6t_1 - 6t_1t_2$, $u_2(t_1, t_2) = 2 - 4t_1 + 2t_2 - 4t_1t_2$.

In this example, d_1 and d_2 have the same sign, which indicates that a coordination between the players causes their deviations from average. A way to ensure that a class of games respects this coordination is to introduce an appropriate basis vector. For instance, replace $\mathbf{d}_1, \mathbf{d}_2$ with $\mathbf{c}_1 = \mathbf{d}_1 + \mathbf{d}_2$ and $\mathbf{c}_2 = \mathbf{d}_1 - \mathbf{d}_2$.[4] Expressed in bimatrix form, this is

$$
\mathbf{c}_1 = \begin{array}{|c c|c c|}
\hline
1 & 1 & -1 & -1 \\
\hline
-1 & -1 & 1 & 1 \\
\hline
\end{array}, \quad
\mathbf{c}_2 = \begin{array}{|c c|c c|}
\hline
1 & -1 & -1 & 1 \\
\hline
-1 & 1 & 1 & -1 \\
\hline
\end{array} \tag{3.34}
$$

Of value for the following discussion of potential games is that $\mathbf{c}_1$ is a basis for coordination games and $\mathbf{c}_2$ is the basis for matching pennies games. For the moment, an analysis of the $\{\mathbf{a}_1, \mathbf{a}_1, \mathbf{c}_1, \mathbf{c}_2\}$ basis is left to the reader. Clearly, an infinite number of other bases—with different consequences—can be generated.

3.4.5 Congestion and Potential Games

Today, game theory is standard fare for many academic departments, but this was not true in the early 1970s. During this era at Northwestern University, the home for some game theorists, including Robert Rosenthal, was the School of Engineering.

Being surrounded by engineering colleagues may have motivated Rosenthal to develop the *congestion game* [21], which captures a phenomenon common to areas as diverse as engineering, traffic control, sociology, and economics. This is where too many people doing the same "good thing" can be bad—or, maybe too many make a "good thing" better. A standard example is the morning commute along either route $\mathcal{A}$ or $\mathcal{B}$. While the shorter route $\mathcal{A}$ attracts more drivers, the growing congestion on $\mathcal{A}$ can make travel times longer than along $\mathcal{B}$. This leads to Rosenthal's congestion game where each player's payoff for a strategic choice depends on how many other players adopt the same strategy. (See Eq. 3.38.) Other settings are where too many people frequenting the same restaurant can dampen its appeal, but more fans at the game cheering for your basketball team adds to the excitement.

With the congestion game (perhaps the commute to work), an action that improves a player's outcome makes it better for all. A driver frustrated with the route $\mathcal{A}$'s traffic jam can find relief along route $\mathcal{B}$: One less $\mathcal{A}$ driver benefits the other $\mathcal{A}$ drivers. With this insight, Rosenthal developed a solution approach by proving that a congestion game is a special case of what now is known as a potential game.

Many congestion games are monotone, where more is either always better or always worse. Here, the d_j coordination values have the same sign (Eq. 3.30); more is better requires $d_j > 0$. But it is easy to envision transition settings where more can be better up to a limit where more becomes worse. (Or the other way round.) Being alone at a restaurant can be lonely, and so more patrons would be welcomed—to a limit. Driving alone on route $\mathcal{A}$ may cause unease with worries of a flat tire where

---

[4] An alternative choice is $\mathbf{c}_1 = \mathbf{d}_1 + \alpha \mathbf{d}_2$ and $\mathbf{c}_2 = \mathbf{d}_1 - \frac{1}{\alpha} \mathbf{d}_2$ to capture differences.

help would be appreciated. Also a voting game where a majority for a preferred issue is desired, but a "no vote" may be personally profitable. (See Sect. 4.3.2.)

Monderer and Shapley [17] introduced the potential game using a title suggesting its resemblance to potential functions in physics. The potential function for the Newton N-body problem (N bodies, such as planets and stars, moving under the Newtonian force) is determined by the positioning of bodies relative to each other; a change in the positioning of any one body changes the potential's value. Similarly, a potential game is where there is a potential function over the space of strategies, $P(t_1, t_2, \ldots, t_n)$, so that the benefit a player gains by changing pure strategies equals the change in the potential. This is where utility functions play a role; using the first player with pure strategies including t_1^* and t_1', this condition is

$$P(t_1^*, t_2, \ldots, t_n) - P(t_1', t_2, \ldots, t_n) = u_1(t_1^*) - u_1(t_1'). \tag{3.35}$$

Monderer and Shapley proved the nice result that a potential game is isomorphic to a congestion game, which intimately connects both classes.

As it is easy to see, the potential's maximum in such a game identifies a pure Nash point. After all, any change from a maximum of P lowers P's value, which, according to Eq. 3.35, lowers the appropriate player's utility. This Nash point property can be preserved with all sorts of generalizations, such as introducing positive weights $w_1, w_2, \ldots, w_n$ whereby Eq. 3.35 becomes $P(t_1^*, t_2, \ldots, s_n) - P(t_1', t_2, \ldots, t_n) = w_1[u_1(t_1^*) - u_1(t_1')]$, or where the sign change in $P(t_1^*, t_2, \ldots, t_n) - P(t_1', t_2, \ldots, t_n)$ agrees with that of $u_1(t_1^*) - u_1(t_1')$.

3.4.5.1 Decomposing Potential Games

A coordinate system for potential games is derived in Sect. 7.7.3, and so an intuitive description is used here. Having a utility function for each player (Eqs. 3.32, 3.33) simplifies the analysis. (Everything extends to any finite number of players where each has a finite number of pure strategies.)

As differences in a utility function u_j are strictly determined by changes in its u_j^N component, the function $P(t_1, t_2)$ is based on $u_1^N(t_1, t_2)$ and $u_2^N(t_1, t_2)$, which converts Eq. 3.35 into (with Eq. 3.32)

$$\begin{aligned} P(t_1^*, t_2) - P(t_1', t_2) &= u_1^N(t_1^*, t_2) - u_1^N(t_1', t_2) = h(t_1^*)2(a_1 + d_1 t_2) \\ P(t_1, t_2^*) - P(t_1, t_2') &= u_2^N(t_1, t_2^*) - u_2^N(t_1, t_2') = h(t_2^*)2(a_2 + d_2 t_1) \end{aligned} \tag{3.36}$$

where $h(1) = 1, h(-1) = -1$. Expressing $P(t_1, t_2)$ in a Taylor series about $t_1 = t_2 = 0$, it follows that $P(t_1, t_2)$ exists if and only if $d_1 = d_2$ and $P(t_1, t_2) = a_1 t_1 + a_2 t_2 + d_1 t_1 t_2$.

Theorem 3.8 *A game* **g** *has a potential function* P *that satisfies Eq. 3.36 if and only if* $[\mathbf{g}, \mathbf{c}_2] = 0$. *Stated more generally, orthogonal to the seven-dimensional space of potential games is the one-dimensional space of matching pennies.*

Proof To show that the space of potential games is a seven-dimensional linear subspace of $\mathbb{G}$, the above condition that $d_1 = d_2$ requires c_1, which is a coordinating game, to be one of the basis vectors. In the four-dimensional Nash subspace, $\mathbb{N}$, the space of potential games has the basis

$$\mathbb{N} = \{a_1, a_2, c_1\}, \mathcal{G}^B. \tag{3.37}$$

As there are no restrictions on the $\mathbb{B}$ and $\mathbb{K}$ components, the space of potential games is seven dimensional.

But c_2 is *not* in Eq. 3.37, and it is orthogonal to a_1, a_2, c_1, so the theorem's first sentence follows. Again, properties of a class of games can be established by properties of the orthogonal class of games.   □

As a first example, a symmetric game always satisfies $[g, c_2] = 0$, so a symmetric game is a potential game. The orthogonality of game g with c_2 is the condition Monderer and Shapley [17] derive with a graph of four arrows. By shifting the emphasis to the structure of games, a different, more understandable interpretation arises.

Corollary 3.5 *The one-dimensional space of "matching pennies" is orthogonal to the seven-dimensional space of potential games. Together they span the space of* 2×2 *games.*

According to Corollary 3.5, the only role played by the Monderer and Shapley cyclic condition is to ensure that the game has no components in the matching penny direction.

Illustrating with two similarly appearing games

4	−3	−2	3
10	3	0	5

4	−5	−2	5
10	3	0	5

the first has $[g, c_2] = (4)(1) + (10)(-1) + (-1)(-3) + (1)(3) + (-2)(-1) + 0(1) + 3(1) + 5(-1) = 0$, while the second has $[g, c_2] = (4)(1) + (10)(-1) + (-1)(-5) + (1)(3) + (-2)(-1) + 0(1) + 5(1) + 5(-1) = 4 \neq 0$. Thus the first game satisfies Theorem 3.8 while the second does not.

Proof of Corollary 3.5. The orthogonality condition is a computation. That the subspace of the potential games has seven dimensions follows from the three basis vectors for the Nash component of the space. The behavioral and kernel terms, which add another four dimensions, do not affect the definition.   □

3.4.5.2 Modeling Congestion

Before moving on, it is worth digressing for a couple of paragraphs to interpret Eq. 3.37 basis of potential games in terms of those congestion settings of selecting the route to drive to work. The Nash and behavioral structures of a game are

$$\mathcal{G}^N = \begin{array}{|cc|cc|} a_1+c & a_2+c & a_1-c & -a_2-c \\ -a_1-c & a_2-c & -a_1+c & -a_2+c \end{array}, \quad \mathcal{G}^B = \begin{array}{|cc|cc|} \beta_1 & \beta_2 & -\beta_1 & \beta_2 \\ \beta_1 & -\beta_2 & -\beta_1 & -\beta_2 \end{array}$$

$$(3.38)$$

Assume that T and L correspond to driving on route $\mathcal{A}$, so B and L represent driving on route $\mathcal{B}$. Then $a_1 > 0$ means that Row prefers route $\mathcal{A}$ independent of what Column prefers; after all, this a_1 value is in both cells in the top row. Conversely, $a_1 < 0$ has route $\mathcal{B}$ as Row's preferred route. A similar analysis applies to Column; the sign of a_2 indicates Column's preferred route independent of what Row does.

Attitudes caused by the presence of other drivers are captured by the c coordinate; in a potential game, they agree. For instance, if Row prefers route $\mathcal{A}$, $a_1 > 0$. Should both drivers *not* want to share the same route, $c < 0$. This means that Row's lower $a_1 + c$ payoff in TL represents Row's irritation of sharing route $\mathcal{A}$ with Column, the higher $a_1 - c$ TR value captures Row's delight of being alone on $\mathcal{A}$. Conversely, $c > 0$ indicates that each agent is rewarded, enjoys the company, should the other take the same route. If the game is not a potential game, the c_1 and c_2 values can reflect the agents' different attitudes.

Finally, the route Row takes imposes an externality on Column (if Row takes $\mathcal{A}$, it is β_2, otherwise $-\beta_2$). The β_2 sign characterizes whether the externality is positive or negative. As an example, suppose Row drives a slow truck with a snowplow. Along the snow-covered route $\mathcal{B}$, this would prove to be an advantage, a positive externality, to Column, but on the snow-free route $\mathcal{A}$, it generates a hinderance for Column, so $\beta_2 < 0$. A similar statement holds for what externality Column imposes on Row (either β_1 or $-\beta_1$). To make a distinction, the c attitude values influence *unilateral decisions*. If $a_1 > 0$ and $a_1 + c < 0$, then (TL cell) Row prefers route $\mathcal{A}$, but having to share the road could lead to Row's *unilateral* (or Nash) decision to hop onto route $\mathcal{B}$. Comments about cooperation, coordination efforts (not attitudes) between the drivers are due to $\mathcal{G}^B$ values.

Following the spirit of Sect. 2.4.3, it is worth showing how the coordinates divvy up each payoff entry to a game's various features. For comparisons, $\mathcal{G}_{Challenge}$ (Eq. 2.1) is used. Here Eq. 2.26 division replaces each $\eta_{i,j}$ with an appropriate bracket ([ ]) of a_i and c_i terms. Because $\eta_{1,1} = a_1 + c_1$ and $\eta_{2,1} = a_1 - c_1$, with similar expressions for Column, the potential coordinates follow from algebra and define

$$\mathcal{G}_{Challenge} = \begin{array}{|cc|cc|} -7=[0.5-2.5]-4-1 & 11=[1.5+0.5]+5+4 & 6=[0.5+2.5]+4-1 & 7=-1.5-0.5]+5+4 \\ -3=[-0.5+2.5]-4-1 & 0=[1.5-0.5]-5+4 & 0=[-0.5-2.5]+4-1 & -2=[-1.5-0.5]-5+4 \end{array}$$

The positive $a_1 = 0.5$ and $a_2 = 1.5$ mean that both drivers prefer route $\mathcal{A}$, while the $c_1 = -2.5$ indicates that Row prefers not to follow Column's choice and (reflecting TL's hyperbolic-row status) could make a unilateral move to $\mathcal{B}$ for a personally better outcome. In contrast, $c_2 = 0.5$ captures Column's individual preference to mimic Row.

3.4.5.3 Only Sign Changes in P

Returning to potential functions, the orthogonality condition $[\mathbf{g}, \mathbf{c}_2] = 0$ (which extends to any number of players and strategies) ensures that potential games reside in a lower dimensional subspace of $\mathbb{G}$. This is not the case with the weaker condition whereby the sign change in P corresponds to the sign change of a player's utility function. Here, at least for 2×2 games, such a potential often exists. Indeed, according to a result developed later (Theorem 3.14), about 87.5% of the 2×2 games have a potential function with this relaxed property. The tool of analysis is the dynamic structure of cells as illustrated in Fig. 1.2. Converting the payoffs of a given game into this structure is as described above.

Start with Fig. 1.2a; there is no Nash cell, so arrows on opposite edges point in different directions to have a clockwise cycle. If a potential P existed, then (starting in the upper left corner) $P(TL) < P(TR)$ because moving from L to R improves Column's utility. Similarly, $P(TR) < P(BR)$ because moving from T to B improves Row's utility. Following the cycle, $P(BR) < P(BL)$ and $P(BL) < P(TR)$, with the contradiction that $P(TR) < P(TR)$.

Proposition 3.1 *A 2×2 game with no Nash cells is not a potential game even for the relaxed condition involving a change in sign. This assertion holds for any $k_1 \times k_2 \times \cdots \times k_n$ game where the best response dynamic creates a closed loop.*[5]

The remaining of Fig. 1.2 diagrams admit a potential function that preserves signs of changes. To prove this assertion, the form of the potential needs to be described, where $P(t_1, t_2) = at_1 + bt_2 + ct_1t_2$. Higher order terms are not necessary because if a higher order term, such as $dt_1^4 t_2^6$, were included, the fact that the t_i's assume only $+1$ and -1 values relegate the term to the role of a constant that can be ignored. Higher order terms with odd exponents change sign, so they can be incorporated into the first three choices.

Thus, it is safe to consider only potentials of $P(t_1, t_2) = at_1 + bt_2 + ct_1t_2$ form. This means that $P(1, 1) = a + b + c$, $P(-1, 1) = -a + b - c$, $P(1, -1) = a - b - c$, $P(-1, -1) = -a - b + c$. The goal is to determine for which games can a, b, c values be found to define a potential for the game. If a Nash cell is at TL, and BR is a repeller, then the sets of inequalities are $P(1, 1) > P(1, -1) > P(-1, -1)$ and $P(1, 1) > P(1, -1) > P(-1, -1)$, which become $a + c > 0$, $b + c > 0$, $b > c$, $a > c$, which are trivially satisfied. Similarly, if the repeller is elsewhere, say BL, then the inequalities become $b + c > 0$, $a > c$, $c > b$, which also can be satisfied. All that remains is if BR also is a Nash cell, where the inequalities become $a + c$, $b + c > 0$, $c > a, b$, which again are consistent.

Proposition 3.2 *If a 2×2 game has a Nash equilibrium, there exists a potential function $P(t_1, t_2)$ whereby if a player changes pure strategies, the sign of change in the player's utility is the same as the change in sign of the potential.*

---

[5]For examples of loops, see Sect. 4.5.2.

As a parting comment, if the purpose of a potential game is to find Nash cells, the answer follows immediately from $\mathcal{G}^N$, which is easier to compute and analyze. For instance, a global maximum of a potential function is a risk dominant cell. After all, with a larger $\eta_{i,j}$ value, the difference in the change of the potential is the largest at a risk dominant setting. But, potential functions are not needed to identify such cells. According to Eq. 2.52, all that is required is to compute $\mathcal{G}^N$ and then select the Nash cell with the largest entries.

3.4.6 Zero-Sum, Identical Play Games

In the above, properties of a class of games are found by using the orthogonal space of games. To further pursue this thought, return to zero-sum and identical play games.

A basis for zero-sum games is obvious; insert $(1, -1)$ in a different cell of four bimatrices to create the subspace $\mathbb{Z}$ spanned by

$$\mathbf{z}_1 = \begin{array}{|cc|cc|} \hline 1 & -1 & 0 & 0 \\ \hline 0 & 0 & 0 & 0 \\ \hline \end{array}, \quad \mathbf{z}_2 = \begin{array}{|cc|cc|} \hline 0 & 0 & 1 & -1 \\ \hline 0 & 0 & 0 & 0 \\ \hline \end{array},$$

$$\mathbf{z}_3 = \begin{array}{|cc|cc|} \hline 0 & 0 & 0 & 0 \\ \hline 1 & -1 & 0 & 0 \\ \hline \end{array}, \quad \mathbf{z}_4 = \begin{array}{|cc|cc|} \hline 0 & 0 & 0 & 0 \\ \hline 0 & 0 & 1 & -1 \\ \hline \end{array} \tag{3.39}$$

Similarly, a basis for the space of identical play games $\mathbb{I}$ is

$$\mathbf{i}_1 = \begin{array}{|cc|cc|} \hline 1 & 1 & 0 & 0 \\ \hline 0 & 0 & 0 & 0 \\ \hline \end{array}, \quad \mathbf{i}_2 = \begin{array}{|cc|cc|} \hline 0 & 0 & 1 & 1 \\ \hline 0 & 0 & 0 & 0 \\ \hline \end{array},$$

$$\mathbf{i}_3 = \begin{array}{|cc|cc|} \hline 0 & 0 & 0 & 0 \\ \hline 1 & 1 & 0 & 0 \\ \hline \end{array}, \quad \mathbf{i}_4 = \begin{array}{|cc|cc|} \hline 0 & 0 & 0 & 0 \\ \hline 0 & 0 & 1 & 1 \\ \hline \end{array} \tag{3.40}$$

By spanning $\mathbb{G}$, these eight vectors form a basis. This observation, coupled with the orthogonality of $\mathbb{Z}$ and $\mathbb{I}$, proves the comment (Eq. 3.5) that any game can be *uniquely* written as the sum of a zero-sum and an identical play game. Even stronger, as suggested with the symmetric and antisymmetric games (Theorem 3.6), this summation assertion should be expected from *any class of games*.

Theorem 3.9 *For $n \geq 2$ players where the ith player has $k_i \geq 2$ pure strategies, suppose a class of games, $\mathbb{CL}$, defines a linear subspace of dimension $0 < c < (k_1 k_2 \cdots k_n)n = kn$. The space normal to $\mathbb{CL}$, of dimension $kn - c$, defines a class of anti-class games, $\mathbb{ACL}$; any game can be uniquely written as the sum of games from these two classes.*

The space of zero-sum games, $\mathbb{Z}$, has dimension $k(n - 1)$, so the class of anti-zero-sum games, which are the identical play games $\mathbb{I}$, has dimension k. Any game can be uniquely expressed as a sum of a zero-sum and an identical play game.

The first assertion generalizes the fact that any vector in the plane can be uniquely expressed as a sum of a vector in the x-direction and in the y-direction.

Proof The bases for $\mathbb{CL}$ and $\mathbb{ACL}$ span the full space, so any game can be represented by the joint bases. As $\mathbb{CL}$ and $\mathbb{ACL}$ are orthogonal, the game's representation is a unique sum of games with one coming from each class.

The proof that $\mathbb{Z}$ has dimension $k(n - 1)$ reflects that the n entries (each player's payoff) in each cell have a zero-sum. This cell requirement has a basis of dimension $n - 1$ where the jth term has 1 for the jth player, -1 for the nth player, and zeros for all others. (For $n = 2$, this is the earlier $(1, -1)$.) The cell basis has dimension $(n - 1)$, so, with k cells, the space of zero-sum games has the advertised dimension of $k(n - 1)$. $\square$

3.4.6.1 Properties of Zero-Sum and Identical Play Games

Thanks to Eqs. 3.39, 3.40, a 2×2 game $\mathbf{g}$ is zero-sum if and only if $[\mathbf{g}, \mathbf{i}_j] = 0$ for $j = 1, \ldots, 4$, and identical play if and only if $[\mathbf{g}, \mathbf{z}_j] = 0$ for $j = 1, \ldots, 4$. While properties of these games follow from these orthogonality relationships (e.g., Theorem 3.8), our emphasis on the importance of selecting appropriate bases suggests finding a suitable $\mathbb{I}$ choice (from the $\{\mathbf{i}_j\}$ vectors) that more transparently describes $\mathbb{Z}$ characteristics. As the goal is to discover kernel, behavioral, and Nash properties, mimic these structures to derive a new basis for $\mathbb{I}$.

The approach is simple: Rewrite the kernel, behavioral, and Nash vectors (Sect. 3.4.2) in terms of $\{\mathbf{i}_j\}$. For instance, $\mathbf{k}_1 + \mathbf{k}_2$ has $(1, 1)$ cells, which leads to the first choice of

$$\mathbf{oz}_1 = \mathbf{k}_1 + \mathbf{k}_2 = \mathbf{i}_1 + \mathbf{i}_2 + \mathbf{i}_3 + \mathbf{i}_4. \tag{3.41}$$

For the behavioral $\mathbf{b}_1$, Row has $+1$ in each L cell and -1 in each R cell. To mimic $\mathbf{b}_1$ with $\mathbb{I}$ vectors, Column must be assigned the same values. This requires adding $\mathbf{n}_{1,2}$ (for the top row) and $\mathbf{n}_{2,2}$ (for the bottom row) to define the vector $\mathbf{oz}_2 = \mathbf{b}_1 + \mathbf{n}_{1,2} + \mathbf{n}_{2,2} = \mathbf{i}_1 - \mathbf{i}_2 + \mathbf{i}_3 - \mathbf{i}_4$. A similar argument involving $\mathbf{b}_2$ leads to the third vector of $\mathbf{oz}_3 = \mathbf{b}_2 + \mathbf{n}_{1,1} + \mathbf{n}_{2,1} = \mathbf{i}_1 + \mathbf{i}_2 - \mathbf{i}_3 - \mathbf{i}_4$.

Only $\{\mathbf{n}_{i,j}\}$ vectors remain. Start with $\mathbf{n}_{1,1}$, which (to satisfy the TL cell requirement of $\mathbb{I}$) must be accompanied by $\mathbf{n}_{1,2}$. The negative entries these vectors leave in other cells mandate adding $-\mathbf{n}_{2,1}$ (for the TR cell) and $-\mathbf{n}_{2,2}$ (for BL) to define $\mathbf{oz}_4 = \mathbf{n}_{1,1} + \mathbf{n}_{1,2} - \mathbf{n}_{2,1} - \mathbf{n}_{22} = \mathbf{i}_1 - \mathbf{i}_2 - \mathbf{i}_3 + \mathbf{i}_4$. (See Eq. 3.42.) The new $\mathbb{I}$ basis is given by Eq. 3.41, and

$$\mathbf{oz}_2 = \begin{array}{|cc|cc|} \hline 1 & 1 & -1 & -1 \\ 1 & 1 & -1 & -1 \\ \hline \end{array}, \quad \mathbf{oz}_3 = \begin{array}{|cc|cc|} \hline 1 & 1 & 1 & 1 \\ -1 & -1 & -1 & -1 \\ \hline \end{array}, \quad \mathbf{oz}_4 = \begin{array}{|cc|cc|} \hline 1 & 1 & -1 & -1 \\ -1 & -1 & 1 & 1 \\ \hline \end{array}$$
$$\tag{3.42}$$

The same constructive approach creates a $\mathbb{Z}$ basis to extract properties of identical play games. They are $\mathbf{oi}_1 = \mathbf{k}_1 - \mathbf{k}_2 = \sum_j \mathbf{z}_j$, $\mathbf{oi}_2 = \mathbf{b}_1 - \mathbf{n}_{1,2} - \mathbf{n}_{2,2} = \mathbf{z}_1 -$

$z_2 + z_3 - z_4$, $oi_3 = b_2 - n_{1,1} - n_{2,1} = -z_1 - z_2 + z_3 + z_4$ and $oi_4 = n_{1,1} - n_{1,2} - n_{2,1} - n_{2,2} = z_1 - z_2 - z_3 + z_4$.

These orthogonality conditions lead to the following result.

Proposition 3.3 *Game* **g** *is zero-sum if and only if* $[\mathbf{g}, \mathbf{oz}_j] = 0$ *for* $j = 1, \ldots, 4$. *The* $j = 1$ *equation requires* $\kappa_1 = -\kappa_2$ *while* $j = 2, 3$ *are Eq. 3.7, and* $j = 4$ *is Eq. 3.6.*

Game **g** *is identical play if and only if* $[\mathbf{g}, \mathbf{oi}_j] = 0$ *for* $j = 1, \ldots, 4$. *The* $j = 1$ *expression requires* $\kappa_1 = \kappa_2$ *while* $j = 2, 3$ *are Eq. 3.11, and* $j = 4$ *is Eq. 3.10.*

It is interesting that both potential and identical play games are orthogonal to matching penny games. This commonality requires these classes of games to be closely related: Identical play games are potential games, and any potential game can be converted into an identical play game without affecting the Nash structures. So, a way to create a potential game is to start with an identical play game and jazz it up with a $\mathcal{G}^B$ choice.

Theorem 3.10 *A* 2×2 *identical play game is a potential game. With a potential game* $\mathcal{G}$, *there exists a unique behavioral term* $\mathcal{G}^B$ *so that* $\mathcal{G} + \mathcal{G}^B$ *is an identical play game. Thus a potential game's* $\mathcal{G}^N$ *component satisfies* $\eta_{1,1} - \eta_{1,2} = \eta_{2,1} - \eta_{2,2}$ *(which is Eq. 3.10).*

Proof According to Eq. 3.37, the $\mathcal{G}^N$ general form of a potential game is

$$\mathcal{G}^N = \begin{array}{|cc|cc|} \hline a_1 + c & a_2 + c & a_1 - c & -a_2 - c \\ \hline -a_1 - c & a_2 - c & -a_1 + c & -a_2 + c \\ \hline \end{array}$$

By selecting $\beta_1 = a_2$ and $\beta_2 = a_1$, it follows that

$$\mathcal{G}^N + \mathcal{G}^B = \begin{array}{|cc|cc|} \hline a_1 + a_2 + c & a_2 + a_1 + c & a_1 - a_2 - c & -a_2 + a_1 - c \\ \hline -a_1 + a_2 - c & a_2 - a_1 - c & -a_1 - a_2 + c & -a_2 - a_1 + c \\ \hline \end{array}$$

which is identical play. Elementary algebra shows that this choice is unique. □

3.5 Missing Terms When Analyzing Beliefs

When playing a game, it helps to understand an opponent's beliefs. Games $\mathcal{G}_8$ and $\mathcal{G}_9$, for instance, have TL and BR Nash cells, so success requires the players to coordinate by making the same choice. How is this done?

$$\mathcal{G}_8 = \begin{array}{|cc|cc|} \hline 16 & 16 & -2 & 8 \\ \hline 8 & -2 & 2 & 2 \\ \hline \end{array}, \quad \mathcal{G}_9 = \begin{array}{|cc|cc|} \hline 16 & 16 & 13 & 12 \\ \hline 12 & 13 & 15 & 15 \\ \hline \end{array} \tag{3.43}$$

Which game do you, the reader, believe is more likely for both players to select TL?

It is reasonable to expect it is more likely for a player to select TL in $\mathcal{G}_8$ than in $\mathcal{G}_9$ because of the vast differences between the two Nash cells. This suggests examining how the games fare in an $EV(T) - EV(B)$ Nash computation. A clever argument developed by Battalio, Samuelson, and Van Huyck [2] goes beyond finding where $EV(T) - EV(B)$ equals zero to express $EV(T) - EV(B)$ as a linear expression in q. Their more general presentation provides added information.

To check out this thought, both Eq. 3.46 games have the same $p^* = q^* = \frac{1}{3}$ Nash mixed strategy. The linear expressions are

$$
\begin{aligned}
EV(T) - EV(B) &= [16q - 2(1-q)] - [8q + 2(1-q)] = 12(q - \tfrac{1}{3}) \\
&\text{and } EV(L) - EV(R) = 12(p - \tfrac{1}{3}) \quad \text{for } \mathcal{G}_8
\end{aligned}
\tag{3.44}
$$

$$
\begin{aligned}
EV(T) - EV(B) &= [16q + 13(1-q)] - [12q + 15(1-q)] = 6(q - \tfrac{1}{3}) \\
&\text{and } EV(L) - EV(R) = 6(p - \tfrac{1}{3}) \quad \text{for } \mathcal{G}_9
\end{aligned}
\tag{3.45}
$$

Perhaps $\mathcal{G}_8$'s larger coefficient, which is double that of $\mathcal{G}_9$, explains a preference for $\mathcal{G}_8$. After all, a larger coefficient supplies an earlier alert as to whether Column is leaning toward L or R. With $q = q^* + \frac{1}{12}$, for instance, Eq. 3.44 value of 1 makes it easier for Row to detect Column's movement to L than with Eq. 3.45 value of $\frac{1}{2}$.

A way to check this conjecture is to compare the following two games

$$
\mathcal{G}_{10} = \begin{array}{|cc|cc|} \hline 28 & 28 & 22 & 20 \\ \hline 20 & 22 & 26 & 26 \\ \hline \end{array}, \quad
\mathcal{G}_{11} = \begin{array}{|cc|cc|} \hline 28 & 28 & 1 & 24 \\ \hline 24 & 1 & 3 & 3 \\ \hline \end{array},
\tag{3.46}
$$

and again ask to select the one for which both players would be more likely to select TL. Applying the earlier argument, a reasonable answer is $\mathcal{G}_{11}$. But this choice contradicts expectations that Eqs. 3.44, 3.45 coefficients explain this behavior. This is because $\mathcal{G}_8$ and $\mathcal{G}_{10}$ share Eq. 3.44 expression while $\mathcal{G}_9$ and $\mathcal{G}_{11}$ satisfy Eq. 3.45. The reason is the Nash equivalency of these two pairs: Remember, these $EV(T) - EV(B)$ comparisons thoroughly wash out *all* $\mathcal{G}^B$ information to *strictly depend on* $\mathcal{G}^N$!

$$
\mathcal{G}_8^N = \mathcal{G}_{10}^N = \begin{array}{|cc|cc|} \hline 4 & 4 & -2 & -4 \\ \hline -4 & -2 & 2 & 2 \\ \hline \end{array}, \quad
\mathcal{G}_9^N = \mathcal{G}_{11}^N = \begin{array}{|cc|cc|} \hline 2 & 2 & -1 & -2 \\ \hline -2 & -1 & 1 & 1 \\ \hline \end{array},
$$

The fact that the first set's entries are twice that of the second completely explains why Eq. 3.44 coefficients are twice that of Eq. 3.45! (See Theorem 3.11, Eq. 3.48.) Incidentally, Eq. 3.49 dependency on β_j captures another role of $\mathcal{G}^B$.

Theorem 3.11 *If a 2×2 game $\mathcal{G}$ has an NME with the probability p^* of playing T and q^* of playing L, then*

$$
p^* = \frac{\eta_{2,2}}{\eta_{2,2} - \eta_{1,2}}, \quad q^* = \frac{\eta_{2,1}}{\eta_{2,1} - \eta_{1,1}}
\tag{3.47}
$$

and

$$EV(T) - EV(B) = 2[\eta_{1,1} - \eta_{2,1}](q - q^*), \quad EV(L) - EV(R) = 2[\eta_{1,2} - \eta_{2,2}](p - p^*).$$
(3.48)

At the NME

$$EV(T) = EV(B) = [2q^* - 1]\beta_1 + \kappa_1, \quad EV(L) = EV(R) = [2p^* - 1]\beta_2 + \kappa_2.$$
(3.49)

Returning to the example, the flaw in expecting Eqs. 3.44, 3.45 to explain differences in selecting between Nash cells is clear.[6] The $\mathcal{G}^B$ component is what spruces up a cell to make it attractive; the $\mathcal{G}^B$ terms, more than Nash terms, capture mutual advantages (and disadvantages) among cells. Consequently, by ignoring β values, Eq. 3.48 disregards a central source of a cell's appeal. This same observation applies to all "best response" arguments that depend on the Nash terms and ignore the crucial $\mathcal{G}^B$ impacts.

What can be done? What makes $\mathcal{G}_8$ a choice over $\mathcal{G}_9$, and $\mathcal{G}_{11}$ over $\mathcal{G}_{10}$, is the difference between the TL and BR entries, which are affected by $\mathcal{G}^B$. A more accurate indicator might be $\beta_j \pm [\eta_{1,j} + \eta_{2,j}]$, which, respectively, measures the jth player's {TL, BR} and {BL, TR} differences. In fact, should the β_j terms have small values (so $\mathcal{G}^B$ does not strongly affect the game), these expressions reduce to the $EV(T) - EV(B)$ method.[7]

Continuing to explore how β values can make a game's entries attractive, the choices for $\mathcal{G}_8$ are $\beta_1 = \beta_2 = 5$, while for $\mathcal{G}_9$ they are $\beta_1 = \beta_2 = 0$. Choosing negative β_j value (to focus attention on a different cell), such as adding $\beta_1 = \beta_2 = -6$ and $\kappa_1 = \kappa_2 = 10$ to $\mathcal{G}_8^N$ leads to $\mathcal{G}_8^*$, where attention focuses on BR, but with the same $EV(T) - EV(B)$ expression of Eq. 3.44 that was used to explain why TL is a natural choice with $\mathcal{G}_8$. Similarly, adding $\mathcal{G}_9^{B*} + \mathcal{G}_9^{K*}$, with $\beta_1 = \beta_2 = -16$, $\kappa_1 = \kappa_2 = 16$, to $\mathcal{G}_9^N$ leads to $\mathcal{G}_9^*$ with the same equation, Eq. 3.45, as $\mathcal{G}_9$, *but* now the emphasis is strongly on BR rather than TL.

$$\mathcal{G}_8^* = \begin{array}{|cc|cc|} \hline 8 & 8 & 14 & 0 \\ \hline 0 & 14 & 18 & 18 \\ \hline \end{array}, \quad \mathcal{G}_9^* = \begin{array}{|cc|cc|} \hline 4 & 4 & 29 & 0 \\ \hline 0 & 29 & 31 & 31 \\ \hline \end{array}$$
(3.50)

Any analysis exploring beliefs of players, coordinated choices of Nash equilibria, and so forth *must* involve $\mathcal{G}^B$; after all, $\mathcal{G}^B$ is what captures these interaction effects. (This observation holds for any number of players and strategies.) The difficulty with best response approaches is that the approaches drop this critical information! Our similar analysis for difficulties of the *quantal response equilibria* is in [14]. The

---

[6]Battalio et al. [2] stress stag hunt games, but by use of the coordinate system, illustrating examples could be made of that type.

[7]To express this argument with vectors, let the $\mathbf{g}(t_1, t_2)$ entries include probability values p, q. This means that $[\mathbf{g}(p, q), \mathbf{a}_1]$ and $[\mathbf{g}(p, q), \mathbf{a}_2]$ (with $\mathbf{a}_j$ from Eq. 3.26) equal, respectively, $EV(T) - EV(B)$ and $EV(L) - EV(R)$. But as $\mathbf{a}_1, \mathbf{a}_2$ reside in $\mathbb{N}$, any approach comparing differences in utility ignores all behavioral components. One way to include the behavioral information is with $\mathbf{b}_j \pm \mathbf{a}_j$.

argument is not repeated here only because it would require developing extensive background material.

Even expressions involving $\mathcal{G}^B$ must be viewed with care. It is reasonable to argue that the likelihood Row would select T is $\frac{EV(T)}{EV(T)+EV(B)}$. While this measure involves β_j values (Eq. 3.49) and the q^* value, it ignores $\eta_{i,j}$ terms. Moreover, at an NME, $EV(T) = EV(B)$ (Eq. 3.48), so $\frac{EV(T)}{EV(T)+EV(B)} = \frac{1}{2}$ for NME games, or Row would play T and B equally likely. Creating examples to cast doubt now involves $\mathcal{G}^N$ terms, such as (with $\eta_{1,1} = \eta_{1,2} = \beta_j = 10, \eta_{2,1} = \eta_{2,2} = -1, q^* = p^* = \frac{1}{11}$)

$$\begin{array}{|cc|cc|} \hline 20 & 20 & -11 & 0 \\ \hline 0 & -11 & -9 & -9 \\ \hline \end{array} \qquad (3.51)$$

It is doubtful that players would be equally likely to play B as T. An appropriate choice requires a mixture of $\mathcal{G}^N$ and $\mathcal{G}^B$ terms.

Proof of Theorem 3.11. At a q^* NME value, $EV(T) = EV(B)$. According to Sect. 2.3, the q^* value can be determined with $\mathcal{G}^N$, which means that

$$EV(T) = \eta_{1,1}q_1 + \eta_{2,1}q_2 = \lambda, \quad EV(B) = -\eta_{1,1}q_1 - \eta_{2,1}q_2 = \lambda. \quad (3.52)$$

Adding these equations proves that $\lambda = 0$. Solving for q_1 and q_2 reduces to $x\eta_{1,1} + y\eta_{2,1} = 0$, with a solution $x = -\eta_{2,1}$ and $y = \eta_{1,1}$. To determine q_j, normalize in the following way: Divide x and then y by $x + y$ to obtain $\mathbf{q} = (\frac{-\eta_{2,1}}{-\eta_{2,1}+\eta_{1,1}}, \frac{\eta_{1,1}}{-\eta_{2,1}+\eta_{1,1}})$, to prove Eq. 3.47.

The proof of Eq. 3.48 is a direct computation. By recognizing (Sect. 2.3) that only $\mathcal{G}^N$ terms are involved, the $EV(T) - EV(B)$ equation becomes

$$EV(T) - EV(B) = [\eta_{1,1}q + \eta_{2,1}(1-q)] - [-\eta_{1,1}q - \eta_{2,1}(1-q)] = 2[\eta_{1,1} - \eta_{2,1}]q + 2\eta_{2,1}.$$

Adding and subtracting $-2[\eta_{1,1} - \eta_{2,1}]q^*$ (which, according to Eq. 3.47, equals $2\eta_{2,1}$) to this expression leads to Eq. 3.48.

Similarly, Eq. 3.49 is a direct computation. Illustrating with $EV(T)$,

$$EV(T) = [q^*\eta_{1,1} + (1-q^*)\eta_{2,1}] + [q^*\beta_1 + (1-q^*)(-\beta_1)] + [q^*\kappa_1 + (1-q^*)\kappa_1]$$

where, as shown with Eq. 3.52, the first bracketed term on the right-hand side is zero. The conclusion follows by collecting terms.   □

3.6 Roadmap of 2 × 2 Games

This decomposition of 2 × 2 games leads to a surprisingly useful map that relates the various games. To see the basic ideas, recall that $\mathcal{G}^N$ is uniquely determined by the four $\eta_{i,j}$ values, $\boldsymbol{\eta}_j = (\eta_{1,j}, \eta_{2,j})$, $j = 1, 2$, which can be plotted as points in $\mathbb{R}^2$. Figure 3.2a, b choices of $\boldsymbol{\eta}_j = (\eta_{1,j}, \eta_{2,j})$ are from $\mathcal{G}_4^N$ (Eq. 2.12) where $\boldsymbol{\eta}_1 = (1, -1)$ and $\boldsymbol{\eta}_2 = (1, -3)$ (with the mixed strategy $(p, q) = (\frac{3}{4}, \frac{1}{2})$). Next plot the $\boldsymbol{\beta} = (\beta_1, \beta_2)$ point, where the $\boldsymbol{\beta} = (2, 3)$ choice in Fig. 3.2c is from $\mathcal{G}_4^B$.

3.6.1 The Nash Portion Is a Square; Ok, A Torus

With this $\mathcal{G}^N$ geometric representation, *all entries* on a ray emanating from the origin and passing through a specified $\boldsymbol{\eta}_j$ (see Fig. 3.2a, b) have the same p and q values. This makes sense; a typical best response analysis (e.g., for Eq. 2.5)

$$EV(T) - EV(B) = [2]q + [-2](1 - q)$$

is to determine whether this expression is positive, zero, or negative. The answer remains the same should the coefficients be divided by the same positive value.

In particular, using the magnitude $\|\boldsymbol{\eta}_j\| = \sqrt{\eta_{1,j}^2 + \eta_{2,j}^2}$ of vector $\boldsymbol{\eta}_j = (\eta_{1,j}, \eta_{2,j})$, the $EV(T) - EV(B) = 2\eta_{1,1}q + 2\eta_{2,1}(1 - q)$ analysis has the same conclusions as the expression

$$\frac{\eta_{1,1}}{\|\boldsymbol{\eta}_1\|}q + \frac{\eta_{2,1}}{\|\boldsymbol{\eta}_1\|}(1 - q). \tag{3.53}$$

This means that $\boldsymbol{\eta}_j$ and $\frac{\boldsymbol{\eta}_j}{\|\boldsymbol{\eta}_j\|}$ always define the same mixed strategy.

Theorem 3.12 *All games on the two rays defined by fixed $\frac{\boldsymbol{\eta}_j}{\|\boldsymbol{\eta}_j\|}$ values, $j = 1, 2$, have the same Nash strategic structure. If the game has an NME, then*

$$p = \frac{\eta_{2,2}}{\eta_{2,2} - \eta_{1,2}}, \qquad q = \frac{\eta_{2,1}}{\eta_{2,1} - \eta_{1,1}} \tag{3.54}$$

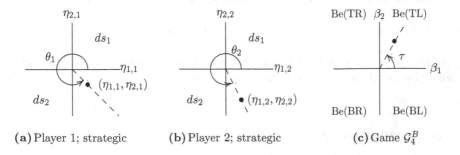

(a) Player 1; strategic   (b) Player 2; strategic   (c) Game $\mathcal{G}_4^B$

Fig. 3.2 Strategic/behavior decomposition of the game $\mathcal{G}_4$

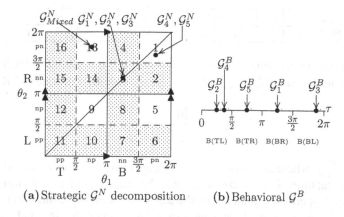

Fig. 3.3 Roadmap of 2 × 2 games

Let θ_j be the angle from the positive $\eta_{1,j}$ axis to the ray passing through $\frac{\eta_j}{||\eta_j||}$. The Nash structure of a game $\mathcal{G}$ is completely determined by the (θ_1, θ_2) values.

Even simpler, a game's basic Nash structure follows from the quadrants in which the (θ_1, θ_2) values reside. To see this, if θ_1 is in the first quadrant (so $0 < \theta_1 < \frac{\pi}{2}$), then $\eta_{1,1}$ and $\eta_{2,1}$ are positive (Fig. 3.2a). Should θ_2 be in the third quadrant (where $\pi < \theta_2 < \frac{3\pi}{2}$), then $\eta_{1,2}$ and $\eta_{2,2}$ are negative (Fig. 3.2b). Thus the $\mathcal{G}^{N,signs}$ structure is

$$\mathcal{G}^{N,signs} = \begin{array}{|cc|cc|} \hline +1 & -1 & +1 & +1 \\ \hline -1 & -1 & -1 & +1 \\ \hline \end{array},$$

where TR is the sole Nash cell and BL is the repelling cell. Similarly, θ_1 in the second quadrant ($\eta_{1,1} < 0$ and $\eta_{2,1} > 0$), and θ_2 in the fourth ($\eta_{1,2} > 0$ and $\eta_{2,2} < 0$) define

$$\mathcal{G}^{N,signs} = \begin{array}{|cc|cc|} \hline -1 & 1 & +1 & -1 \\ \hline 1 & -1 & -1 & +1 \\ \hline \end{array}$$

with no Nash cells, but a mixed equilibrium (Eq. 3.54) of $p = \frac{\eta_{2,2}}{\eta_{2,2}-\eta_{1,2}}, q = \frac{\eta_{2,1}}{\eta_{2,1}-\eta_{1,1}}$.

The promised roadmap is the $[0, 2\pi] \times [0, 2\pi]$ square in Fig. 3.3a, where the horizontal and vertical axes represent all θ values. Each axis is divided into four parts; one segment for each quadrant. The entries along Fig. 3.3a axes, such as *pn*, identify the quadrant signs of $(\eta_{1,j}, \eta_{2,j})$; e.g., *pn* on the horizontal axis means that $\eta_{1,1} > 0$ and $\eta_{2,1} < 0$.

Each of the $4 \times 4 = 16$ smaller squares defines a particular Nash structure. Square 1 (upper right corner) with a (pn, pn) designation has $\mathcal{G}^{N,signs}_{Square\,1}$ (Eq. 3.55) with two Nash cells and one NME. In contrast, the neighboring square 4 with (nn, pn) has $\mathcal{G}^{N,signs}_{Square\,4}$ with a BR Nash cell and a TR repelling cell.

$$\mathcal{G}_{Square\,1}^{N,signs} = \begin{array}{|cc|cc|} \hline 1 & 1 & -1 & -1 \\ \hline -1 & -1 & 1 & 1 \\ \hline \end{array}, \quad \mathcal{G}_{Square\,4}^{N,signs} = \begin{array}{|cc|cc|} \hline -1 & 1 & -1 & -1 \\ \hline 1 & -1 & 1 & 1 \\ \hline \end{array} \tag{3.55}$$

The line segment separating these two squares is where $\eta_{1,1}$ changes sign; it is where $\eta_{1,1} = 0$, which creates $\mathcal{G}_{Transition}^{N,signs}$ with a TL weak Nash, BR Nash, and a TR repeller.

$$\mathcal{G}_{Transition}^{N,signs} = \begin{array}{|cc|cc|} \hline 0 & 1 & -1 & -1 \\ \hline 0 & -1 & 1 & 1 \\ \hline \end{array} \tag{3.56}$$

This $\mathcal{G}_{Transition}^{N,signs}$, where the weak Nash cell changes from making the TL cell Nash or hyperbolic, captures Sect. 1.5 comments that the properties of a weak Nash equilibrium reflect properties of games from the associated generic settings. For instance, TL is a Nash cell in square 1, demoted to a weak Nash on the line, and emerges as hyperbolic in square 4. (The added zeros for vertices in Fig. 3.3a mean that these weak Nash cells reflect properties of more games.)

A game's Fig. 3.3a location is determined by the θ_js as computed by $\boldsymbol{\eta}_j$ terms. For instance, a symmetric game's Nash portion is on the diagonal line connecting the bottom-left and top-right corners. The Nash equivalent $\mathcal{G}_1, \mathcal{G}_2, \mathcal{G}_3$ (Eqs. 2.16, 2.17, 2.18) have $\boldsymbol{\eta}_1 = \boldsymbol{\eta}_2 = (-1, -1)$, so $\boldsymbol{\eta}_1$ and $\boldsymbol{\eta}_2$ are on rays bisecting the fourth quadrant. Thus the $\theta_1 = \theta_2 = \frac{7\pi}{4}$ places their common $\mathcal{G}^N$ games at the center of square 3. The positioning of $\mathcal{G}_4^N$, $\mathcal{G}_5^N$ (Eq. 2.3), and $\mathcal{G}_{Mixed}^N$ (Eq. 2.37) also are in Fig. 3.3a.

Theorem 3.13 *Only games in the four unshaded squares of Fig. 3.3a have an NME. Only games in squares 1 and 9 have two Nash cells. Games in cells 5 and 13 have no Nash cells.*

All games in a shaded square have one Nash cell. If a shaded square shares an edge with a blank square, the repelling cell for all games in this square is adjacent to the Nash cell. If the shaded square shares only a vertex with a blank cell, the game's repelling cell is diametrically opposite the Nash cell.

If games in a square have Nash cells, one is in the cell indicated by the B or T on the horizontal axis and T or L on the vertical axis.

All games on the boundaries between labeled squares have a weak Nash equilibrium.

The theorem's last line is immediate because boundaries, which identify the passage between one generic class of games to another, require some $\eta_{i,j} = 0$. To illustrate the penultimate comment, consider Fig. 3.3a quadrant consisting of squares 1, 2, 3, 4. Below the horizontal axis and to the right of the vertical one are the B and R designations. Because games in all four squares have a Nash cell, these B and R labels require one Nash cell to be in BR. Only games in square 1 (of the BR quadrant) have two Nash cells, so one is at BR and the other at TL.

An elementary but messy proof of Theorem 3.13 is to compute the sixteen $\mathcal{G}^{N,signs}$. A more satisfying method uses the following tool.

Proposition 3.4 *If a change in a game affects the signs in two cells of $\mathcal{G}^{N,signs}$, then the sum of the two cell values before a change is their sum after the change.*

A change in the sign of one entry in a cell of $\mathcal{G}^{N,signs}$ changes the cell value by ± 2.

To illustrate the first sentence with the transition in Eq. 3.55, the sum of the two cell values of $\mathcal{G}^{N,signs}_{Square\,1}$ (of $2 + (-2) = 0$) is the same as the sum of the two cell values in $\mathcal{G}^{N,signs}_{Square\,4}$ (of $0 + 0 = 0$). To illustrate the second comment, in the transition, the TL cell changes the sign of Row's entry, and the cell value drops from 2 to 0.

Proof The sum of all cell values is zero (Theorem 2.2), so if two cells are not changed, then the sum of the two altered cell values must remain the same as those before any changes were made. Changing the sign of a cell entry in $\mathcal{G}^{N,signs}$ changes the entry value by 2.   □

To apply Proposition 3.4, start with the Nash structure of the square 1 games (given by Eq. 3.55). Going from square 1 to square 2 changes Column player's entries (vertical axis) from pn to nn, which changes the $\eta_{1,2}$ sign: Only TL and TR cells are affected. In square 1, TL is a Nash cell and TR is a repeller, so the sum of the two cell values is $2 + (-2) = 0$. With the move to square 2, TL loses 2 in the cell value, so TL becomes a hyperbolic cell. As the change is in the column direction, TL is hyperbolic-column. The sum of the changed two cells must equal zero, so TR also is hyperbolic; it is hyperbolic-row. Thus, BR remains a Nash cell and BL remains a repeller.

Moving from square 2 to square 3 changes Row's $\eta_{1,1}$ sign, which involves cells TL and BL. In square 2, the sum of these cell values is $-2 + 0 = -2$, which means that change moves the repeller status from BL to TL, while BR and TR remain, respectively, Nash and hyperbolic-row. Similarly, moving from square 3 to 4 again affects Column's $\eta_{1,2}$ sign, which affects cells TL and TR. The same analysis shows that this change moves the repelling cell in square 4 to TR. More generally, the shaded squares in each quadrant of four squares rotate the location of the repelling cell.

Adventurous readers can discover symmetry arguments of Theorem 3.2 type. Games in each of the squares 3, 7, 11, and 15, for instance, have a Nash cell diametrically opposite the repeller. Going from square 3 to 7 can be identified with interchanging a game's columns.

As an aside, as angle $\theta = 0$ is equated with $\theta = 2\pi$, the left and right edge of Fig. 3.3a square can be identified to create a cylinder (Fig. 3.5). Equating the square's top and bottom edges glues the cylinder's top circle with the bottom one. In this manner, Fig. 3.3a square is identified with a *torus* T^2, which is the surface of a donut. In this manner, square 1 is adjacent to squares 6 and 16, where Proposition 3.4 analysis applies.

3.6.2 Likelihood of Different Strategic Structures

How about applications of the roadmap? It is reasonable to wonder what portion of all 2×2 games has a single Nash cell. What about no Nash cells? A previous obstacle in finding answers is that $\mathbb{G}$, the space of games, is eight dimensional, so games with a single Nash equilibrium define an eight-dimensional region! As it is difficult to envision a four-dimensional object, trying to analyze an *eight-dimensional* object can be daunting. The roadmap, however, radically simplifies the analysis.

Our answers could use spherical probability measures such as Gaussian or exponential distributions. But to keep the analysis simple and intuitive, we use a uniform distribution. Unfortunately, it is not defined over $\mathbb{R}^8$, *but* a standard way to circumvent this difficulty is to use compact approximations. That is, let the $\mathbb{R}^8$ orthogonal basis be given by (η_1, η_2), $\beta = (\beta_1, \beta_2)$, $\kappa = (\kappa_1, \kappa_2)$. If B_ρ^8 is the ball of radius ρ with center at the $\mathbb{R}^8$ origin, then as $\rho \to \infty$; the ball encompasses the full $\mathbb{R}^8$.

Theorem 3.14 *For any $\rho > 0$ and with a uniform distribution over B_ρ^8,*

- *One-fourth of all games have an NME with the division:*

  – *One-eighth of all games have no Nash cells.*
  – *One-eighth of all games have two Nash cells.*

- *Three-fourths of all games have a single Nash cell with the division:*

  – *One-fourth of all games have a single Nash cell with a diametrically opposite repelling cell.*
  – *One half of all games have a single Nash wih an adjacent repelling cell.*

Proof According to Figs. 3.2a, b, with a uniform distribution, each Fig. 3.3a square is equally likely in the four-dimensional subspace of (η_1, η_2). For this space, the above assertions hold. Because the β and κ variables play no role in the Nash structure, the assertions follow.   □

3.6.3 Extending a Monderer–Shapley Observation

The roadmap can be used to extend the Monderer and Shapley [16] comment (end of Sect. 3.3) that a non-degenerate game is best response equivalent in the Nash mixed strategies to either a zero-sum or an identical play game. According to Fig. 3.3a, this result applies only to the four unshaded squares, which constitute just one-fourth of the space of games. An issue is to determine which games in these four unshaded squares are best response equivalent to a zero-sum or an identical play game. Then, what about the other 12 squares?

To respond, our extension goes beyond games with NME to include almost all 2×2 games. As a hint of what we do, the dashed edges of squares are weak Nash points and the diagonal line in Fig. 3.3a locates the Nash component for symmetric games. Our approach is to determine the Nash locations of all zero-sum and identical play games (Fig. 3.4a).

A game's Nash structure is determined by its $\mathcal{G}^N$ component. Arguments similar to that used in Eq. 3.53 show that the dynamic structure remains intact even after scaling the players' $\mathcal{G}^N$ entries. For instance,

$$
\begin{array}{|cc|cc|}
\hline
1 & 1 & -1 & -1 \\
\hline
-1 & -1 & 1 & 1 \\
\hline
\end{array}
\quad \text{and} \quad
\begin{array}{|cc|cc|}
\hline
3 & 2 & -3 & -2 \\
\hline
-3 & -2 & 3 & 2 \\
\hline
\end{array}
$$

have identical Nash structures even though, in the second game, Row's payoffs are tripled and Column's are doubled. Both games are best response equivalent in Nash mixed strategy because the scaling of the agents' entries does not affect the analysis (Eq. 3.53). This leads to the following.

Proposition 3.5 *Two 2 × 2 games with an NME are best response equivalent in Nash mixed strategy if and only if they are represented by the same point in Fig. 3.3a.*

As an example, the choices of $\eta_{1,1} = 10$, $\eta_{1,2} = -2$, $\eta_{2,1} = \eta_{2,2} = 8$ do not satisfy Eq. 3.6 (zero-sum) nor Eq. 3.10 (identical play), so this $\mathcal{G}^N$ is not the Nash component of a zero-sum nor an identical play game. *But,* after scaling one of the agent's entries (divide the second agent's choices by 2 to have $\eta_{1,1} = 10$, $\eta_{1,2} = -2$, $\eta_{2,1}^* = \eta_{2,2}^* = 4$), Eq. 3.6 now is satisfied. Thus, this class of games are best response equivalent to a zero-sum game. Moreover, the specific zero-sum game can be constructed by using Theorem 3.3 with the $\eta_{1,1} = 10$, $\eta_{1,2} = -2$, $\eta_{2,1}^* = \eta_{2,2}^* = 4$ information.

At this stage we move beyond best response equivalence, which only focuses on the four roadmap's squares with an NME, to consider the *full Nash strategic structure.* That is, best response equivalent refers only to NME; the following "strategically equivalent" includes all NME, Nash, hyperbolic, and repelling information.

Definition 3.2 Two 2 × 2 games are said to be *strategically equivalent* if a scaling of the $\eta_{i,j}$ entries for players in one game makes the two games Nash equivalent.

The following connects Definition 3.2 with Fig. 3.3a roadmap.

Proposition 3.6 *Two 2 × 2 games are strategically equivalent if and only if they define the same point in Fig. 3.3a.*

The proof follows from the construction of Fig. 3.3a using Eq. 3.53.

Thanks to Proposition 3.6, the Monderer–Shapley result can be extended beyond NME to now include all 2 × 2 games. This is accomplished by determining the positioning of zero-sum and identical play games in the roadmap as given by Fig. 3.4a.

Theorem 3.15 *If a 2 × 2 game has $\eta_{1,j} \neq \eta_{2,j}$ for both players, then the game is strategically equivalent to either a zero-sum or an identical play game. The choice is given by the game's location in Fig. 3.4a, where a game in the shaded regions (where $\eta_{1,1} - \eta_{2,1}$ and $\eta_{1,2} - \eta_{2,2}$ have opposite signs) is strategically equivalent to a zero-sum game, while a game in an unshaded region (where $\eta_{1,1} - \eta_{2,1}$ and $\eta_{1,2} - \eta_{2,2}$ have the same sign) is strategically equivalent to an identical play game.*

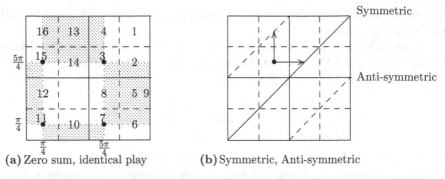

(a) Zero sum, identical play   (b) Symmetric, Anti-symmetric

Fig. 3.4 Two orthogonal classes of games

The boundaries between the shaded and unshaded regions are where $\eta_{1,j} = \eta_{2,j}$ for one player; no zero-sum nor identical play game can have this property. But if $\eta_{1,j} = \eta_{2,j}$ holds for both players (the four Fig. 3.4a bullets), the game is strategically equivalent to a zero-sum and an identical play game.

Figure 3.4a identifies which games are strategically equivalent to a zero-sum or an identical play game. The only squares with NME are 1 and 9, which are in unshaded regions, and 5 and 13, which are in shaded regions. Thus this geometry identifies the NME games that are best response equivalent to a zero-sum or an identical play game.

Corollary 3.6 *A 2×2 game with two Nash cells is best response equivalent to an identical play game; a 2×2 game with zero Nash cells is best response equivalent to a zero-sum game.*

To indicate how to use Fig. 3.4a, choose a point, say in square 6, that is in the shaded region. Rather than the $\frac{3\pi}{2} < \theta_1 < 2\pi$ and $\frac{\pi}{4} < \theta_2 < \frac{\pi}{2}$ form, use Fig. 3.2a, b geometry to select a $\boldsymbol{\eta_1}$ in the fourth quadrant, such as $\eta_{1,1} = 1$, $\eta_{2,1} = -2$, and a $\boldsymbol{\eta_2}$ in the first quadrant above the $x = y$ diagonal, such as $\eta_{1,2} = 3$ and $\eta_{2,2} = 4$.

The goal is to find whether, say, the second agent's $\boldsymbol{\eta_2}$ can be scaled to satisfy the $\eta_{1,1} - \eta_{2,1} = \eta_{2,2} - \eta_{1,2}$ version of Eq. 3.6. Namely, find whether a $\lambda > 0$ exists where

$$[\eta_{1,1} - \eta_{2,1}] = \lambda[\eta_{2,2} - \eta_{1,2}], \tag{3.57}$$

or $[1 + 2] = \lambda[4 - 3]$, which is $\lambda = 3$. Thus, an associated strategically equivalent zero-sum game is defined by $\eta_{1,1}^* = \eta_{1,1} = 1$, $\eta_{1,2}^* = \eta_{2,1} = -2$, $\eta_{1,2}^* = 3\eta_{1,2} = 9$, $\eta_{2,2}^* = 3\eta_{2,2} = 12$ with corresponding β_j values of $\beta_1 = \frac{1}{2}[9 + 12]$, $\beta_2 = \frac{1}{2}[1 - 2]$.

To show there does not exist a scaling whereby the identical play Eq. 3.10 can be satisfied, determine whether a $\lambda > 0$ exists where $[\eta_{1,1} - \eta_{2,1}] = \lambda[\eta_{2,1} - \eta_{2,2}]$. But the right-hand side of this expression is the negative of the right-hand side of Eq. 3.57, which ensures the expression cannot be satisfied.

Finally, with the exception of the four Fig. 3.4a bullets, the boundary between shaded and clean regions of Fig. 3.4a cannot represent zero-sum nor identical play games. For an example, modify the above example so the square 6 point is on the boundary where $\eta_{1,2} = \eta_{2,2}$; one choice is $\eta_{1,2} = \eta_{2,2} = 3$. Thus the right-hand side of Eq. 3.57 is zero, which can be satisfied only if $\eta_{1,1} = \eta_{2,1}$. A similar statement holds identical play games.

Proof of Theorem 3.15. The proof mimics the above discussion: If $\eta_{1,1} - \eta_{2,1}$ and $\eta_{1,2} - \eta_{2,2}$ have the same sign, there is a $\lambda > 0$ value where $[\eta_{1,1} - \eta_{2,1}] = \lambda[\eta_{1,2} - \eta_{2,2}]$, so the game is strategically equivalent to a zero-sum game as the scaled values satisfy Eq. 3.6. It they have opposite signs, there is a $\lambda > 0$ scale so that $[\eta_{1,1} - \eta_{2,1}] = \lambda[\eta_{2,2} - \eta_{1,2}]$, so the game is strategically equivalent to an identical play game; the scaled values satisfy Eq. 3.11.

If only one player has $\eta_{1,j} = \eta_{2,j}$, the game cannot satisfy Eq. 3.6 nor 3.11, so a scaled version cannot satisfy these conditions. These equations form the boundaries between the shaded and unshaded regions of Fig. 3.4a. If $\eta_{1,j} = \eta_{2,j}$ for both players, the game satisfies Eqs. 3.6 and 3.11.   □

Here is a natural question: It is well-known that a game $\mathcal{G}$ can be expressed as the sum of an identical play and zero-sum game $\mathcal{G} = \mathcal{G}_I + \mathcal{G}_Z$. So, if $\mathcal{G}$ is strategically equivalent to a game of one of these types, is $\mathcal{G}_I$ or $\mathcal{G}_Z$ the strategically equivalent game? For instance,

$$
\mathcal{G} = \begin{array}{|cc|cc|} \hline -4 & 4 & 2 & 2 \\ \hline 8 & 8 & 6 & -6 \\ \hline \end{array} = \begin{array}{|cc|cc|} \hline 0 & 0 & 2 & 2 \\ \hline 8 & 8 & 0 & 0 \\ \hline \end{array} + \begin{array}{|cc|cc|} \hline -4 & 4 & 0 & 0 \\ \hline 0 & 0 & 6 & -6 \\ \hline \end{array} = \mathcal{G}_I + \mathcal{G}_Z
$$

has a BL Nash cell and TR repeller. Because $\mathcal{G}_I$, its identical play component, has Nash cells at BL and TR, the $\mathcal{G}_I$ and $\mathcal{G}$ Nash structures differ significantly. Both $\mathcal{G}_Z$ and $\mathcal{G}$, however, have BL as a sole Nash cell and TR as a repeller, so they *do* share similar Nash structures. If the $\mathcal{G} = \mathcal{G}_I + \mathcal{G}_Z$ division has anything to do with Theorem 3.15, it must be expected that the strategic equivalence of $\mathcal{G}$ is a version of $\mathcal{G}_Z$. It is not!

To prove this, compute $\mathcal{G}$'s $\eta_{i,j}$ choices with $[\mathbf{g}, \mathbf{n}_{i,j}]$ (Sect. 3.4.2); they are $\eta_{1,1} = -6, \eta_{2,1} = -2, \eta_{1,2} = 1, \eta_{2,2} = 7$. Because $\eta_{1,1} - \eta_{2,1} = -4$ and $\eta_{1,2} - \eta_{2,2} = -6$ have the same sign, $\mathcal{G}$ is strategically equivalent to an identical play game—*not a zero-sum game!*

To create a strategically equivalent identical play game, the scaling argument shows that $\lambda = \frac{2}{3}$ satisfies the identical play requirement $[\eta_{1,1} - \eta_{2,1}] = \lambda[\eta_{2,2} - \eta_{1,2}]$. Thus, multiply each $\eta_{j,1}$ by $\frac{3}{2}$ to obtain $\eta_{1,1}^* = -9, \eta_{2,1}^* = -3, \eta_{1,2}^* = 1, \eta_{2,1}^* = 7$. Therefore (Eq. 3.11) $\beta_1^* = 4, \beta_2^* = -6$ define the identical play game that is strategically equivalent to $\mathcal{G}$

$$
\mathcal{G}^* = \begin{array}{|cc|cc|} \hline -9 & 1 & -3 & -1 \\ \hline 9 & 7 & 3 & -7 \\ \hline \end{array} + \begin{array}{|cc|cc|} \hline 4 & -6 & -4 & -6 \\ \hline 4 & 6 & -4 & 6 \\ \hline \end{array} = \begin{array}{|cc|cc|} \hline -5 & -5 & -7 & -7 \\ \hline 13 & 13 & -1 & -1 \\ \hline \end{array}
$$

where $\mathcal{G}^*$ differs significantly from $\mathcal{G}_I$ and $\mathcal{G}_Z$.

Of course, Definition 3.2 could be weakened to "strategically resembling" where two games have the same Nash structure if and only if their Nash structures belong to the same figure, Fig. 3.3a, square: Weaker than sharing the same Fig. 3.2a point, they share the same square. Answers for questions follow from Figs. 3.3a and 3.4a.

3.6.4 Why Not Symmetric, Antisymmetric?

The Monderer and Shapley comment about a non-degenerate game being best response equivalent in Nash mixed strategies to a zero-sum or an identical play game is partly motivated by the fact a game is the (orthogonal) sum of a zero-sum and identical play game. As this assertion holds for wide classes of games (Theorem 3.9), perhaps our extension of the Monderer and Shapley result (Theorem 3.15) holds for several other choices.

Well, not quite. Instead of the strong "either …or," in many cases the claim reduces to the weaker "this *and* that." The following result is intended to clarify this comment and identify what is needed for an extension.

Theorem 3.16 *A* 2×2 *game without weak Nash cells is strategically equivalent to the sum of a symmetric and an antisymmetric game. In general, this sum does* not *consist of only one type of game.*

As the symmetric and antisymmetric games span $\mathbb{G}$, each Fig. 3.3a point is a sum of these games, so the first assertion follows (Proposition 3.6). But rarely is a game strategically equivalent to *one* of the two types. Indeed, the Nash structure of symmetric games are on Fig. 3.4b diagonal line where (Theorem 3.5) $(\eta_{1,1}, \eta_{2,1}) = (\eta_{1,2}, \eta_{2,2})$; antisymmetric games are on the two slanted dashed lines (Theorem 3.6) where $(\eta_{1,1}, \eta_{2,1}) = -(\eta_{1,2}, \eta_{2,2})$. Games not on one of these lines *cannot* be strategically equivalent to only one type!

The {zero-sum, identical play} and {symmetric, antisymmetric} pairs differ in that the Nash part for the first pair almost covers Fig. 3.3a, while the second pair covers only two lines. More generally, this {zero-sum, identical play} assertion extends to other classes of games only if their Nash structures project to cover most of Fig. 3.3a.

Incidentally, by twisting the square into a torus (Fig. 3.5), Fig. 3.4b dashed lines connect. To see the sum of a symmetric and antisymmetric game, consider the bullet in Fig. 3.4b. The horizontal arrow hits the solid diagonal precisely at the relevant symmetric game while the horizontal arrow identifies the appropriate antisymmetric game.

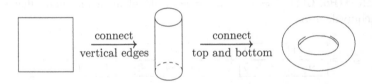

Fig. 3.5 From square to cylinder to donut (torus)

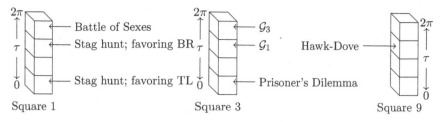

Fig. 3.6 Positioning of games

3.7 The Behavioral Portion and a Circle

The $\mathcal{G}^B$ behavioral terms are handled similar to the above. Central aspects of $\beta = (\beta_1, \beta_2)$ are the Pareto superior and inferior $\mathcal{G}^B$ cells, which are located by the unit vector $\frac{\beta}{\|\beta\|}$ and (Fig. 3.2c) the angle $\tau \in [0, 2\pi]$. Also relevant is the size of β terms relative to η values.

If τ is in the first quadrant, $\mathcal{G}^B$'s Pareto superior point is at TL; all possible positions of this point are represented in Fig. 3.2c. Thus a 2×2 game's relevant information, the Nash and Pareto superior structures, collapse into a point in the three-dimensional cube $[0, 2\pi] \times [0, 2\pi] \times [0, 2\pi]$. (By equating angle zero with angle 2π, the cube is a *three-torus* T^3.) In this way, Fig. 3.6 captures the relationships among all 2×2 games.

Figure 3.6 starts with square 1 (Fig. 3.3a) with TL and BR Nash cells. If TL is $\mathcal{G}^B$'s Pareto superior cell (τ is in the first quadrant), $\mathcal{G}^B$ terms enhance the TL Nash cell, as indicated by the "Stag Hunt" arrow. A reversed stag hunt (stag hunting is in the bottom row) is where BR is $\mathcal{G}^B$'s Pareto superior cell. This, again, demonstrates the critical role played by $\mathcal{G}^B$. Moving $\mathcal{G}^B$'s Pareto superior cell to BL identifies the home for the Battle of Sexes. In this manner, all 2×2 games can be located.

Chapter 4
Two-Person Multi-strategy Games

Much of what has been developed in previous chapters holds for games with more players and/or pure strategies. Rather than carrying out fairly obvious extensions, we identify attributes that differ from 2×2 games. This includes explaining why certain 2×2 assertions collapse with more strategies, differences in $\mathcal{G}^B$ features, a mystery associated with the Ultimatum game, and the Nash structure of $k_1 \times k_2$ games (including possible hiding places for NME). In Chap. 5, this line of thinking is extended to extensive games. Results where three or more player games differ from two-player games are in Chap. 6.

4.1 Introduction

Expect more options to usher in added intrigue. As a Battle of the Sexes (Eq. 1.7, Sect. 2.5.5) sampler, imagine the wife's growing disgust over the bickering whether to go to the opera or the football game. At a certain point, she may prefer to clean.

With modification to Eq. 1.7, the interactions are modeled by the 3×2 game

$$
\mathcal{G}_{Sexes,1} = \quad
\begin{array}{c}
\\
\text{Opera} \\
\text{Football} \\
\text{Cleaning}
\end{array}
\begin{array}{c}
\text{Opera} \quad \text{Football} \\
\begin{array}{|cc|cc|}
\hline
6 & 2 & 0 & 0 \\
\hline
0 & 0 & 2 & 6 \\
\hline
1.8 & 1.8 & 1.8 & 1.8 \\
\hline
\end{array}
\end{array}
\qquad (4.1)
$$

The cleaning rewards—which are the smallest Eq. 4.1 positive entries (hyperbolic-row cells that the wife would gladly abandon)—are insignificant enough to expect that the Sects. 1.4.4, 2.5.5 descriptions hold: The two Nash cells (at TL and MR, which is Middle-R) are accompanied with an NME at $\mathbf{p} = (\frac{3}{4}, \frac{1}{4}, 0)$ for the wife and $\mathbf{q} = (\frac{1}{4}, \frac{3}{4})$ for the husband. While this $(\mathbf{p}, \mathbf{q})$ defines an NME if the husband and

D. T. Jessie and D. G. Saari, *Coordinate Systems for Games*, Static & Dynamic Game Theory: Foundations & Applications, https://doi.org/10.1007/978-3-030-35847-1_4

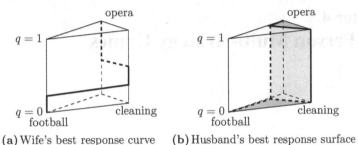

(a) Wife's best response curve   (b) Husband's best response surface

Fig. 4.1 The difference a dimension makes

wife concentrate on football or opera, it is arguable that it is *not* a Nash equilibrium for the full game! The reason is that with added strategies, the NME can drift off to, well, another dimension.

The problem is that the NME expected value is 1.5 for each player (e.g., use Eq. 3.49), while the cleaning option rewards each player an improved 1.8 payoff. To see where this crops up in the wife's best response curve (Fig. 4.1a), if her husband (Column) plays L, or tends to play L, her best response is to play T—large q values lead to her best response of $p = 1$. Conversely, should he tend toward football, which is a small q value, her best response is $p = 0$, or football. Should he oscillates with values near $q = \frac{1}{4}$, the mixed strategy expected payoff is 1.5. She can do better with a payoff of 1.8 by cleaning.

The NME, which is the intersection of the wife's best response curve and the husband's best response surface (Fig. 4.1), is an *interval*. Each point in this interval corresponds to the wife's pure strategy of cleaning and the husband's mixed strategy of agonizing over the opera or football. To compute when she prefers cleaning, set

$$EV_{Wife}(Opera) = EV_{Row}(T) = 6q + 0(1 - q) = EV_{Wife}(Cleaning) = 1.8$$

while

$$EV_{Wife}(Football) = EV_{Row}(B) = 0q + 2(1 - q) = EV_{Wife}(Cleaning) = 1.8.$$

She is indifferent between opera and cleaning for $q = 0.3$ and between football and cleaning for $1 - q = 0.9$ or $q = 0.1$. Her best response is to clean for $0.1 < q < 0.3$. The derivation of the husband's best response surface mimics the earlier (Sect. 1.4.4) analysis.

An alternative description uses the reality that, unless the probabilities are actually used to make choices, an NME is *not* an action, it is a mathematical construct that provides guidance: An NME resembles a car's GPS spouting the annoying "Recalculating" while supplying updated recommendations about which way to go. It is the driver's actions—not the GPS—that constitute the activity. Similarly, the husband's indecision introduces a new option—vacillate, which converts Eq. 4.1 into the 3×3

game

$$\mathcal{G}_{Sexes,2} = \begin{array}{c|cc|cc|cc} & \text{Opera} & & \text{FB} & & \text{Indec.} & \\ \hline \text{Opera} & 6 & 2 & 0 & 0 & 0 & 0 \\ \text{FB} & 0 & 0 & 2 & 6 & 0 & 0 \\ \text{Cleaning} & 0 & 0 & 0 & 0 & 1.8 & 1.8 \end{array} \qquad (4.2)$$

The Nash structure of this new game has $2^3 - 1 = 7$ Nash equilibria consisting of

1. the (Opera, Opera) and (Football, Football) Nash cells plus a new one at (Clean, Indecision),
2. the standard NME at $p_1 = q_2 = \frac{3}{4}$, $q_1 = p_2 = \frac{1}{4}$,
3. plus new NME at $\{p_1 = \frac{9}{19}, p_2 = q_2 = 0, p_3 = \frac{10}{19}, q_1 = \frac{3}{13}, q_3 = \frac{10}{13}\}$, $\{p_1 = q_1 = 0, p_2 = \frac{3}{13}, p_3 = \frac{10}{13}, q_2 = \frac{9}{19}, q_3 = \frac{10}{19}\}$, and $\{q_1 = p_2 = \frac{3}{22}, p_1 = q_2 = \frac{9}{22}, p_3 = q_3 = \frac{10}{22}\}$, which, respectively, serve as guideposts for the selection of the first and third, second and third, and among all three Nash cells.

The point is that reality can introduce options into the modeling. As another example, perhaps the husband, irked by the deliberations, threatens to install a door in the family room. The door benefits both, but, trust us, cleaning and carpentry are not compatible. This situation explodes the original 2×2 game into a 4×4 game captured by

$$\mathcal{G}_{Sexes,3} = \begin{array}{c|cc|cc|cc|cc} & \text{Opera} & & \text{FB} & & \text{Indec} & & \text{Door} & \\ \hline \text{Opera} & 6 & 2 & 0 & 0 & 0 & 0 & 0 & 0 \\ \text{FB} & 0 & 0 & 2 & 6 & 0 & 0 & 0 & 0 \\ \text{Indec.} & 0 & 0 & 0 & 0 & 0 & 0 & 1.8 & 1.8 \\ \text{Cleaning} & 0 & 0 & 0 & 0 & 1.8 & 1.8 & 0 & 0 \end{array} \qquad (4.3)$$

where there are $2^4 - 1 = 15$ Nash equilibria consisting of

1. four Nash cells (Opera, Opera), (FB, FB), (Indecision, Door) and (Cleaning, Indecision),
2. and at least eleven NME providing guidance in selecting between any of the six pairs, the four triplets, and the set of all four Nash cells. (Some of these NME have other interpretations; e.g., for a triplet, it could provide guidance between selecting an NME for a pair and the remaining Nash cell.)

Mixed strategy settings can motivate adding alternatives or strategies. It is arguable that a Hawk–Dove example of this type arose in August 2017 when North Korea threatened to launch a missile toward Guam—a highly provocative threat because of an American base on the island. The continual saber-rattling from the United States and North Korea, complete with "fire and fury" threats coupled with insults that sunk 'diplomacy' to levels below that found on playgrounds,[1] resembled the expected tit-for-tat retaliatory strategies in a brinkmanship version of the Hawk–Dove game. No resolution seemed to be available, which caused consternation across the world.

---

[1] Yet, this resulted in a June 10, 2018 summit meeting between Tump and Kim in Singapore! Figure that out if you can.

A mixed strategy that arises during multistage negotiations and manifested with words, threats, and insults differs significantly from the actual adoption of a mixed strategy where, randomly, a pure strategy is selected. What can be done? A resolution emerged on August 28, 2017, when North Korea hurled a ballistic missile over *Japan*, not *Guam*: This action can be viewed as creating a new alternative, similar to the wife deciding to "clean," in reaction to circumstances.

An example without Nash cells has Bob and Sue out for dinner where the cost is $51, or, with the tip, $60 each. After dinner, Bob discovers he does not have enough money. Aha! perhaps he can outsmart Sue in a "Heads or Tails" game, so he suggests "let's match pennies to see who pays." If the faces agree, Sue (Row) pays, otherwise Bob (Column) does, which defines the first Eq. 4.4 bimatrix. But when Bob sees Sue preparing to *flip a coin*, which he did not anticipate, Bob suggests, "Tell you what, I have to leave, so I'll pick up the tip, you can get the rest." By paying $18 and saving $42, this converts the usual

$$
\begin{array}{|cc|cc|}
\hline
-60 & 60 & 60 & -60 \\
\hline
60 & -60 & -60 & 60 \\
\hline
\end{array}
\quad \text{into} \quad
\begin{array}{|cc|cc|cc|}
\hline
-60 & 60 & 60 & -60 & -42 & 42 \\
\hline
0 & 0 & 0 & 0 & -42 & 42 \\
\hline
60 & -60 & -60 & 60 & -42 & 42 \\
\hline
\end{array}.
\tag{4.4}
$$

Both games have a 50–50 NME, but the second one has a Nash cell at [Flip, Pay Tip]. But there is *not* an NME between the NME block and the Nash cell. Surprises caused by admitting more pure strategies and agents are partly alleviated with our decomposition.

4.2 Decomposition

The decomposition for general games, with any finite number of players where each has a finite number of pure strategies, follows the lead of Sect. 2.3. To define the jth player's $\mathcal{G}^N$, or Nash entries, start by selecting pure strategies for each of the other $n - 1$ players; this defines an array of $\mathcal{G}$ cells. Replace the jth player's entry in each cell of the array by how it differs from the jth player's average payoff over the array. Doing so for each player defines $\mathcal{G}^N$; the "average array values" define $\mathcal{G}^{Averages}$; i.e., $\mathcal{G}^{Averages} = \mathcal{G} - \mathcal{G}^N$.

If κ_j is the jth player's average payoff in $\mathcal{G}$, the kernel component, $\mathcal{G}^K = \mathcal{G}^K(\kappa_1, \ldots, \kappa_n)$, replaces the jth player's entry in each $\mathcal{G}$ cell with κ_j. The behavioral $\mathcal{G}^B$ is where the jth player's entry in a $\mathcal{G}^{Averages}$ cell is replaced by how the entry differs from κ_j.

The following assertion should be expected from the 2×2 discussion in Chap. 2.

Theorem 4.1 *For $n \geq 2$ players, let $\mathcal{G}$ be a $k_1 \times k_2 \times \cdots \times k_n$ game. Then*

$$
\mathcal{G} = \mathcal{G}^N + \mathcal{G}^B + \mathcal{G}^K.
\tag{4.5}
$$

The sum of the jth player's payoffs in any $\mathcal{G}^N$ array, defined by pure strategy choices of the other players, is zero. Similarly, any $k_1 \times k_2 \times \cdots \times k_n$ game with this property for each player overall arrays is a Nash component for a class of games.

The $\mathcal{G}^N$ component has all $\mathcal{G}$ information needed to determine the game's Nash structure. The sum of the jth agent's $\mathcal{G}^B$ entries is zero; conversely, a matrix where a player's entries in each array is the same and their sum equals zero is a $\mathcal{G}^B$ for games.

An annoying cost of adding options is the need to tweak the notation. In 2×2 games, $\eta_{i,j}$ adequately captured $\mathcal{G}^N$ values, but it does not suffice for three or more players and/or options. The adjusted notation is $\eta_{i,a,j}$ meaning, from left to right, the ith item in the ath array (determined by the pure strategies selected for each of the other players) for the jth player. Similarly, $\beta_{a,j}$ is the common β value in the ath array for the jth player. The following corollary uses this notation to restate a portion of Theorem 4.1.

Corollary 4.1 *For $n \geq 2$ players, and each array a and player j,*

$$\sum_{i=1}^{k_j} \eta_{i,a,j} = 0 \tag{4.6}$$

Similarly, for each player j,

$$\sum_{a=1}^{J} \beta_{a,j} = 0 \text{ where } J = [k_1 k_2 \ldots k_n]/k_j. \tag{4.7}$$

4.2.1 More on Risk and Payoff Dominance

A way to illustrate Corollary 4.1 is to use it to expose risk dominance features that arise for games with three or more Nash cells (see Sect. 2.6). According to Eq. 4.6,

$$
\begin{array}{|cc|cc|cc|}
\hline
2 & 2 & v & x & w & -2-x \\
\hline
u & y & 4 & 4 & -6-w & -4-y \\
\hline
-2-u & z & -4-v & -6-z & 6 & 6 \\
\hline
\end{array}
\rightarrow
\begin{array}{|cc|cc|cc|}
\hline
2 & 2 & v & 1 & 2 & -3 \\
\hline
1 & y & 4 & 4 & -8 & -4-y \\
\hline
-3 & 2 & -4-v & -8 & 6 & 6 \\
\hline
\end{array}
\tag{4.8}
$$

Equation 4.8 game on the left is a $\mathcal{G}^N$ component. This is because the sum of Row's entries in each column, and the sum of Column's entries in each row, equals zero. If the numbers without variables are the maximum in respective rows and columns (e.g., in the first column, $2 > u$, $-2 - u$ and in the second row $4 > y$, $-4 - y$), then the three cells on the diagonal are Nash cells. Label them as N_1, N_2, N_3.

Pairs of these Nash cells can be ranked with the risk dominance criteria. That is, $N_i \succ_{RD} N_j$ (N_i is risk dominant over N_j) iff Eq. 2.49 criterion (Sect. 2.6) is satisfied in the corresponding 2×2 subgame. A concern is whether these paired comparisons always define a transitive ranking. That is, could there be u, v, w, x, y, z choices that define the cycle $N_3 \succ_{RD} N_2$, $N_2 \succ_{RD} N_1$, but $N_1 \succ_{RD} N_3$?

Thanks to the decomposition, designing such a cycle involves only simple algebra.

- $N_3 \succ_{RD} N_2$ iff $(10 + v)(10 + y) > (8 + v)(8 + y)$, which is satisfied,
- $N_2 \succ_{RD} N_1$ iff $(4 - u)(4 - x) > (2 - u)(2 - x)$, which also is true, and
- $N_1 \succ_{RD} N_3$ iff $(4 + u)(4 + x) > (6 - w)(6 - z)$.

These three inequalities are satisfied (to create a cycle) with $u = x = 1$ and $w = z = 2$; these values define the $\mathcal{G}^N$ on the right in Eq. 4.8.

Thus, anticipate "$\succ_{RD}$" cycles. A reason is that, in general, restricting a game's $\mathcal{G}^N$ component to a subgame does *not* define the subgame's $\mathcal{G}^N$ term. Yes, a projection of a Nash cell always is a Nash cell, but, in general, the projection of $\mathcal{G}^N$ includes some of the subgame's behavioral and kernel terms. With Eq. 4.8, for instance, concentrating on the subgame defined by T, B, and L, R leads to

2	2	2	−3
−3	2	6	6

with Nash component

2.5	2.5	−2	−2.5
−2.5	−2	2	2

.

Here, the earlier Sect. 2.6 result applies; the product of the TL $\mathcal{G}^N$ entries (both 2.5) is larger than the product of BR entries (both 2), so TL $\succ_{RD}$ BR.

As described in Sect. 2.6, *if* there is an NME (p_1, p_2, p_3) and (q_1, q_2, q_3) relating the three Nash cells, then a risk dominance ranking can be determined by ranking the values of the three products $p_1 q_1$, $p_2 q_2$, *and* $p_3 q_3$. A natural question is whether the ranking of the triplet is related to the above rankings the pairs. As this example already illustrates, expect differences: This is a common phenomenon in the social sciences such as where the plurality vote ranking of three candidates need not have anything to do with their paired majority vote rankings (e.g., [24, 25]). A further complication, as shown later, is that the second game in Eq. 4.8 does *not* have an NME relating the three cells should $v \geq 0$.

This analysis underscores the fact that the content of a 3×3 (or larger) game need not be reflected by that of a 2×2 subgame. This assertion is another example of a social science phenomenon where the sum of the parts (the smaller games) need not reflect the whole. (As argued in [31], appropriate connecting information is required.)

$$\mathcal{G}^B = \begin{array}{|ccc|ccc|ccc|} \hline 8 & & 9 & 2 & & 9 & -10 & & 9 \\ 8 & & 3 & 2 & & 3 & -10 & & 3 \\ 8 & & -12 & 2 & & -12 & -10 & & -12 \\ \hline \end{array} \qquad (4.9)$$

Payoff dominant outcomes (see Sect. 2.6), on the other hand, require sufficiently large β values. It follows from the independence of $\mathcal{G}^N$ and $\mathcal{G}^B$ that a payoff dominant ranking need not have anything to do with a risk dominant ranking (if it exists). Arguments and extensions are essentially those of Sect. 2.6, so only an example is given. Equation 4.9 game is a $\mathcal{G}^B$ component because whatever column Column selects, Row's payoffs are the same; whatever row Row selects, Column's payoffs are the same. Finally, Eq. 4.7 is satisfied because all of Row's β values sum to zero as do Column's β values. Adding this choice to Eq. 4.8 games leads to the "payoff dominant ranking" $N_1 \succ_{PD} N_2 \succ_{PD} N_3$.

4.2.2   More on Zero-Sum and Identical Play Games

Treat the Corollary 4.1 notation in the same manner as a fire extinguisher: Use only when really needed. But, to prepare for emergency, it should be understood. A way to encourage familiarity is to prove other results. Readers who are more interested in game structures, or who suffer allergies from prolonged exposure to analysis, could jump to Sect. 4.3.

For continuing readers, we selected a topic to explain why adding options can change conclusions; it is to determine whether we can expand upong our extension (Theorem 4.3) of the Monderer and Shapley [16] result. Recall, we proved that almost all 2×2 games are strategically equivalent to either a zero-sum or an identical play game. The exceptions are where $\eta_{1,j} = \eta_{2,j}$ for one player but not the other.

As shown next, this result *fails to hold* with more strategies and players. The problem is that the extra options increase the dimension of the space of games. In doing so, the Nash structures of zero-sum games are separated from those of identical play games in a manner similar to Theorem 3.16 description. (Recall, Theorem 3.16 explained why the "either—or" assertion fails to hold for {symmetric, antisymmetric} games.) More specifically, the added dimensions create open gaps of $k_1 \times k_2$ games, $k_1, k_2 \geq 3$, where the assertion *fails*.

The key step in the proof is the following decomposition of zero-sum and identical play games. (Rather than using vectors, an alternative proof is provided.) Essentially the same proof extends the assertion to three and more players.

Theorem 4.2 *If $\mathcal{G}$ is a $k_1 \times k_2$ zero-sum game, $k_1, k_2 \geq 3$, then $\kappa_1 = -\kappa_2$. If $\mathcal{G}$ is identical play, then $\kappa_1 = \kappa_2$. For zero-sum, the β values are*

$$\beta_{i,1} = -\frac{1}{k_1}\sum_{i=1}^{k_1}\eta_{j,i,2} \text{ for } j = 1, \ldots, k_1 - 1, \ \beta_{i,2} = -\frac{1}{k_2}\sum_{i=1}^{k_2}\eta_{j,i,1} \text{ for } j = 1, \ldots, k_2 - 1.$$

(4.10)

and, for each i and j,

$$k_1[k_2\eta_{i,j,1} - \sum_{a=1}^{k_2}\eta_{i,a,1}] = -k_2[k_1\eta_{j,i,2} - \sum_{s=1}^{k_1}\eta_{j,s,2}]$$

(4.11)

For identical play, signs of Eq. 4.10 are changed to have

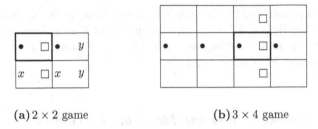

(a) 2×2 game   (b) 3×4 game

Fig. 4.2 Geometry of Eq. 4.11

$$\beta_{i,1} = \frac{1}{k_1} \sum_{i=1}^{k_1} \eta_{j,i,2} \, for \, j = 1, \ldots, k_1 - 1, \ \beta_{i,2} = \frac{1}{k_2} \sum_{i=1}^{k_2} \eta_{j,i,1} \, for \, j = 1, \ldots, k_2 - 1,$$

(4.12)

and Eq. 4.11 is replaced with

$$k_1[k_2 \eta_{i,j,1} - \sum_{a=1}^{k_2} \eta_{i,a,1}] = k_2[k_1 \eta_{j,i,2} - \sum_{s=1}^{k_1} \eta_{j,s,2}]$$

(4.13)

Equations 4.11 and 4.13 appear to be inscrutable. To add understanding, examples are carried out. Start with a 2×2 game and the TL cell, where the Eq. 4.11 form of $2[2\eta_{1,1} - [\eta_{1,1} + \eta_{2,1}] = -2[2\eta_{1,2} - [\eta_{1,2} + \eta_{2,2}]$ reduces to $\eta_{1,1} - \eta_{2,1} = \eta_{2,2} - \eta_{1,2}$, which is Eq. 3.6. In Fig. 4.2a, Row's $\mathcal{G}^N$ terms are given by bullets that lie in the T row, while Column's are specified by squares that are in the L column. Of importance for our argument is that these four Figs. 4.2a values completely determine $\mathcal{G}^N$. (The. x and y values are, respectively, the negative of the associated bullet and square values.)

The 3×4 game depicted in Fig. 4.2b illustrates which $\mathcal{G}^N$ terms to compute for the $(2, 3)$ cell. Row's η values are given by the bullets in the second row, while Column's η values, designated by squares, are in the third column. An important difference from Fig. 4.2a is that *after selecting values for the bullets and squares, still other entries remain to be chosen to define $\mathcal{G}^N$.* The flexibility of having added options is what causes difficulties.

$$\mathcal{G}^N = \begin{array}{|c|c|c|c|c|c|} \hline 2 & 2 & 3 & \mathbf{5} & 5 & 4 \\ \hline \mathbf{1} & \mathbf{3} & \mathbf{2} & \mathbf{3} & \mathbf{-2} & \mathbf{1} \\ \hline x & x & x & \mathbf{4} & x & 1 \\ \hline \end{array}$$

(4.14)

To illustrate with numbers, the bold entries of Eq. 4.14 define a partially completed $\mathcal{G}^N$; as in Fig. 4.2b, the designated cell is $(2, 3)$ where the seven relevant entries are in bold. Both Eqs. 4.11, 4.13 require computing $k_1[k_2 \eta_{i,j,1} - \sum_{a=1}^{k_2} \eta_{i,a,1}]$ for Row and $k_2[k_1 \eta_{j,i,2} - \sum_{s=1}^{k_1} \eta_{j,s,2}]$ for Column; here, $k_1 = 3$, $k_2 = 4$. To start, add Row's entries in the middle row to obtain $1 + 3 + 2 - 2 = 4$ and subtract it from

$k_2(2) = (4)(2)$, for a value of $8 - 4 = 4$. Because $k_1 = 3$, Row's contribution to either of the two equations is 12.

Similarly, the sum of Column's entries in the third column is 12; subtract this value from $3k_1 = 9$ for the outcome of -3, and multiply -3 by $k_2 = 4$ for Column's contribution of -12. As Row's and Column's contributions agree in magnitude but differ in signs, the zero-sum Eq. 4.11 is satisfied.

So far, four of Row's and three of Column's $\mathcal{G}^N$ entries are specified, which leaves four more to be selected for each player (actually eight, but this reduces to four because of Eq. 4.6); the choices are independent—they need not respect the zero-sum structure. Indeed, with the non-bold numbers associated with the TR cell, Row's contribution is $3[(4)5 - 12] = 24$ while Column's contribution is $4[3(4) - 6] = 24$. The values agree, so the *identical play* Eq. 4.13 (rather than the zero-sum choice) is satisfied. Consequently, the partially completed $\mathcal{G}^N$ of Eq. 4.14 *cannot* be the Nash component for a zero-sum nor an identical play game! Incidentally, the x's are where values are determined by entries already in Eq. 4.13. According to Eq. 4.6, Row's entries in any column must sum to zero. So, going from left to right in the bottom row, the x values are $-3, -5, -5, -3$.

Proof For Row, the (i, j) cell is the ith item in the jth column; for Column, it is the jth entry in the ith row. As the sum of the players' entries in a zero-sum game equals zero,

$$[\eta_{i,j,1} + \beta_{j,1}] + [\eta_{j,i,2} + \beta_{i,2}] = -[\kappa_1 + \kappa_2] = K. \qquad (4.15)$$

Summing over $i = 1, \ldots, k_1$, $j = 1, \ldots, k_2$, yields (with Eqs. 4.6, 4.7) $[k_1 + k_2]K = 0$, or $\kappa_1 = -\kappa_2$. Setting $K = 0$, holding j fixed and summing Eq. 4.15 over-all i leads to $k_1 \beta_{j,1} = -\sum_{i=1}^{k_1} \eta_{j,i,2}$, which is the first term in Eq. 4.10; the second term is found in the same way. Equation 4.11 emerges by substituting these β values into Eq. 4.15. For identical play games, Eq. 4.15 becomes $[\eta_{i,j,1} + \beta_{j,1}] - [\eta_{j,i,2} + \beta_{i,2}] = -[\kappa_1 - \kappa_2] = K$.

In a similar manner, appropriate conditions can be derived for the $\mathcal{G}^N, \mathcal{G}^B, \mathcal{G}^K$ terms for any $k_1 \times \ldots \times k_n$ zero-sum or identical play game. $\qquad \square$

Equation 4.14 example indicates why Theorem 3.15 holds for 2×2 games, but not with more pure strategies. The following theorem considers $k_1 \times k_2$ games with $k, k_2 \geq 3$; the $k_1 \geq 3$, $k_2 = 2$ case is left to the reader. (Hint: Create a diagram similar to Fig. 4.2.)

Theorem 4.3 *For $k_1 \times k_2$ games where $k_1, k_2 \geq 3$, there are open sets of games that are not strategically equivalent to either a zero-sum or an identical play game.*

Proof The central argument for 2×2 games (Theorem 3.15), was to scale a player's Nash entries so that the resulting $\mathcal{G}^N$ satisfies either the Nash structure of a zero-sum (Eq. 3.6) or identical play (Eq. 3.10) game: a positive or negative λ value exists where

$$[\eta_{1,1} - \eta_{2,1}] = \lambda[\eta_{1,2} - \eta_{2,2}].$$

Here is where the geometry associated with the number of pure strategies arises. With 2×2 games, the $\eta_{1,1} = \eta_{2,1}$ line divides the space of η_1 values into two half-spaces: one for $[\eta_{1,1} - \eta_{2,1}] > 0$ and the other for $[\eta_{1,1} - \eta_{2,1}] < 0$. The same holds for η_2 with $\eta_{1,2} = \eta_{2,2}$. These four subregions are characterized by whether the $\eta_{1,j} - \eta_{2,j}$ value is positive or negative. For any η_1 and η_2 not on the boundary lines, such a λ can be found.

The situation changes for $k_1, k_2 \geq 3$; the corresponding division, created by setting each Eq. 4.11 expression equal to zero, divides the η_1 and η_2 spaces into $2^{(k_1-1)(k_2-1)}$ open regions corresponding to the signs of the required expressions. For a $\mathcal{G}^N$ to have an identical play game in its equivalence class, both η_j's must be on the same side of *each* hyperplane (Eq. 4.11); for a zero-sum, both η_j's must be on opposite sides. But the geometry admits open sets of η_j that are on the same side of some hyperplanes and on the opposite side of other planes (i.e., the sum of terms in Eq. 4.11 have different signs).

This is precisely what Eq. 4.14 illustrates: Added options increase the dimensionality of the $\mathcal{G}^N$ space. Two different designated cells defines arrays that have nothing to do with each other, which permits one array to respect the zero-sum convention while the other belongs to the identical play setting. $\qquad\square$

4.3 Examples

A civilized way to illustrate the coordinate system is with examples. For instance, the husband's cooperative attitude about the cleaning option in Eq. 4.2 can be computed with Theorem 4.1 and Corollary 4.1, where the answer comes from $\mathcal{G}^B_{Sexes,2}$.

Each player's average payoff is $\kappa_j = \frac{9.8}{9}$, which defines $\mathcal{G}^K_{Sexes,2}$. The wife's average payoff in the first, second, and third columns are, respectively, 2, $\frac{2}{3}$, and 0.6. Thus, the wife's $\mathcal{G}^N_{Sexes,2}$ entries for the first column are, respectively, $6 - 2 = 4$, -2, and -2; the $\mathcal{G}^N$ entries for the second column are similarly computed. For the husband, the average in the top, middle, and bottom are, respectively, 1, 3, and 1.8, which leads to his $\mathcal{G}^N_{Sexes,2}$ entries in the three rows. Then, $\mathcal{G}^B = \mathcal{G} - \mathcal{G}^N - \mathcal{G}^K$, which leads to

$$
\begin{array}{|cc|cc|cc|}
\hline
4 & \tfrac{4}{3} & -\tfrac{2}{3} & -\tfrac{1}{3} & -\tfrac{3}{3} & -\tfrac{1}{3} \\
\hline
-2 & -2 & \tfrac{4}{3} & 4 & -\tfrac{3}{5} & -2 \\
\hline
-2 & -\tfrac{3}{5} & -\tfrac{2}{3} & -\tfrac{3}{5} & \tfrac{6}{5} & \tfrac{6}{5} \\
\hline
\end{array}
+
\begin{array}{|cc|cc|cc|}
\hline
\tfrac{41}{45} & -\tfrac{19}{45} & -\tfrac{19}{45} & -\tfrac{19}{45} & -\tfrac{22}{45} & -\tfrac{19}{45} \\
\hline
\tfrac{41}{45} & \tfrac{41}{45} & -\tfrac{19}{45} & \tfrac{41}{45} & -\tfrac{22}{45} & \tfrac{41}{45} \\
\hline
\tfrac{41}{45} & -\tfrac{22}{45} & -\tfrac{19}{45} & -\tfrac{22}{45} & -\tfrac{22}{45} & -\tfrac{22}{45} \\
\hline
\end{array}
+ \mathcal{G}^K\left(\frac{49}{45}, \frac{49}{45}\right).
$$

Properties of this game (and recognizing that multiplying all payoffs by 45 eliminate unsightly fractions) become apparent. The three Nash cells are along the $\mathcal{G}^N_{Sexes,2}$ diagonal. According to the $\mathcal{G}^B_{Sexes,2}$ Pareto superior cell at ML (M = Middle), the game structure continues its cooperative intent of having both spouses reach agreement. This comment is further reflected by the $\mathcal{G}^B_{Sexes,2}$ Pareto inferior cell at BR, suggesting a strong group dislike for the (Indecision, Cleaning) option.

4.3.1 Ultimatum Game

A $\mathcal{G}^B$ role is nicely captured with the ultimatum game. This is where each of the two players knows everything about the game, including who is doing what. The only unknown is the other player's identify.

Player 1 is offered a fixed sum, say ten one-dollar bills, with the following conditions:

- Player 1 is to offer the unknown Player 2 a portion of the dollar bills denoted by $X.
- If Player 2 accepts, Player 1 has $(10-X) and Player 2 has $X.
- If Player 2 rejects the offer, then neither player receives anything.

The challenge is to determine how much to offer, the X value.

As computed below, the Nash equilibrium is $X = 1$. This makes sense; if the first player offers nothing, the second player has nothing to lose by rejecting. Sure, accepting the offer of zero is a weak Nash equilibrium, but rather than being indecisive between Yes or No, Player 2 most surely would reject. But if the first player offers a dollar, the second player has at least something—not much, but something. And so, expect Player 1 to offer Player 2 a single dollar.

The mystery associated with this game is that in experiments involving actual money, a common choice is *not* $X = 1$; it is to evenly divide the amount to offer $X = 5$. This unexpected phenomenon was continuously supported in an extensive study involving cultures across the world [9]. Of course, rather than a 50/50 split, the value could reflect cultural aspects of different societies. But do not expect the Nash equilibrium to be the choice.

Among the several explanations, Brian Skyrms [35] used replicator dynamics with evolutionary game theory to show that, with repeated interactions, a 50/50 social norm could emerge. Using a wider class of dynamics, Saari [30, Chap. 1] obtained a broader list of possibilities leading to the 50/50 split. An alternative explanation is given next.

Modeling this game depends on the culture. With European/North American societies, it is reasonable to argue that offering more than half is unthinkable, so $X \leq 5$. So the game is as follows where Column is Player 1. The form for other cultures might differ.

$$\mathcal{G}_U = \begin{array}{c|cc|cc|cc|cc|cc|cc} \text{Offer:} & \multicolumn{2}{c}{0} & \multicolumn{2}{c}{1} & \multicolumn{2}{c}{2} & \multicolumn{2}{c}{3} & \multicolumn{2}{c}{4} & \multicolumn{2}{c}{5} \\ \hline \text{Yes} & 0 & 10 & 1 & 9 & 2 & 8 & 3 & 7 & 4 & 6 & 5 & 5 \\ \text{No} & 0 & 0 & 0 & 0 & 0 & 0 & 0 & 0 & 0 & 0 & 0 & 0 \end{array} \qquad (4.16)$$

Row's Nash components are the average of the amount offered and the zeros in the No row of $\mathcal{G}_U$. Column's $\mathcal{G}^N$'s entries are equally easy to find: Column's cell value in the No row is zero. In the Yes row, the entries are the differences of Column's $\mathcal{G}_U$ entry from Column's row average of $\frac{45}{6} = 7.5$.

$$\mathcal{G}_U^N = \begin{array}{l|rr|rr|rr|rr|rr|rr|} \text{Yes} & 0 & 2.5 & .5 & 1.5 & 1 & .5 & 1.5 & -.5 & 2 & -1.5 & 2.5 & -2.5 \\ \hline \text{No} & 0 & 0 & -.5 & 0 & -1 & 0 & -1.5 & 0 & -2 & 0 & -2.5 & 0 \end{array}$$

Offering zero is a weak Nash cell, which was discussed and dismissed. Only two $\mathcal{G}_U^N$ cells have positive entries; both are in the top row where Column decides. So the strict Nash cell is where $X = 1$.

As this use of coordinates for games has shown, expect games to experience a tension between the $\mathcal{G}^N$ me-choice and the $\mathcal{G}^B$ we-option. To compute $\mathcal{G}_U^B$, subtract $\mathcal{G}_U^K(1.25, 3.75)$ from $\mathcal{G}_U^{Averages}$ to obtain

$$\mathcal{G}_U^B = \begin{array}{l|rr|rr|rr|rr|rr|rr|} \text{Yes} & -\frac{5}{4} & 3.75 & -\frac{3}{4} & 3.75 & -\frac{1}{4} & 3.75 & \frac{1}{4} & 3.75 & \frac{3}{4} & 3.75 & \frac{5}{4} & 3.75 \\ \hline \text{No} & -\frac{5}{4} & -3.75 & -\frac{3}{4} & -3.75 & -\frac{1}{4} & -3.75 & \frac{1}{4} & -3.75 & \frac{3}{4} & -3.75 & \frac{5}{4} & -3.75 \end{array}$$

where the Pareto superior cell is the 50/50 split!

Not really a surprise.
In examining this tension between the we-cooperative equal split ($X = 5$) and a me-Nash choice of $X = 1$, recall that this game mandates cooperation: For Player 1 to receive *any money*, Player 2 must cooperate. With this framing mind set, rather than the Nash cell serving as the focal starting place, expect Player 1's analysis to begin with the we-cooperative cell of an equal split. The me-Nash force could, of course, reduce the offer. With moderate amounts of money, it is reasonable to expect the cooperative force to prevail. (If each dollar bill is replaced with a million dollars, it is not clear whether a conservative cooperative approach, or the Nash term would have more influence.)

These comments counter a standard assumption that a Nash point should dominate. Why should it? As the experimentalist Katri Sieberg pointed out in a private conversation, it is difficult in an experiment to describe to a participant what to do without imposing some sense of expected cooperation. As she pointed out, a person finding $10 on the sidewalk need not have even a fleeting thought about sharing, but should the same person be involved in an Ultimatum or Dictator[2] game experiment, the game's instructions already convey a sense of joint action. Common manifestations of the Dictator game, such as leaving a tip, support Sieberg: At a restaurant, cooperation is a norm where tips are standard (and actions somewhat monitored); this is more common than leaving a tip for an hotel maid.

However, instructions for the following two-player games, where each is to select an integer between 2 and 6, can spark competitiveness rather than cooperation. Here, if both select the same number, each receives that amount of money. But, should different amounts be selected, the person choosing the *smaller* number is rewarded with the larger value *plus* two, while the person who selected the larger number is punished by being assigned the lower value. With (3, 4), for instance, by selecting

---

[2]Think of this as the Ultimatum game where whether the second person accepts or rejects has no impact on what the first player receives.

the smaller 3, Row receives $4 + 2 = 6$ while Column only gets 3. (All payoffs are in Eq. 4.17.) What outcome should be expected?

$$
\mathcal{G} = \begin{array}{c|ccccc}
 & 2 & 3 & 4 & 5 & 6 \\
\hline
2 & 2\ 2 & 5\ 2 & 6\ 2 & 7\ 2 & 8\ 2 \\
3 & 2\ 5 & 3\ 3 & 6\ 3 & 7\ 3 & 8\ 3 \\
4 & 2\ 6 & 3\ 6 & 4\ 4 & 7\ 4 & 8\ 4 \\
5 & 2\ 7 & 3\ 7 & 4\ 7 & 5\ 5 & 8\ 5 \\
6 & 2\ 8 & 3\ 8 & 4\ 8 & 5\ 8 & 6\ 6
\end{array} \quad \text{with} \quad \mathcal{G}^K(4.8, 4.8) \tag{4.17}
$$

The $\mathcal{G}^B$ Pareto superior point (Eq. 4.19) captures the cooperative choice of $(6, 6)$. But if the game is framed in a competitive manner, the counterproductive weak Nash cell of $(2, 2)$ (Eq. 4.18) probably plays a more dominate role.

$$
\mathcal{G}^N = \begin{array}{c|ccccc}
 & 2 & 3 & 4 & 5 & 6 \\
\hline
2 & 0\ \ \ \ 0 & 1.6\ \ \ \ 0 & 1.2\ \ \ \ 0 & .8\ \ \ \ 0 & .4\ \ \ \ 0 \\
3 & 0\ \ 1.6 & -.4\ -.4 & 1.2\ -.4 & .8\ -.4 & .4\ -.4 \\
4 & 0\ \ 1.2 & -.4\ \ 1.2 & -.8\ -.8 & .8\ -.8 & .4\ -.8 \\
5 & 0\ \ \ .8 & -.4\ \ \ .8 & -.8\ \ \ .8 & -1.2\ -1.2 & .4\ -1.2 \\
6 & 0\ \ \ .4 & -.4\ \ \ .4 & -.8\ \ \ .4 & -1.2\ \ \ .4 & -1.6\ -1.6
\end{array} \tag{4.18}
$$

$$
\mathcal{G}^B = \begin{array}{|ccccc|}
\hline
-2.8\ \ -2.8 & -1.4\ \ -2.8 & 0\ \ -2.8 & 1.4\ \ -2.8 & 2.8\ \ -2.8 \\
-2.8\ \ -1.4 & -1.4\ \ -1.4 & 0\ \ -1.4 & 1.4\ \ -1.4 & 2.8\ \ -1.4 \\
-2.8\ \ \ \ \ 0 & -1.4\ \ \ \ \ 0 & 0\ \ \ \ \ 0 & 1.4\ \ \ \ \ 0 & 2.8\ \ \ \ \ 0 \\
-2.8\ \ \ 1.4 & -1.4\ \ \ 1.4 & 0\ \ \ 1.4 & 1.4\ \ \ 1.4 & 2.8\ \ \ 1.4 \\
-2.8\ \ \ 2.8 & -1.4\ \ \ 2.8 & 0\ \ \ 2.8 & 1.4\ \ \ 2.8 & 2.8\ \ \ 2.8 \\
\hline
\end{array} \tag{4.19}
$$

4.3.2 Legislator Pay Raise Game

Another game exhibiting a me-we tension is the three-player "legislator game." Here, each legislator votes either for or against a bill to increase their salary. Everyone wants more money, so each wants the bill to pass. On the other hand, this bill will not be popular with constituents who loudly complain about public expenditures, so voting for the bill carries a cost. This setting, where each legislator wants the bill to pass while being able to vote against it, forms a threshold-congestion game (Sect. 3.4.5).

In designing the game, let {vote against the bill, bill passes} be worth 4 units. Assign 2 units for {vote for the bill, bill passes}. But {vote against the bill, bill loses} is worth zero units. The nightmare scenario {vote for the bill, bill loses} is worth -2 units. This three-person games are

$$
\mathcal{G}_{Front} = \begin{array}{c|cc}
 & \text{Yes} & \text{No} \\
\hline
\text{Yes} & 2\ \ 2\ \ 2 & 2\ \ 4\ \ \ \ \ 2 \\
\text{No} & 4\ \ 2\ \ 2 & 0\ \ 0\ \ -2
\end{array} \qquad
\mathcal{G}_{Back} = \begin{array}{c|cc}
 & \text{Yes} & \text{No} \\
\hline
\text{Yes} & 2\ \ \ \ \ 2 & 4\ -2\ \ 0\ \ 0 \\
\text{No} & 0\ -2\ \ 0 & 0\ \ 0\ \ 0
\end{array} \tag{4.20}
$$

where "Front" represents the third player's vote of "Yes." As an example, the TLF (Top-Left-Front) cell is where the bill passes with an unanimous vote.

The description, where each legislator prefers to vote no while the other two votes yes signals a personal vs. group tension, which the coordinates nicely extract. The sum of each player's payoffs over the eight cells is eight with an average of one, so the kernel is $G^K(1, 1, 1)$. The Nash component, which follows, has four Nash cells.

$$
G^N_{Front} = \begin{array}{c|ccc|ccc}
 & \multicolumn{3}{c}{\text{Yes}} & \multicolumn{3}{c}{\text{No}} \\
\hline
\text{Yes} & -1 & -1 & -1 & 1 & 1 & 1 \\
\text{No} & 1 & 1 & 1 & -1 & -1 & -1
\end{array}, \quad
G^N_{Back} = \begin{array}{c|ccc|ccc}
 & \multicolumn{3}{c}{\text{Yes}} & \multicolumn{3}{c}{\text{No}} \\
\hline
\text{Yes} & 1 & 1 & 1 & -1 & -1 & -1 \\
\text{No} & -1 & -1 & -1 & 1 & 1 & 1
\end{array}
$$

$$\tag{4.21}$$

Three of them are expected; they are where the bill passes while a free-riding legislator snares a personal benefit by voting "no." Of interest is the fourth (BRBa = Bottom, Right, Back) Nash cell where the bill is *unanimously defeated*!

These four G^N Nash cells are indistinguishable, so any differentiating characteristics *must* reflect the game's G^B component. Although the TLF G^N cell's Eq. 4.20 outcome (the bill passes) would be welcomed by each legislator, TLF is a *repelling cell*, whereby any of the three could be personally rewarded by voting no. The G^B TLF component, which is its Pareto superior cell, explicitly exposes this me-we tension.

$$
G^B_{Front} = \begin{array}{c|ccc|ccc}
 & \multicolumn{3}{c}{\text{Yes}} & \multicolumn{3}{c}{\text{No}} \\
\hline
\text{Yes} & 2 & 2 & 2 & 0 & 2 & 0 \\
\text{No} & 2 & 0 & 0 & 0 & 0 & -2
\end{array} \quad
G^B_{Back} = \begin{array}{c|ccc|ccc}
 & \multicolumn{3}{c}{\text{Yes}} & \multicolumn{3}{c}{\text{No}} \\
\hline
\text{Yes} & 0 & 0 & 2 & -2 & 0 & 0 \\
\text{No} & 0 & -2 & 0 & -2 & -2 & -2
\end{array}
$$

$$\tag{4.22}$$

Notice how G^B's Pareto inferior cell (BRBa) coincides with the G^N Nash cell of an unanimous vote *against* the bill. *It is this G^B entry that makes BRBa a Pareto inferior Nash point.* In contrast, the G^B Pareto superior cell (TLF) is positioned over a G^N repelling cell: cooperatively (G^B terms) all three want the bill to pass; privately (G^N factors), each wants the bill to pass while voting against it.

This positioning of G^B Pareto superior and inferior cells relative to G^N Nash and repelling cells creates a multiplayer Prisoner's Dilemma. The two "yes" voters may not want the remaining voter to free ride at their expense; they may prefer a cooperative unanimous vote (as indicated by G^B). With a single shot game, not much can be done; in a repeated setting (repetition might reflect negotiations over this particular bill, or the passing of multiple bills), retaliatory actions (e.g., tit-for-tat, grim trigger) become feasible.

Following the lead of Theorem 2.4, the first step is to determine whether the reneging player can be punished. Fortunately, we know where to find punishment opportunities; they are in the arrays emanating from the G^B Pareto inferior cell. In this game, they represent the hard-love "If you want the bill passed, you better vote for it, or else both of us will vote against it" type of party discipline that can be seen in practice. A computation similar to that in the proof of Theorem 2.4 shows that cooperation can be expected if $\delta > \frac{1}{2}$.

4.4 Modeling and Creating Games

Using these coordinates to analyze specified games offers fresh insights. Of more interest and fun is to use the decomposition to create games with novel features. As discovered with 2×2 games, what makes the coordinate system a powerful tool is that it separates the design of individual opportunities ($\mathcal{G}^N$) from what it takes to make certain cells attractive to both players ($\mathcal{G}^B$). The properties of $\mathcal{G}^B$ are examined in Sects. 4.4.1 and 4.5.

4.4.1 Differences in the $\mathcal{G}^B$ Structure

The $\mathcal{G}^B$ structure gains interest with more pure strategies. Using examples to illustrate, the $\mathcal{G}^B$ Pareto superior cell in a 2×2 game can be located wherever desired. With $\beta_1 = 6$, $\beta_2 = 5$, a $\mathcal{G}^B$ with TL as the Pareto superior cell is $\mathcal{G}_2^B$ in Eq. 4.23. This choice requires Row's first column $\mathcal{G}_2^B$ entries to be 6, and Column's first row $\mathcal{G}_2^B$ entries to be 5. The x and y values of $x = -6$ and $y = -5$ are determined by Eq. 4.7.

$$
\mathcal{G}_2^B = \begin{array}{|c|c|} \hline 6\ \ 5 & x\ \ 5 \\ \hline 6\ \ y & x\ \ y \\ \hline \end{array}, \quad
\mathcal{G}_3^B = \begin{array}{|c|c|c|} \hline 6\ \ 5 & x_1\ \ 5 & x_2\ \ 5 \\ \hline 6\ \ y_1 & x_1\ \ y_1 & x_2\ \ y_1 \\ \hline 6\ \ y_2 & x_1\ \ y_2 & x_2\ \ y_2 \\ \hline \end{array}
\tag{4.23}
$$

In 2×2 games, the Pareto superior position uniquely determines the location of the $\mathcal{G}_2^B$ Pareto inferior cell; here it *must* be BR. As discussed (e.g., right after Eq. 2.22), $\mathcal{G}^B$'s Pareto superior and inferior cells for 2×2 games *always are* diametrically opposite. This mandated structure is what causes the behavior for several classes of games. With two Nash cells, for instance, placing the $\mathcal{G}^B$ Pareto superior choice over one of them (as in the stag hunt) makes that cell more attractive, but the *required* location of $\mathcal{G}^B$'s Pareto inferior cell is over the other Nash cell, which reduces its appeal to make one Nash cell preferred over the other.

Alternatively, placing $\mathcal{G}^B$'s Pareto superior choice over a repelling cell can introduce coordination issues (e.g., Battle of the Sexes), or it can make the repelling cell more inviting while (with the compulsory placement of $\mathcal{G}^B$'s Pareto inferior cell) forcing the other repelling cell into a reprehensible status (as in Hawk–Dove). Even more, the only way to elevate a non-Nash cell to a group preferred cell (Theorem 2.3) is with the positioning of the $\mathcal{G}^B$ Pareto superior cell. Accompanying the cell's promotion is that the $\mathcal{G}^B$ Pareto inferior entries (see Eq. 2.38) make repeated game punishment (e.g., tit-for-tat or grim trigger) feasible by ensuring there is an array of outcomes to punish a reneging player.

These properties can change with added options. To assist with comparisons, the $\mathcal{G}_3^B$ Pareto superior cell in Eq. 4.23 also is at TL. The only constraints (Eq. 4.7) on the remaining variables are $x_1 + x_2 = -6$, $y_1 + y_2 = -5$, so the $\mathcal{G}_3^B$ Pareto inferior cell can be *any* of the four choices outside of the first row and first column. While one of

x_j's and y_i's must be negative (Eq. 4.7), the others could be positive, zero, or negative, so $\mathcal{G}^B$ could enhance the payoffs for *several cells*. (If the x_j, y_j are all negative, the $\mathcal{G}_3^B$ Pareto inferior cell need not guarantee the availability of a punishment strategy. See Theorem 6.3.)

$$
\begin{array}{|cc|cc|cc|}
\hline
-1 & -1 & 0 & 0 & 0 & 0 \\
\hline
0 & 0 & -1 & 2 & 0 & 0 \\
\hline
0 & 0 & 0 & 0 & 2 & 2 \\
\hline
\end{array}
\rightarrow
\begin{array}{|cc|cc|cc|}
\hline
5 & 4 & 5 & 5 & -11 & 5 \\
\hline
6 & 4 & 4 & 6 & -11 & 4 \\
\hline
6 & -9 & 5 & -9 & -9 & -7 \\
\hline
\end{array}
\rightarrow
\begin{array}{|cc|cc|cc|}
\hline
5 & 4 & -11 & 5 & 5 & 5 \\
\hline
6 & 4 & -12 & 6 & 5 & 4 \\
\hline
6 & -9 & -11 & -9 & 7 & -7 \\
\hline
\end{array}
$$

$$(4.24)$$

To illustrate, the first Eq. 4.24 bimatrix has a repelling (1, 1) cell, a hyperbolic-row (2, 2) cell, and a (3, 3) Nash cell. Adding the $\mathcal{G}_3^B$ (defined by $x_1 = 5$, $y_1 = 4$, $x_2 = -11$, $y_2 = -9$) to the first bimatrix defines the second one with *four* group preferred cells. What advances these cells to a preferred status are the positive values for one of the x_i's and one of the y_j's. Equation 4.7 now *guarantees* a large negative value for the remaining x_i and y_j, which ensures that any of these four cells can be cooperatively enforced in a repeated situation. The reason is that the Pareto inferior cell ensures there are arrays of punishment possibilities for a reneging player—here the third column for Row and the bottom row for Column. The same holds for the last bimatrix; it only changes the location of the $\mathcal{G}_3^B$ Pareto inferior cell. (Here $x_2 = 5$, $y_1 = 4$, $x_1 = -11$, $y_2 = -9$.)

Armed with these observations, a theorem, similar to Theorem 2.4, can be fashioned asserting that if $\mathcal{G}^B$ makes some non-Nash cells group preferred along with certain properties, then possibilities are created to penalize a non-cooperating player in a repeated setting. An obvious condition is if $\mathcal{G}^B$ has only one cell with negative entries.

Most $\mathcal{G}^B$ properties become evident from the mandated structure: Row's entries for each column remain the same, Column's entries for each row are fixed.[3] These conditions box in options of assisting some $\mathcal{G}^N$ cell, while not others; the main constraint is Eq. 4.7. And so, while simple, the $\mathcal{G}^B$ structure can generate complicated games.

It remains to describe the structure of $\mathcal{G}^N$, which begins in Sect. 4.5. But first, some needed technical comments about the dimensions of the component subspaces are specified.

4.4.2 Dimension Counting

Recall that $\mathbb{N}$, $\mathbb{B}$, and $\mathbb{K}$ are, respectively, the subspace of Nash, Behavioral, and Kernel entries in the space of games $\mathbb{G}$. Each subspace is identified, respectively,

---

[3]There is an interesting comparison with a $\mathcal{G}^B$ structure and how a cell is evaluated to determine whether it is Nash. To illustrate with $\mathcal{G}_3^B$ (Eq. 4.23), to check if TL is a Nash cell, Row compares his entries in the L column; these are his fixed $\beta = 6$ values. Similarly, Column compares her values in the T row, and her $\beta = 5$ values are in this row.

with the entries of $\mathcal{G}^N$, $\mathcal{G}^B$, and $\mathcal{G}^K$. Computing the dimensions of these subspaces is the same as finding the number of admissible entries for these three $\mathcal{G}$ components.

First, find the number of possible entries for an n-person $k_1 \times k_2 \times \cdots \times k_n$ game, where, for convenience, assume that $k_1 \geq k_2 \geq \cdots \geq k_n$. In a 2×2 game, the number of cells is $(2)(2) = 4$, where each cell has two entries, so the total number is $(4)2 = 8$. In a general n-person game, the number of cells is the product $k = k_1 k_2 \ldots k_n$. Each cell has the payoff for each of the n players, so the total number of entries is kn. This means that a $6 \times 5 \times 4 \times 3$ game has $(360)(4) = 1440$ entries; the space of such games can be identified with the 1440-dimensional space $\mathbb{R}^{1440}$.

What a monstrous object! It is difficult to envision a space with more than three dimensions, which dashes any hope of understanding a 1440-dimensional one. No wonder game theory rapidly becomes so complicated by adding options. To have a slightly more manageable object, the component subspaces are described. The simplest choice is $\mathcal{G}^K$, where its entries are the averages of each player's payoffs. Thus the dimension of $\mathbb{K}$ is n, so for the $6 \times 5 \times 4 \times 3$ game, the dimension of $\mathbb{K}$ is four.

A player's array of values in $\mathcal{G}^B$ is determined by the pure strategy choices of the other $(n-1)$ players. Thus the jth player confronts $\frac{k}{k_j}$ such arrays. A player's $\mathcal{G}^B$ entries in a specified array have the same value, so the jth player has $\frac{k}{k_j}$ choices of β values. But Eq. 4.7 constraint means that $\frac{k}{k_j} - 1$ values are independent and determine that last value. Thus, overall n players, there are $\sum_{j=1}^{n}[\frac{k}{k_j} - 1] = k\left[\sum_{j=1}^{n}\frac{1}{k_j}\right] - n$ independent choices, so the dimension of $\mathbb{B}$ is $k\left[\sum_{j=1}^{n}\frac{1}{k_j}\right] - n$. For the space of $6 \times 5 \times 4 \times 3$ games, $\mathbb{B}$'s dimension is $360(\frac{1}{6} + \frac{1}{5} + \frac{1}{4} + \frac{1}{3}) - 4 = 338$.

To compute the dimension of $\mathbb{N}$, the number of arrays for the jth player is $\frac{k}{k_j}$ and each array has k_j cells. According to Eq. 4.6, only $k_j - 1$ of them are independent, so the dimension of $\mathbb{N}$ is $\sum_{j=1}^{n}\frac{k}{k_j}(k_j - 1) = k\sum_{j=1}^{n}\frac{k_j - 1}{k_j}$. With the space of $6 \times 5 \times 4 \times 3$ games, the dimension of $\mathbb{N}$ is $360[\frac{5}{6} + \frac{4}{5} + \frac{3}{4} + \frac{2}{3}] = 1098$. As it must be, the sum of the dimensions of $\mathbb{N}$, $\mathbb{B}$, and $\mathbb{K}$ equals the dimension of $\mathbb{G}$—the space of these games.[4] For the space of $6 \times 5 \times 4 \times 3$ games, this is $1098 + 338 + 4 = 1440$.

The next statement catalogues these dimensional statements along with a count on the number of possible Nash cells.

Theorem 4.4 *For the space of $k_1 \times k_2 \times \ldots \times k_n$ games $\mathbb{G}$, where $k_1 \geq k_2 \geq \ldots \geq k_n$, let $k = k_1 k_2 \ldots k_n$. The dimension of $\mathbb{G}$ is kn, so it can be identified with the Euclidean space $\mathbb{R}^{kn}$. $\mathbb{G}$ can be orthogonally divided into the subspaces $\mathbb{N}$, $\mathbb{B}$, and $\mathbb{K}$ with subspace dimensions of, respectively, $k\sum_{j=1}^{n}\frac{k_j - 1}{k_j}$, $k\left[\sum_{j=1}^{n}\frac{1}{k_j}\right] - n$, and n.*

For integer α satisfying $0 \leq \alpha \leq k_2 k_3 \ldots k_n = \frac{k}{k_1}$, there is a game with precisely α strict Nash cells. No game can have $\alpha > \frac{k}{k_1}$ strict Nash cells.

---

[4] An easy proof is that the dimension of $\mathbb{N}$ is $k\sum_{j=1}^{n}\frac{k_j - 1}{k_j} = kn - k\sum_{j=1}^{n}\frac{1}{k_j}$, which is the dimension of $\mathbb{G}$ minus the sum of the dimensions of $\mathbb{B}$ and $\mathbb{K}$.

If each player has a nonzero $\mathcal{G}^B$ entry, then for a two-player games, $\mathcal{G}^B$ has a Pareto superior cell with all positive entries, and a Pareto inferior cell with all negative entries. For three or more players, $\mathcal{G}^B$ need not have such cells. In 2×2 games the $\mathcal{G}^B$ Pareto superior and Pareto inferior cells are diametrically opposite one another; this need not be true with more pure strategies.

Thus there are $6 \times 5 \times 4 \times 3$ games with $5 \times 4 \times 3 = 60$ Nash cells. The $\mathcal{G}^B$ structures can further modify these cells to introduce all sorts of unexpected behavior.

Proof Constructing games with the maximum number of strict Nash cells makes it clear how to construct games with any smaller number. To show that a $k_1 \times k_2$ game can have k_2 strict pure Nash equilibria, it suffices to create a $\mathcal{G}^N$ with these properties; the only constraint on selecting $\eta_{i,s,j}$ terms is Eq. 4.6. To be strict, for each j and array s, one $\eta_{i,s,j}$ term is the largest. Only ordinal information is being used, so it suffices to assume that one value in the array is positive and the rest are negative.

There are k_1 first column cells to place player one's positive $\eta_{i,1;1}$ entry; let it be $\eta_{i^*,1;1}$. To make $(i^*, 1)$ a Nash cell, let it have player two's positive entry for that row. These choices specify positive signs of player one's first column entry and player two's first row.

For the induction step, assume up to player two's strategy j, $j < k_2$ that j Nash points are created. This identifies signs for player one's entries for the first j columns (positive in all Nash cells, negative otherwise), and signs for player two's entries for j rows. For player two's strategy $(j + 1)$, there remain $k_1 - j > 0$ rows where the sign of player one's entry is not specified. Select one of these $k_1 - j$ rows to be the Nash cell by containing both player's positive entry; this completes this induction step. By construction, it is impossible to have more than k_2 Nash points because all of player two's positive entries have been assigned.

For the second induction step, assume that $\alpha = k_2 k_3 \ldots k_j$ strict Nash equilibria can be created for $k_1 \times \ldots \times k_j$ games, but no games can have more equilibria. It must be shown that this assertion extends to $k_1 \times \ldots \times k_j \times k_{j+1}$ games. For the $(j + 1)$th player's first strategy, select one of these $k_1 \times \ldots \times k_j$ games with the maximum number of Nash points. Assign the $(j + 1)$th agent's positive value for each Nash point, which makes it a Nash point for this $k_1 \times \ldots \times k_{j+1}$ game. Other choices of the $(j + 1)$th agent are negative. But, for the $k_{j+1} - 1$ the assignment of signs is not made. So for the kth strategy, construct one of the $k_1 \times \ldots \times k_j$ games with the maximum number of Nash points. This can only be done for each strategy, which leads to the conclusion.

To prove the two-player comment about $\mathcal{G}^B$ Pareto superior and inferior cells, the $\mathcal{G}^B$ structure requires Row's maximum $\mathcal{G}^B$ entry to be repeated as Row's entry in some column, while Column's maximum $\mathcal{G}^B$ value is repeated as Column's entry in some row. The $\mathcal{G}^B$ Pareto superior cell is where the column and row intersect. A similar statement holds for each player's $\mathcal{G}^B$ minimum value. (There can be several Pareto superior and/or inferior cells should several β values agree.)

To prove that $\mathcal{G}^B$ need not have a Pareto superior or inferior cell with more than two players, consider the following $2 \times 2 \times 2$ game:

$$\mathcal{G}^B_{Front} = \begin{array}{|ccc|ccc|} \hline 5 & 4 & -8 & -5 & 4 & -2 \\ 5 & -7 & 4 & -5 & -7 & 6 \\ \hline \end{array}, \quad \mathcal{G}^B_{Back} = \begin{array}{|ccc|ccc|} \hline 3 & 4 & -8 & -3 & 4 & -2 \\ 3 & -1 & 4 & -3 & -1 & 6 \\ \hline \end{array},$$

$$(4.25)$$

This is a $\mathcal{G}^B$ component because entries in appropriate arrays are the same and the sum of each player's β values equals zero (Theorem 4.1). The maximum β value for Row, Column, and Face are, respectively, 5, 4, and 6. As no $\mathcal{G}^B$ cell has all three values (actually, no cell has all positive values), no Pareto superior cell exists. The same comment holds for the minimum β values for the three players. The proof for two players, then, is that in a rectangle, a vertical and a horizontal line must meet. In three dimensions, vertical, East-West, and North-South lines need not intersect. □

4.5 Structure of Two-Person Games

Generically, $k \times k$ games can have no more than k Nash cells (Theorem 4.4), but the number of Nash equilibria can reach the exponentially large $2^k - 1$. With $k = 40$, for instance, the 1,099,511,627,775 possible Nash equilibria equal the number of seconds of time since the last glacial period 34,865 years ago—an era where leading game theory concerns involved confronting a mammoth or mastodon. As k grows in value, even a bank of computers just trying to *count* the trillions of trillions of ...equilibria becomes insurmountable.

To expand upon this mind-boggling number, suppose a model in defense or economics is stripped to a 100×100 game, which for practical purposes can be a significant reduction. Here[5] there can be over a nonillion (a million times a trillion times a trillion, or 10^{30}) equilibria, which is more than a billion times the number of seconds of time since the Big Bang. Forget $1,000 \times 1,000$ games if only because it difficult to concoct an understandable "Gee Whiz" comparison. Although the number of equilibria quickly becomes impossible to count, such games are easy to construct.

Theorem 4.5 *A $k \times k$ game can have $2^k - 1$ Nash equilibria, which consists of k strict Nash equilibria and $2^k - (k + 1)$ NME. If k is an even integer, then a $k \times k$ game with zero Nash cells can have $2^{\frac{k}{2}} - 1$ NME.*

The theorem is proved by creating examples. The $k \times k$ game $\mathcal{G}$ with $(1, 1)$ entries along the main diagonal and zeros elsewhere has precisely k Nash cells. Accompanying each of the $\binom{k}{2} = \frac{k(k-1)}{2}$ pairs of Nash cells is an NME, with the $(p, q) = (\frac{1}{2}, \frac{1}{2})$ probability of selecting one or the other pure Nash choice. Each of the $\binom{k}{3} = \frac{k(k-1)(k-2)}{3!}$ triplets of Nash cells admits another NME (where $(p, q, r) = (\frac{1}{3}, \frac{1}{3}, \frac{1}{3})$) providing guidance. Continuing in this manner, the number of NME is

$$\sum_{j=2}^{k} \binom{k}{j} = \sum_{j=0}^{n} \binom{n}{j} - (k + 1) = 2^k - (k + 1)$$

---

[5]Recall, $2^k \approx 10^{0.301k}$.

An example without Nash cells but an overabundance of NME is similarly created by replacing the (1, 1) entries of the first example with 2×2 matching penny bimatrices; all other cells have zeros. To attach a story to a $k = 8$ game, suppose there are four types of coins: pennies, nickels, dimes, and quarters. Each player selects a coin and then selects either Heads or Tails. If the players selected different coins, the payoff for both is zero. If both players selected the same coin, Bob wins if the faces agree, Sue wins if they disagree.

For each 2×2 bimatrix, the NME is $p = q = \frac{1}{2}$. Associated with each pair of matching coin games is an NME essentially describing which of the two coins the players select, which is 50–50. Thus the NME for each of the four options is $\frac{1}{4}$. Everything continues as above, with triplets, etc. The computations for the number of NME is as above.

Other mix-and-match examples are similarly constructed. Again along the main diagonal, let some terms be Nash cells, others 2×2 matching coin cells, and maybe throw in some 3×3 games. Of importance for what is described next, the primary units (e.g., Nash cells, matching penny cells, etc.) should be first analyzed. Then consider secondary units consisting of combinations of these primary ones; e.g., find the Nash equilibria of selecting one of these primary terms. Quickly the number of equilibria become incredibly large. A partial way out of this headache, which is developed next, is based on determining the structure (i.e., Nash, hyperbolic, and repelling cell arrangements) of two-person games.

4.5.1 Where's Waldo?

Knowing a game's $\mathcal{G}^N$ term simplifies the search for Nash cells. The stumbling block is to identify the Nash dynamic, which is needed to appreciate possible modeling possibilities. Compare this search problem with the child's game of finding a character called Waldo, who is hiding somewhere in a detailed illustration. An efficient way to solve the Waldo challenge is to know where to look. As developed starting in Sect. 4.5.2, the same holds for games. Namely, the structure developed in Sect. 4.5.2 identifies where to look in our "Waldo" search for a game's Nash structure, which includes its NME.

In particular, the Nash dynamic suggests where some NME are hiding in $k_1 \times k_2$ games. To explain with a 2×2 game's NME of $(p, q) = (\frac{1}{3}, \frac{3}{4})$, a $q < \frac{3}{4}$ indicates that Row's best response is to move in one direction, while $q > \frac{3}{4}$ dictates the opposite best response shift. Thus, treat a mixed strategy as a road sign at a fork providing advice about which way to move depending on the circumstances. This intimate connection between NME and best response choices suggests that a way to unearth NME potential hiding places is to discover where the best response dynamic creates twisting shifts.

The approach is similar to what was done for 2×2 games (Sect. 2.4.1) with the Definition 1.2 and Theorem 1.1 characterization of game cells. Figure 1.2 dynamic, with arrows along the edges aimed at each agent's best response cell, arises in larger games to define what we call *irreducible* $j \times j$ *blocks*. As examples, the Nash cells in the first example for the proof of Theorem 4.5 are 1×1 irreducible blocks. In contrast, each of the 2×2 matching coin games in the second example corresponds to 2×2 irreducible blocks.

Secondary blocks combine irreducible blocks. Similar to how q on one side or the other of $\frac{3}{4}$ moves Row's best response in different directions, the NME of these secondary blocks, should they exist (and they may not), define forks in the road to determine which irreducible unit should be considered. With Theorem 4.5 examples, these secondary equilibria identify whether the best response should move toward particular Nash cells or 2×2 games.

A strategy to capture the Nash dynamics is obvious: fill in the arrows. Once the approach is described and the structure of irreducible blocks is found, a modeling tool emerges: It becomes possible to dream up a coveted best response dynamic for $k_1 \times k_2$ games and, with imaginative $\mathcal{G}^B$ choices, to construct all sorts of games.

4.5.2 Next Response Dynamic

To avoid being smothered in a swamp of special cases, adopt the generic assumption that each agent has a unique best response in reaction to whatever is the other player's pure strategy.[6] This means that each column of a $k_1 \times k_2$ game has a unique cell with Row's best response, and each row has a unique cell with Column's best response. To depict these choices, place a bullet (a "•") on the left or right side of an identified cell if it represents, respectively, Row's or Column's best choice (see Fig. 4.3a). Different positioning of these bullets lead to different equilibria structures.

The "next response dynamic"[7] starts with a cell that has a bullet and moves in a column-row or row-column alternating fashion to the next cell with a bullet. The dynamic is uniquely defined: Should the starting point be Row's best response in some column, it defines a row in which Column has a unique maximum. This cell selected by Column defines a column in which Row has a maximum entry. The dynamic continues.

To illustrate with Fig. 4.3a (the (i, j) cell is in the ith row from the top and jth column from the left), start at cell $(1, 1)$, which is Row's best response in the first column. In this first row, Column's next response is to move (top arrow in Fig. 4.3b) to cell $(1, 3)$. Here in column 3, it now is Row's turn. So Row's next response is to move to cell $(2, 3)$ as indicated by the right-side downward arrow in Fig. 4.3b. From this $(2, 3)$ cell in row 2, Column's next response is cell $(2, 1)$, as given by the horizontal leftward pointing arrow. From $(1, 2)$, Row moves the dynamic to cell $(1, 1)$, which completes the loop. This loop, consisting of the top two cells in the first and third column, or cells $\{(1, 1), (2, 1), (1, 3), (2, 3)\}$, defines a 2×2 irreducible block. With

---

[6]With modifications, games not satisfying this assumption can be similarly analyzed.

[7]"Next response dynamic" was coined to avoid confusion with "best response curves" as in Fig. 1.2.

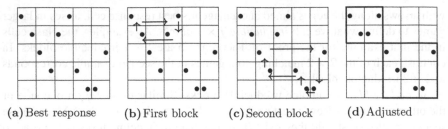

(a) Best response   (b) First block   (c) Second block   (d) Adjusted

Fig. 4.3 Finding mixed equilibria boxes

both players experiencing twisting best response shifts—requiring a switch from one strategy to the other—one must anticipate that the block has an NME.

There are remaining bullets, so start with one of them such as (3, 5), which is Column's best response in column 5. Row starts the next response dynamic by moving to (5, 5) (Fig. 4.3c), and Column next moves to (5, 4). Row moves to (4, 4), Column changes to (4, 2), Row moves to (3, 2), and Column competes the cycle by moving to starting cell of (3, 5). This next response dynamic defines a 3×3 irreducible block consisting of the bottom three cells in the third, fourth, and fifth columns; with each player experiencing several best response shifts, with twisting behavior, it is reasonable to expect an NME is this block.

All blocks with a bullet have been handled. The *twelve* remaining cells are repellers. Illustrating with (5, 1) (at the bottom of the first column), the bullet placements show that Row would do better by moving to (1, 1), and Column would find cell (5, 4) of personal advantage.

What a designer's delight! Just put a bullet in each column for Row and one in each row for Column, and, in this simple manner, a desired Nash dynamic can be created.

- A cell with two bullets is a Nash cell.
- A cell with one bullet is a hyperbolic cell. If the bullet represents Row's entry, then the cell is hyperbolic-column. Otherwise, it is hyperbolic-row.
- A cell with no bullets is a repelling cell.

Adding $\mathcal{G}^B$ components to this Nash structure generates an overflowing treasure chest of complicated behaviors.

An example
Figure 4.3 represents the strategic behavior of a continuum of games (values are only compared), so examples of the $\mathcal{G}^N$ structure of a game follow immediately. (Attached to each $\mathcal{G}^N$ is a continuum of $\mathcal{G}^B$ choices.) Remember (Theorem 4.1), the sum of Row's entries in any column and Column's entries in any row is zero. In each column, place Row's unique maximum entry in the designated cell; in each row, enter Column's unique maximum entry in the appropriate cell. An example illustrating Fig. 4.3, with maxima in boldface, is

$$\mathcal{G}^N_{Mixed} = \begin{array}{|cc|cc|cc|cc|cc|} \mathbf{5} & 1 & 3 & -4 & 4 & \mathbf{5} & 4 & -2 & -4 & 0 \\ -3 & \mathbf{3} & -6 & 0 & \mathbf{7} & -3 & 0 & 1 & -1 & -1 \\ 1 & 0 & \mathbf{6} & -3 & 1 & 4 & -6 & -6 & 3 & \mathbf{7} \\ -4 & -4 & 1 & \mathbf{9} & -8 & 5 & \mathbf{7} & 0 & -2 & -10 \\ 1 & -5 & -4 & 3 & -4 & -1 & -5 & \mathbf{5} & \mathbf{4} & -2 \end{array} \qquad (4.26)$$

To simplify the figures, notice how the first 2×2 block involves the first and third columns but not the second. Interchanging the second and third columns does not affect the game, but, as indicated in Fig. 4.3d, it creates a block of abutting lines and columns. The corresponding Eq. 4.26 adjustment is

$$\mathcal{G}^N_{Adjusted\ Mixed} = \begin{array}{|cc|cc||cc|cc|cc|} \mathbf{5} & 1 & 4 & \mathbf{5} & 3 & -4 & 4 & -2 & -4 & 0 \\ -3 & \mathbf{3} & \mathbf{7} & -3 & -6 & 0 & 0 & 1 & -1 & -1 \\ 1 & 0 & 1 & 4 & \mathbf{6} & -5 & -6 & -6 & 3 & \mathbf{7} \\ -4 & -4 & -8 & 5 & 1 & \mathbf{9} & \mathbf{7} & 0 & -2 & -10 \\ 1 & -5 & -4 & -1 & -4 & 3 & -5 & \mathbf{5} & \mathbf{4} & -2 \end{array} \qquad (4.27)$$

Figure 4.3d dark lines identify two irreducible blocks caused by the Nash dynamic of the hyperbolic cells. Each block experiences up-down and left-right shifts of best responses, so it is reasonable to anticipate that each block has an NME serving as a traffic cop indicating which way to move. Cells without bullets are repellers: It is clear from the figure that in the 3×2 collection of repelling cells, Row wishes to move to the 2×2 irreducible block, while Column finds the 3×3 block to be attractive. Thus, *repellers play a role in whether secondary NME exist;* if such an NME does exist, it directs the best response traffic toward one or the other irreducible block.

As an alert to what is discussed later: The analysis uses only each player's top choice in each column or row. Can other values have an impact? The answer is an emphatic *yes*, which is illustrated with the cautionary "anticipate" rather than "there exists an NME." The problem is captured by the Eq. 4.1 Battle of the Sexes example, which was constructed to show how the wife's second best entry in each column changed the location (and structure) of certain Nash equilibria. As described in Sect. 4.7.3, these other values are needed to analyze the governing structure among the repellers.

4.6 Structure of the Next Response Dynamic

Figure 4.3d identifies the Nash behavior of one 5×5 game. To find *all possible* $k_1 \times k_2$ structures, notice that a cell with an agent's best response for that row or column is either a Nash or a hyperbolic cell (Definition 1.2). If it is a hyperbolic cell defined by Row's best response, only Column can defect, so the cell is hyperbolic-column. The next response dynamic, then, moves from one type of hyperbolic point to the other.

4.6.1   Number of Hyperbolic Cells

Hyperbolic cells are the building blocks for the next response dynamic, so a first step is to identify their availability. This is done by generalizing the approach used for 2×2 games, where the central tool (Theorem 2.2) was that the sum of $\mathcal{G}^{N,signs}$ cell values is zero. The next definition is an extension.

Definition 4.1 The cell entries of a $k_1 \times k_2$ game $\mathcal{G}$ determine the entries of associated $\mathcal{G}^{N,signs}$ cell. If Row's maximum $\mathcal{G}$ value for the cell's column is in the cell, then Row's entry in this $\mathcal{G}^{N,signs}$ cell is $(k_1 - 1)$; otherwise Row's $\mathcal{G}^{N,signs}$ entry is (-1). Similarly, Column's value for this $\mathcal{G}^{N,signs}$ cell is $(k_2 - 1)$ if the $\mathcal{G}$ cell has Column's maximum value for the row, otherwise it is (-1).[8] A $\mathcal{G}^{N,signs}$ cell's value is the sum of the cell's entries.

To motivate these choices, consider a column in a 4×4 game. If a specific cell in this column does not have Row's maximum, then Row prefers to move to the cell that has the maximum entry. Thus, for three of these cells, the motion is to move away, which is captured by the -1 value. The cell with Row's column maximum has three arrows pointing toward it, which explains the $4 - 1 = 3$ value. The following generalizes Theorem 2.2.

Theorem 4.6 *Generically, for a $k_1 \times k_2$ game $\mathcal{G}^{N,signs}$, the sum of its cell values is zero.*

The number of hyperbolic-row and hyperbolic-column cells differ by $(k_1 - k_2)$. If $k_1 = k_2$, the game has the same number of both kinds of hyperbolic cells.

Proof The sum of Row's and of Column's entries in, respectively, a $\mathcal{G}^{N,signs}$ column and row is zero. As the sum of the cell values merely reorders the summations overall rows and columns, the first assertion follows.

In each column of $\mathcal{G}$, Row selects the cell with her maximum value. Only Column might have a reason to defect; if so, this cell is hyperbolic-column. Consequently, there are k_2 possible hyperbolic-column cells. Similarly, there are k_1 possible hyperbolic-row cells. A potential hyperbolic cell is not hyperbolic only if it is Nash. Each Nash cell reduces the number of possible cells of each type by one, which proves the assertion. $\Box$

And so $\mathcal{G}^{N,signs}$ reduces the complexity of finding Nash cells. This is because a necessary condition for a cell to be a (strict) Nash cell is for all of its $\mathcal{G}^N$ entries to be positive, but it is not a sufficient condition. It is necessary and sufficient requirement for $\mathcal{G}^{N,signs}$.

---

[8]To capture subtle aspects about repelling cells, which will be mentioned (Sect. 4.7.3) but not fully developed, for each cell in an array, attach an arrow going toward each array cell with a larger value. For each cell, count the number of arrows pointing toward it. Generically, the values for a column, from maximum to minimum, are $k_1 - 1, k_1 - 3, \ldots, -(k_1 - 1)$. The zero-sum part of Theorem 4.6 holds.

Theorem 4.7 *For an n player* $2 \times 2 \times \cdots \times 2$ *game, a* $\mathcal{G}$ *cell is a strict Nash cell if and only if all of its* $\mathcal{G}^N$ *entries are positive. If players have three or more strategies, this condition is necessary, but not sufficient, for the cell to be strict Nash. With* $\mathcal{G}^{N,signs}$, *the positivity of all entries in a cell is a necessary and sufficient condition for it to be a strict Nash.*

Proof The only item that needs to be proved is that a $\mathcal{G}^N$ cell with positive entries need not be a Nash cell. This is a consequence of Eq. 4.6, where, for three or more pure strategies, two of the η values for $\mathcal{G}^N$ can be positive; this cannot happen with $\mathcal{G}^{N,signs}$. A 3×3 example with η entries $1, 2, -3$ is

$$
\mathcal{G}^N =
\begin{array}{|cc|cc|cc|}
\hline
1 & 1 & -3 & 2 & 2 & -3 \\
\hline
2 & -3 & 1 & 1 & -3 & 2 \\
\hline
-3 & 2 & 2 & -3 & 1 & 1 \\
\hline
\end{array}, \quad
\mathcal{G}^{N,signs} =
\begin{array}{|cc|cc|cc|}
\hline
-1 & -1 & -1 & 2 & 2 & -1 \\
\hline
2 & -1 & -1 & -1 & -1 & 2 \\
\hline
-1 & 2 & 2 & -1 & -1 & -1 \\
\hline
\end{array}
$$

$$(4.28)$$

The only $\mathcal{G}^N$ cells with positive entries are along the diagonal.[9] Each, however, is a *repelling cell;* for each cell, each player can move to a cell with the value 2, so this game has no Nash cells. Theses features are exposed with $\mathcal{G}^{N,signs}$. □

The number of available hyperbolic cells (Theorem 4.6) provides information about the game's Nash structure. To illustrate with 3×3 games, a $\mathcal{G}^{N,signs}$ Nash cell value is $2 + 2 = 4$, a $\mathcal{G}^{N,signs}$ hyperbolic cell value is $2 - 1 = 1$, and a $\mathcal{G}^{N,signs}$ repelling cell's value is -2.

Corollary 4.2 *The associated* $\mathcal{G}^{N,signs}$ *of a* 3×3 *game* $\mathcal{G}$ *has between three and six repelling cells.*

- *If* $\mathcal{G}^{N,signs}$ *has three repelling cells,* $\mathcal{G}$ *has no Nash cells, three hyperbolic-column, and three hyperbolic-row cells.*
- *If* $\mathcal{G}^{N,signs}$ *has four repelling cells,* $\mathcal{G}$ *has one Nash cell, two hyperbolic-column, and two hyperbolic-row cells.*
- *If* $\mathcal{G}^{N,signs}$ *has five repelling cells,* $\mathcal{G}$ *has two Nash cells, one hyperbolic-column, and one hyperbolic-row cells.*
- *If* $\mathcal{G}^{N,signs}$ *has six repelling cells,* $\mathcal{G}$ *has three Nash cells.*

And so a 3×3 game with one Nash cell has four repelling and two pairs of hyperbolic cells. Imagine how selecting appropriate $\mathcal{G}^B$ choices can generate a rich smorgasbord of Prisoner's Dilemma flavored games!

Proof The positive cell sums of Nash and hyperbolic cells require a sufficient number of repelling cells to satisfy Theorem 4.6. For instance, two repelling cells contribute the cell value of -4 to the summation, so the cell sum for the remaining seven cells

---

[9]Notice the rotating symmetry construction. Row's first column entries are $1, 2, -3$, the second column has $2, -3, 1$ and the third column has $-3, 1, 2$, with a similar but opposite rotation for Column's three rows. This symmetry structure, described as a Z_3 orbit in the symmetry literature, arises in Chap. 6 (where it is introduced with a "rotating wheel") when considering more than two players.

must equal $+4$. But each of the seven remaining cells has a cell value of at least 1, so their sum is at least 7, which prohibits such a setting.

Three repelling cells have a cell value summation of -6, so the sum of cell values for the remaining six cells must be $+6$. This can happen only if each cell's value is unity, which requires six hyperbolic cells. It follows from the second assertion of Theorem 4.6 that half of these hyperbolic cells are hyperbolic-column and the other half are hyperbolic-row.

The cell value sum for four repelling cells is -8, so the cell value summation of the remaining five cells equals 8. The only way to do so with cell values of 4 or 1 is to have one Nash and four hyperbolic cells. Of the hyperbolic cells, two are hyperbolic-column and two are hyperbolic-row (Theorem 4.6).

The two remaining cases are similarly proved. $\square$

4.6.2   The Next Response Dynamic

Here is a problem: We know how many cells are of each type, but how are they connected? What is the Nash dynamic? As a next response dynamic can start anywhere, the general structure of the associated curves needs to be described. A similar snag arose when creating a qualitative approach for dynamics and evolutionary game theory in the social sciences [30, Chap. 1], where the difficulty was handled by using fundamental components of dynamical systems such as cycles, attracting limit cycles, and attracting limit points.

Each of these units from dynamics has a parallel in next response motion. (What everything means will become clear with examples and the description.) Representing the attracting limit point is where the next response curve ends at a Nash cell. What replaces the cycles are the $j \times j$ irreducible blocks. Standing in for the attracting limit cycle is where a next response dynamic ends at a $j \times j$ irreducible block.

Theorem 4.8 *Generically, for a two-person games, next response curves can define*

- *a Nash cell,*
- *a $j \times j$ loop among hyperbolic cells, which is a $j \times j$ irreducible block,*
- *a next response curve that terminates at a Nash cell, and*
- *a next response curve that terminates at a $j \times j$ loop.*

Proof Each iterate in the next response dynamic is unique. If an iterate is at a hyperbolic cell, there is a next iterate. If it is a Nash cell, the dynamic stops.

With a finite number of rows and columns, the next response dynamic eventually must return to a column or row. The argument is essentially the same in either case, so assume it is a row. A row has precisely one hyperbolic-column cell, but it can have several hyperbolic-row cells; one of these cells was involved in an earlier stage of the dynamic. If the dynamic returns to the original hyperbolic-row cell, a loop is defined. If it hits a different hyperbolic-row cell, the next iterate will be the row's hyperbolic-column cell, which was in an earlier stage of the dynamic. Thus this new hyperbolic-row cell can be viewed as a starting cell of a loop.

For some integer value $j > 0$, the loop is $j \times j$. This is because the hyperbolic-column, hyperbolic-row cells alternative and the dynamic cannot hit a column or row twice except at a starting cell. For the same reason, the loop is irreducible.

The same argument shows that any dynamic not defining a loop must terminate in a Nash point or loop.   □

It is interesting how Theorem 4.8 and Corollary 4.2 identify *all possible* next response curves for 3×3 games. By understanding the next response dynamic, all that remains are the repelling cells. Thus, a first cut at understanding (and, for modeling, designing) the Nash dynamic is known.

1. With three repelling cells, there are six hyperbolic cells.

   (a) One possibility is a 3×3 loop. A best response for the three repelling cells is to move to the loop.
   (b) A second possibility is a 2×2 loop, which uses four hyperbolic cells. The remaining two hyperbolic cells generate a next response dynamic that terminates at the loop.

2. Four repelling cells requires four hyperbolic cells and a Nash cell. The possible arrangements are:

   (a) The hyperbolic cells form a next response curve that terminates at the Nash cell.
   (b) The hyperbolic cells form a 2×2 irreducible block that is independent of the Nash cell,

3. Five repelling cells are accompanied with two Nash cells and two hyperbolic cells. There are not enough hyperbolic cells to create a loop, so a next response dynamic terminates at one of the Nash cells.
4. Six repelling cells require three Nash cells and no hyperbolic cells. All dynamic moves toward some Nash cell.

Illustrating examples for zero or one Nash cells are in Fig. 4.4. Figure 4.4a has a 3×3 irreducible block, Fig. 4.4b has the next response dynamic ending at a Nash cell, Fig. 4.4c has a next response dynamic terminating at a 2×2 irreducible block, and Fig. 4.4d has a separate 2×2 block and a Nash cell. Notice how this geometry

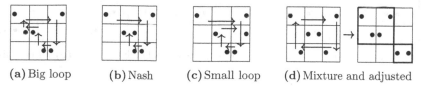

(a) Big loop   (b) Nash   (c) Small loop   (d) Mixture and adjusted

Fig. 4.4 Different equilibria boxes

shifts so slightly; in all figures, nothing changes for Row. For the first three figures, only Column's best choice in the second row changes. Figure 4.4b and d differ only in Column's choice in the bottom row. It is clear from the figures what Row and Column would prefer in repelling cells.

Each Fig. 4.4 diagram represents a class of games, so creating examples with the entries 1, 0, −1 is immediate. In Eq. 4.29, game $\mathcal{G}_j$ is a representative of Fig. 4.4j, $j =$ a, b, c, d. The Nash structures differ significantly, yet the four games share similarities.

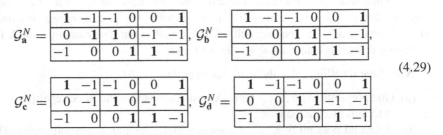

$$
\mathcal{G}_a^N =
\begin{array}{|cc|cc|cc|}
\hline
1 & -1 & -1 & 0 & 0 & 1 \\
\hline
0 & 1 & 1 & 0 & -1 & -1 \\
\hline
-1 & 0 & 0 & 1 & 1 & -1 \\
\hline
\end{array}, \quad
\mathcal{G}_b^N =
\begin{array}{|cc|cc|cc|}
\hline
1 & -1 & -1 & 0 & 0 & 1 \\
\hline
0 & 0 & 1 & 1 & -1 & -1 \\
\hline
-1 & 0 & 0 & 1 & 1 & -1 \\
\hline
\end{array},
$$

$$
(4.29)
$$

$$
\mathcal{G}_c^N =
\begin{array}{|cc|cc|cc|}
\hline
1 & -1 & -1 & 0 & 0 & 1 \\
\hline
0 & -1 & 1 & 0 & -1 & 1 \\
\hline
-1 & 0 & 0 & 1 & 1 & -1 \\
\hline
\end{array}, \quad
\mathcal{G}_d^N =
\begin{array}{|cc|cc|cc|}
\hline
1 & -1 & -1 & 0 & 0 & 1 \\
\hline
0 & 0 & 1 & 1 & -1 & -1 \\
\hline
-1 & 1 & 0 & 0 & 1 & -1 \\
\hline
\end{array}
$$

To complete the 3×3 story, the structures for games with two or three Nash cells are in Fig. 4.5. Figure 4.5a has two Nash cells with numbers representing where to position the two remaining hyperbolic cells. In Fig. 4.5b, two next response curves end at the same Nash cell, while Fig. 4.5c has two next response curves ending at different Nash cells. Figure 4.5d has a two-step next response line ending at a Nash cell.

Imagine the fun that can triggered with appropriately selected $\mathcal{G}^B$ terms! Fig. 4.5b, for instance, suggests an affinity for the TL Nash cell. But selecting a $\mathcal{G}^B$ with a strong Pareto superior cell at $(2, 2)$ and a heavy TL Pareto inferior cell immediately complicates the analysis. Indeed, $\mathcal{G}^B$ could even be selected to convert the BR repeller into a cell of mutual benefit for both players.

Examining 2×3 games
This approach, which holds for any $k_1 \times k_2$ game, is further illustrated by determining the Nash structures of 2×3 games. Column's $\mathcal{G}^{N,signs}$ row entries (Definition 4.1) are 2, −1, −1 while Row's column entries are 1, −1. Consequently, a Nash cell's value is 3, a hyperbolic-row cell's is $2 - 1 = 1$, a hyperbolic-column cell is $1 - 1 = 0$, and a repelling cell is −2.

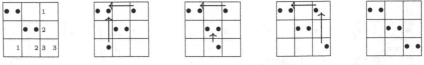

(a) Two cells   (b) Toward one   (c) Toward both   (d) The long way   (e) Three cells

Fig. 4.5 Nash structures

A single repelling cell, with cell value of -2, requires the sum of the remaining five cells to equal 2. While there can be no Nash cells, there are two hyperbolic-row and three hyperbolic-column cells. The next response dynamic defines one 2×2 irreducible block and a next response line that terminates at this loop.

Two repelling cells (cell value sum of -4) requires the other three cells to consist of one Nash, one hyperbolic-row, and two hyperbolic-column cells. All next response dynamics terminate at the Nash cell.

With three repelling cells (cell sum of -6), the remaining three are two Nash cells and one hyperbolic-column. The next response dynamic terminates at a Nash cell.

Diagrams and examples of games representing each of these cases are immediate. Preferred opportunities for repelling cells follow from the diagrams.

The Nash structure of 4×4 games

With 4×4 games, the column and row entries for $\mathcal{G}^{N,signs}$ are 3 for the maximum value and -1's for the three cells without the maximum entry. Thus $\mathcal{G}^{N,signs}$ Nash cells have value 6, hyperbolic cells have value 2, and repelling cells have value -2.

The simplest admissible $\mathcal{G}^{N,signs}$ has eight repelling cells (cell sum of -16), which must be accompanied by four hyperbolic-column and four hyperbolic-row cells. The possible dynamics have a 4×4 irreducible block, a 3×3 irreducible block at which all other next response dynamics terminate, two 2×2 irreducible blocks, and a single 2×2 irreducible block at which all next response dynamics terminate.

With nine $\mathcal{G}^{N,signs}$ repelling cells (cell sum of -18), the remaining seven must have one Nash and six hyperbolic cells. The hyperbolic cells could have a next response curve ending at the Nash cell, a 3×3 irreducible block, or curves terminating at a 2×2 block.

The remaining cases are found in the same manner by using the following theorem that identifies the number of hyperbolic, repelling, and Nash cells in a $k \times k$ game. A similar assertion can be computed for $k_1 \times k_2$ games.

Theorem 4.9 *Generically (i.e., in each column, Row has a unique maximum, in each row, Column has a unique maximum),*[10] *if a $k \times k$ game ($k \geq 2$), has j Nash cells, $0 \leq j \leq k$, then there are precisely $(k - j)$ hyperbolic-column cells, $(k - j)$ hyperbolic-row cells, and $k^2 - 2k + j$ repelling cells.*

Proof A Nash cell's value is $2(k - 1)$, a hyperbolic cell's value is $k - 2$, and a repelling cell has value -2. The sum of nonnegative entries—the hyperbolic and Nash cell values—equals $j(2k - 2) + 2(k - j)(k - 2) = 2k^2 - 4k + 2j$. The sum of the repelling cell values is $-2k^2 + 4k - 2j$, which yields the required zero-sum.

Increasing the number of hyperbolic cells would decrease the number of repelling cells, which would violate the zero-sum requirement. Similarly, decreasing the number hyperbolic cells requires more repelling cells, which increases the sum of negative cell values while keeping (for $k = 2$) or decreasing (for $k \geq 3$) the sum of the positive cell values. This violates the zero-sum condition. □

---

[10]To obtain a similar result with t equal maximum values in a column or row, $1 \leq t < k$, assign $k - t$ to a maximum entry and $-t$ to all others.

4.7 Some Games

To illustrate these notions, consider the Colonel Blotto game, which includes weak Nash equilibria, and Rock-Scissors-Paper that often is played on elementary school playgrounds.

4.7.1 Colonel Blotto

About a century ago, the French mathematician and politician Émile Borel [4] (1871–1956) introduced the Colonel Blotto game. A way to think of this construct is that a war strategist (or a football coach, or …) has limited resources but several fronts (or, several football positions) to handle. Victory requires winning a majority of the skirmishes. For a simple $n = 3$ case, restrict the choices for each of the three fronts to positive integers with a total of 6. This allows Colonel Blotto to allocate $(1, 1, 4)$, or $(1, 2, 3)$, or $(2, 2, 2)$ units among the three fronts. If Blotto decided upon $(1, 1, 4)$ and his opponent, with an equal number of resources, allocated $(1, 2, 3)$, the outcome would be a tie on the 1 and 1 front, the opponent would win on the 1 and 2 front, and Blotto would win on the 4 and 3 front, which leads to a stalemate.

An immediate goal is to extract the Nash structure of this game, where, to avoid fractions, a win is worth 3 units.

$$
\mathcal{G}_{Blotto} =
\begin{array}{c}
\\ (1,1,4) \\ (1,2,3) \\ (2,2,2)
\end{array}
\begin{array}{|cc|cc|cc|}
\multicolumn{2}{c}{(1,1,4)} & \multicolumn{2}{c}{(1,2,3)} & \multicolumn{2}{c}{(2,2,2)} \\
\hline
0 & 0 & 0 & 0 & -3 & 3 \\
0 & 0 & 0 & 0 & 0 & 0 \\
3 & -3 & 0 & 0 & 0 & 0 \\
\hline
\end{array}
\text{ and } \mathcal{G}^N_{Blotto} =
\begin{array}{|cc|cc|cc|}
\hline
-1 & -1 & 0 & -1 & -2 & 2 \\
-1 & 0 & 0 & 0 & 1 & 0 \\
2 & -2 & 0 & 1 & 1 & 1 \\
\hline
\end{array}
\quad (4.30)
$$

The Nash structure becomes apparent with $\mathcal{G}^{N,signs}_{Blotto}$ (Fig. 4.6a). $\mathcal{G}^N_{Blotto}$ identifies BR as the sole strict Nash cell (a fact that is not immediately obvious from $\mathcal{G}_{Blotto}$); this is where each player selects $(2, 2, 2)$. The Blotto game does not satisfy the generic

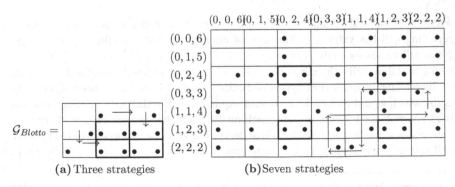

(a) Three strategies   (b) Seven strategies

Fig. 4.6 Blotto games

assumption of a single maximum in rows and columns, so, as Fig. 4.6a makes clear, both players have more than one maximum in certain arrays. This structure introduces weak Nash cells—a row or column can have more than one Nash cell. The four Nash cells, where each player has a bullet, are indicated in Fig. 4.6a with thicker boundary lines. The action associated with the one repelling cell (TL) is that Row prefers BL while Column prefers TR.

Figure 4.6a arrows identify the next response dynamic. Both terminate in a Nash cell, so there are no other irreducible blocks. Thus, the only possible NME would be to indicate among Nash cells. A computation shows there are no distinctions among the four.

Extended game

Now extend the game so that the strategist can concede certain fronts by putting forth only token resistance. This is done in order to assign added strength to other fronts. Here a player selects triplets of nonnegative integers that sum to six. The choice of $(0, 1, 5)$, for instance, concedes one front with the expectation of a clear victory on a second front with a possible tie on the third leading to a stalemate. This happens should the opponent select $(0, 2, 4)$ (with the expectation that strengths for fronts of both players align) where the two token conflicts (both select 0) lead to a tie, in the 2 and 1 conflict, the opponent wins, but in the 4 and 5 conflict, Blotto wins.

It is reasonable to expect $(2, 2, 2)$ to repeat as a Nash cell along with, perhaps, $(0, 3, 3)$. But Fig. 4.6b $\mathcal{G}_{Blotto}^{N,signs}$ graph shows this is not the case; both cells are demoted to being hyperbolic! Instead, the component parts of a Nash cell are $(1, 2, 3)$ and $(0, 2, 4)$. The only irreducible block, which is 3×3, is indicated by Fig. 4.6b arrows. Here $(0, 3, 3)$ and $(2, 2, 2)$ are represented along with $(1, 1, 4)$. It is fairly immediate to show that any other next response dynamic—from a hyperbolic or repelling cell—terminates either at a Nash cell, or on this 3×3 block.

The 3×3 irreducible block defines an NME where a computation shows that $q_4 = q_5 = q_7 = p_4 = p_5 = p_7 = \frac{1}{3}$. That is, with probability $\frac{1}{3}$, each player plays one of $(0, 3, 3)$, $(2, 2, 2)$ or $(1, 1, 4)$. As for selecting among the four Nash cells, because Column's entries for each relevant row, and Row's entries for each relevant column, are the same, the equations become degenerate so any q_3, q_6 and p_3, p_6 suffice—a distinct NME does not exist here. At least for us, geometry (e.g., Fig. 4.6) simplifies the analysis.

4.7.2 Rock-Paper-Scissors

One of the better known 3×3 games is rock-paper-scissors where each player selects one of these objects. Here a rock can break scissors, scissors can cut up a paper, and paper can cover a rock. The bimatrix for this symmetric game is

Fig. 4.7 Rock, scissors, paper

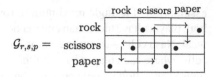

$$\mathcal{G}_{\text{r, s, p}} = \begin{array}{c} \\ \\ \text{rock} \\ \text{scissors} \\ \text{paper} \end{array} \begin{array}{cc} \text{rock} & \text{scissors} \quad \text{paper} \\ \begin{array}{|cc|cc|cc|} \hline 0 & 0 & 1 & -1 & -1 & 1 \\ \hline -1 & 1 & 0 & 0 & 1 & -1 \\ \hline 1 & -1 & -1 & 1 & 0 & 0 \\ \hline \end{array} \end{array} \qquad (4.31)$$

which equals its Nash component; e.g., $\mathcal{G}_{\text{r, s, p}} = \mathcal{G}^N_{\text{r, s, p}}$.

As well understood, there are no Nash cells, and the NME involves all three strategies where the next response dynamic is

$$(1, 2) \rightarrow (1, 3) \rightarrow (2, 3) \rightarrow (2, 1) \rightarrow (3, 1) \rightarrow (3, 2) \rightarrow (1, 2)$$

to define a 3×3 irreducible block as indicated in Fig. 4.7.[11] An immediate computation shows that the NME is $\mathbf{p} = \mathbf{q} = (\frac{1}{3}, \frac{1}{3}, \frac{1}{3})$.

This game illustrates how NME serve as signposts to direct Best response traffic. Figure 4.8a indicates the various $\mathbf{q} = (q_1, q_2, q_3)$ values adopted by Column where Fig. 4.8b has Row's best reaction. Figure 4.8a, an equilateral triangle given by $\{\mathbf{q} \mid q_1 + q_2 + q_3 = 1,\ q_i \geq 0\}$, is divided into six smaller triangles by passing the three diagonals $q_i = q_j$ through the region. The center point is the mixed strategy $q_1 = q_2 = q_3 = \frac{1}{3}$.

Should Column stray from the mixed strategy, $\mathbf{q}$ is in one of the regions; e.g., if $\mathbf{q}$ is in the lower-left triangle, then (according to the dividing lines) $q_1 > q_2 > q_3$, which shows Column's predilection for Rock over Scissors over Paper (as indicated in Fig. 4.8a by rsp). According to the signpost, Row's pure strategy reaction is given is the corresponding small Fig. 4.8b triangle of emphasizing Paper, followed by Rock and then Scissors. Or, if $\mathbf{q}$ is on the line separating the Fig. 4.8a rps and prs triangles, then Row should adopt the mixed strategy of Paper and Scissors. All possibilities are listed in Fig. 4.8.

4.7.3 Those Repulsive Cells

At a repelling cell, both players have a personal incentive to move elsewhere, where an optimal move is toward a cell in the next response curve—a dynamic that eventually terminates in one of the irreducible blocks. (Such a move always exists because Row and Column have, respectively, a wiser selection in each column and each row.)

---

[11] Figure 4.7 also describes the next response dynamic for Eq. 4.28. This is not an accident; both are constructed with the rotating symmetry construction.

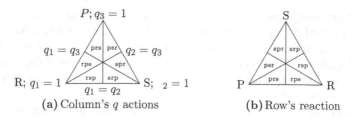

Fig. 4.8 Row's signpost

This suggests that a best response process quickly abandons the repelling cells to eventually focus attention on the $j \times j$ irreducible blocks.

Theorem 4.10 *A $j \times j$ irreducible block for $j = 1$ is a Nash cell. For $j \geq 2$ and where each player is restricted to the strategies identified by this block, the block contains an NME.*

Proof The fact a $j \times j$ irreducible block for $j = 1$ is a Nash cell follows from the above. For $j \geq 2$, the block, and the restriction of players' strategies, defines a game, which, according to Nash's existence theorem, has a Nash equilibrium. No cell in this block is a Nash cell, so the equilibrium must be an NME.   □

But Theorem 4.10 is not the end of the story: The reason is that some repelling cells are more repulsive than others. This distinction need not play a role in understanding the $\mathcal{G}^N$ dynamic as needed for modeling, but it can complicate the search for NME.

$$
\mathcal{G} =
\begin{array}{|cc|cc||cc|cc|}
\hline
5 & -5 & -5 & 5 & 4 & 2 & 2 & 4 \\
-5 & 5 & 5 & -5 & 2 & 4 & 4 & 2 \\
\hline
4 & 4 & 2 & 2 & 5 & -5 & -5 & 5 \\
2 & 2 & 4 & 4 & -5 & 5 & 5 & -5 \\
\hline
\end{array}
\rightarrow
\begin{array}{|cc|cc||cc|cc|}
\hline
\bullet & - & - & \bullet & - & - & - & - \\
- & \bullet & \bullet & - & - & - & - & - \\
\hline
- & - & - & - & \bullet & - & - & \bullet \\
- & - & - & - & - & \bullet & \bullet & - \\
\hline
\end{array}
$$

$$(4.32)$$

To explain, the next response dynamic for Eq. 4.32 defines two 2×2 irreducible blocks in the upper-left and lower right corners. Each block is a matching coin game, so the NMEs assured by Theorem 4.32 are $p_1 = p_2 = q_1 = q_2 = \frac{1}{2}$ and $p_3 = p_4 = q_3 = q_4 = \frac{1}{2}$. Each player's expected winning in each block is zero, which reflects the fairness of this decision process.

Here is where the repelling terms come in; these are the cells in the 2×2 blocks in the upper-right and lower-left corners. As discussed at the beginning of this section, their repeller status assures both players a preferred outcome by moving elsewhere, where a best response for this game is to move to a cell in one of the two irreducible blocks. In this manner, the best response motion quickly abandons the repelling sectors of Eq. 4.32 to move toward an irreducible block, where Row prefers one choice and Column the other.[12]

---

[12]This action is a source of the NME for secondary blocks in Theorem 4.5.

Rather than individual cells, these repellers can be viewed as a unit of cells glued together by mixed strategies. Here something different happens with a best response mentality. To explain, suppose Column is resigned to play $q_3 = q_4 = \frac{1}{2}$, a choice that ties together the third and fourth columns. Row will ignore the bottom two rows because of the expected outcome of zero. Instead her mixed strategy is to play $p_1 = p_2 = \frac{1}{2}$ *rather than* $p_3 = p_4 = \frac{1}{2}$.

This selection, which returns the focus to the upper-right block of *repellers*, rewards each player with an improved expected value of three rather than zero! The behavior, which is essentially equivalent to the cleaning option in Eq. 4.1 version of the Battle of the Sexes, introduces two improved NME of $p_1 = p_2 = q_3 = q_4 = \frac{1}{2}$ and $p_3 = p_4 = q_1 = q_2 = \frac{1}{2}$. The first one captures the NME in a 2×2 game of Fig. 1.2a form, while the second is reflected by Fig. 1.2d that connects two Nash cells. Strange, because Eq. 4.32 has *no Nash cells*. But by restricting attention to the lower-left 2×2 block, cells (3, 1) and (4, 2) become *relative* Nash cells. Thus the dynamic generating these NME is defined by the differing levels of repulsiveness; a player in a cell with payoff 3 might prefer the improved payoff of 4, even though a better choice of 5 is available.

This more refined analysis can be picked up by replacing $\mathcal{G}^{N,signs}$ with $\mathcal{G}^{N.rank}$, where cells are ranked according to payoff values. With different values in a k array, assign the jth largest value $k - (2j - 1)$ points (to allow a Theorem 4.6 type result). With Eq. 4.32, this leads to

$$
\mathcal{G}^{N.rank} =
\begin{array}{|cc|cc||cc|cc|}
3 & -3 & -3 & 3 & 1 & -1 & -1 & 1 \\
-3 & 3 & 3 & -3 & -1 & 1 & 1 & -1 \\
\hline
1 & 1 & -1 & -1 & 3 & -3 & -3 & 3 \\
-1 & -1 & 1 & 1 & -3 & 3 & 3 & -3 \\
\end{array}
\rightarrow
\begin{array}{|cc|cc||cc|cc|}
\bullet & - & - & \bullet & s & - & - & s \\
- & \bullet & \bullet & - & - & s & s & - \\
\hline
s & s & - & - & \bullet & - & - & \bullet \\
- & - & s & s & - & \bullet & \bullet & - \\
\end{array}
$$

$$(4.33)$$

The s's in the adjacent bimatrix are "second ranked" terms, which capture the described secondary dynamic structure of the repelling cells.

There are many ways this can be handled; one is by imposing a definition asserting that blocks of repellers do not deserve an NME. As this approach ignores the basic theme of maximization, we need to appreciate when the above behavior stressing repellers can arise. The answer is as given with Eq. 4.1 Battle of the Sexes; it is when the expected value of a block of repellers exceeds the NME expected values of the appropriate $j \times j$ irreducible blocks. In the Battle of the Sexes, demoting the payoff of cleaning to 1 would restore the NME between the two cultural options of opera or football. But, is this demotion appropriate? Replacing 4 and 3 with, respectively, 0 and -2 in Eq. 4.32 retains Eq. 4.33 structure while lowering the repulsive cells' expected values to be negative, which allows the irreducible blocks to recover ownership of the NME. But, does doing so distort what is being modeled? By playing games with payoff values in any diagram, such as in the right side of Eq. 4.33, other examples follow immediately. Ah, such fun!

4.8 Computing Mixed Equilibria

The previous section concluded by discussing NME's. For continuity, it is reasonable to turn to the ugly arithmetic, which nobody with a semblance of a life relishes, of finding the NME probability values. As shown next, the decomposition reduces some of the agony.

To review approaches, return to $\mathcal{G}_4$ (Eq. 2.3) given by

$$\mathcal{G}_4 = \begin{array}{|cc|cc|} \hline 6 & 6 & 0 & 4 \\ 4 & -4 & 2 & 2 \\ \hline \end{array} \text{ leading to } \mathcal{G}_4^N = \begin{array}{|cc|cc|} \hline 1 & 1 & -1 & -1 \\ -1 & -3 & 1 & 3 \\ \hline \end{array}.$$

What simplifies the analysis is that only $\mathcal{G}_4^N$ is needed to compute the NME. With $q_1, q_2 = 1 - q_1$, the $EV(T) - EV(B) = 0$ expression for $\mathcal{G}_4^N$ becomes

$$q_1 - q_2 = -q_1 + q_2 \text{ subject to } q_1 + q_2 = 1, \tag{4.34}$$

or $2q_1 = 2q_2$, which leads to $q_1 = q_2 = \frac{1}{2}$.

An equivalent approach is to set $EV(T) = EV(B) = \lambda$, which defines the related system of $q_1 - q_2 = \lambda$ and $-q_1 + q_2 = \lambda$, or the linear algebra matrix expression of

$$\begin{pmatrix} 1 & -1 \\ -1 & 1 \end{pmatrix} \begin{pmatrix} q_1 \\ q_2 \end{pmatrix} = \begin{pmatrix} \lambda \\ \lambda \end{pmatrix}, \tag{4.35}$$

where the λ value is not known. If matrix $\mathcal{A}$ is non-singular, this matrix equation of

$$\mathcal{A} \begin{pmatrix} q_1 \\ q_1 \end{pmatrix} = \begin{pmatrix} \lambda \\ \lambda \end{pmatrix} \text{ has the solution } \begin{pmatrix} q_1 \\ q_1 \end{pmatrix} = \mathcal{A}^{-1} \begin{pmatrix} \lambda \\ \lambda \end{pmatrix}, \tag{4.36}$$

where if

$$\mathcal{A} = \begin{pmatrix} a & b \\ c & d \end{pmatrix}, \text{ then matrix } \mathcal{A}^{-1} = \frac{1}{ad - bc} \begin{pmatrix} d & -b \\ -c & a \end{pmatrix}. \tag{4.37}$$

Unfortunately the $\mathcal{A}$ for Eq. 4.35 is singular because $ad - bc = (1)(1) - (-1)(-1) = 0$, which means that Eq. 4.36 is not applicable.

4.8.1 Help from the Decomposition

Linear systems permit a scaling whereby if $\mathcal{A}(\mathbf{a}) = \mathbf{b}$, then for any scalar μ it also is true that $\mathcal{A}(\mu \mathbf{a}) = \mu(\mathbf{b})$. Consequently, should $\mathcal{A}$ be non-singular in Eq. 4.36, set $\lambda = 1$ and modify the resulting x_1, x_2 solution to obtain $q_1 = \frac{x_1}{x_1 + x_2}$, $q_2 = \frac{x_2}{x_1 + x_2}$ where $q_1 + q_2 = 1$. But if $\mathcal{A}$ is singular, which often is the case, the *decomposition* can convert the system into an equivalent (with respect to finding NME) non-singular one! This is because *the decomposition permits violating certain rules of linear algebra.*

Namely, modify an $\mathcal{A}$ column by adding the same value to each column entry. Doing so merely adds an "average value," which the decomposition proves does not affect the outcome. But, this change can create a non-singular matrix,[13] so Eq. 4.37 now applies!

To illustrate, adding 4 to each entry in the first column converts Eq. 4.35 to

$$\begin{pmatrix} 5 & -1 \\ 3 & 1 \end{pmatrix} \begin{pmatrix} x_1 \\ x_2 \end{pmatrix} = \begin{pmatrix} 1 \\ 1 \end{pmatrix},$$

where, using Eq. 4.36, the solution is

$$\begin{pmatrix} x_1 \\ x_2 \end{pmatrix} = \mathcal{A}^{-1} \begin{pmatrix} 1 \\ 1 \end{pmatrix} = \frac{1}{8} \begin{pmatrix} 1 & 1 \\ -3 & 5 \end{pmatrix} \begin{pmatrix} 1 \\ 1 \end{pmatrix} = \begin{pmatrix} \frac{1}{4} \\ \frac{1}{4} \end{pmatrix}.$$

Therefore $q_1 = \frac{x_1}{x_1+x_2} = \frac{1}{2}$ and $q_2 = \frac{x_2}{x_1+x_2} = \frac{1}{2}$.

A wiser modification of Eq. 4.35 is to add 1 to the first and the second columns reducing the system to $2x_1 = 1$, $2x_2 = 1$, which trivially leads to the $x_1 = x_2 = \frac{1}{2}$, or $q_1 = q_2 = \frac{1}{2}$ solution. Using this approach with Eq. 4.27, the 2×2 block of

$$\begin{matrix} 5q_1 + 4q_2 = \lambda \\ -3q_1 + 7q_2 = \lambda \end{matrix} \text{ or } \begin{pmatrix} 5 & 4 \\ -3 & 7 \end{pmatrix} \begin{pmatrix} q_1 \\ q_2 \end{pmatrix} = \begin{pmatrix} \lambda \\ \lambda \end{pmatrix}$$

can be converted (by adding 3 to first column entries and -4 to second column entries) into $8x_1 = 1$, $3x_2 = 1$, with the solution $x_1 = \frac{1}{8}$ and $x_2 = \frac{1}{3}$, to define $q_1 = \frac{\frac{1}{8}}{\frac{1}{3}+\frac{1}{8}} = \frac{3}{11}$ and $q_2 = \frac{\frac{1}{3}}{\frac{1}{3}+\frac{1}{8}} = \frac{8}{11}$.

Similarly, to find the p_1 and p_2 values of this Eq. 4.27 block, convert

$$\begin{pmatrix} 1 & 3 \\ 5 & -3 \end{pmatrix} \begin{pmatrix} p_1 \\ p_2 \end{pmatrix} = \begin{pmatrix} \lambda \\ \lambda \end{pmatrix}$$

(by adding -1 to the first column and 3 to the second) into

$$\begin{pmatrix} 0 & 6 \\ 4 & 0 \end{pmatrix} \begin{pmatrix} y_1 \\ y_2 \end{pmatrix} = \begin{pmatrix} 1 \\ 1 \end{pmatrix}$$

or $6y_2 = 1$ and $4y_1 = 1$, which leads to $p_1 = \frac{\frac{1}{4}}{\frac{1}{4}+\frac{1}{6}} = \frac{3}{5}$ and $p_2 = \frac{\frac{1}{6}}{\frac{1}{4}+\frac{1}{6}} = \frac{2}{5}$.

---

[13] This follows from a dimension counting and gradient argument. For instance, with a 2×2 matrix, a singular value is the three-dimensional surface $ad - bc = 0$. This surface's normal (gradient of the equation) is the vector $(d, -c, -b, a)$. Adding a value to the first column makes the matrix non-singular (take the scalar product of the gradient with $(1, 1, 0, 0)$) if and only if $d \neq c$; similarly adding a value to the second column makes the matrix non-singular if and only if $a \neq b$. The approach fails only if all $\mathcal{A}$ entries are zero.

Notice, this strategy of modifying each column can be directly applied to *any* $k \times k$ game. This is because doing so just adds $\mathcal{G}^B$ terms to the Nash component.

4.8.2 Moving to 3 × 3 Blocks

In a similar way, the decomposition can be applied to $j \times j$ blocks, $j \geq 3$. Illustrating with the 3×3 blocks of Eq. 4.27, the expressions are

$$\begin{pmatrix} 6 & -6 & 3 \\ 1 & 7 & -2 \\ -4 & -5 & 4 \end{pmatrix} \begin{pmatrix} q_3 \\ q_4 \\ q_5 \end{pmatrix} = \begin{pmatrix} \lambda_1 \\ \lambda_1 \\ \lambda_1 \end{pmatrix} \text{ and } \begin{pmatrix} -5 & 9 & 3 \\ -6 & 0 & 5 \\ 7 & -10 & -2 \end{pmatrix} \begin{pmatrix} p_3 \\ p_4 \\ p_5 \end{pmatrix} = \begin{pmatrix} \lambda_2 \\ \lambda_2 \\ \lambda_2 \end{pmatrix}$$

Using the first matrix and the relaxed rules allowed by the decomposition, add 4 to the first column, 5 to the second, and -3 to the third to obtain

$$\begin{pmatrix} 10 & -1 & 0 \\ 5 & 12 & -5 \\ 0 & 0 & 1 \end{pmatrix} \begin{pmatrix} x_3 \\ x_4 \\ x_5 \end{pmatrix} = \begin{pmatrix} 1 \\ 1 \\ 1 \end{pmatrix},$$

from which it follows (bottom row) that $x_5 = 1$. This value reduces the expression to

$$\begin{pmatrix} 10 & -1 \\ 5 & 12 \end{pmatrix} \begin{pmatrix} x_3 \\ x_4 \end{pmatrix} + \begin{pmatrix} 0 \\ -5 \end{pmatrix} = \begin{pmatrix} 1 \\ 1 \end{pmatrix} \text{ or } \begin{pmatrix} 10 & -1 \\ 5 & 12 \end{pmatrix} \begin{pmatrix} x_3 \\ x_4 \end{pmatrix} = \begin{pmatrix} 1 \\ 6 \end{pmatrix}.$$

The answer (from Eq. 4.37) is

$$\begin{pmatrix} x_3 \\ x_4 \end{pmatrix} = A^{-1} \begin{pmatrix} 1 \\ 6 \end{pmatrix} = \frac{1}{125} \begin{pmatrix} 12 & 1 \\ -5 & 10 \end{pmatrix} \begin{pmatrix} 1 \\ 6 \end{pmatrix} = \begin{pmatrix} \frac{18}{125} \\ \frac{55}{125} \end{pmatrix}$$

Thus $q_3 = \frac{x_3}{x_3+x_4+x_5} = \frac{18}{198}$, $q_4 = \frac{55}{198}$, and $q_5 = \frac{125}{198}$. Ugly, but nowhere near as messy as it would be using standard algebraic approaches.

The $\mathcal{G}^N$ matrices *always are singular* because the sum of Row's entries in any column and Column's entries in any row equals zero. To exploit this singularity, the three $\mathcal{G}_a^N$ (Eq. 4.29) equations for **q** are

$$q_1 - q_2 + 0q_3 = \lambda, \quad 0q_1 + q_2 - q_3 = \lambda, \quad -q_1 + 0q_2 + q_3 = \lambda.$$

The sum of these three equations is $0 = 3\lambda$, or $\lambda = 0$; this $\lambda = 0$ expression always holds for a Nash component. Consequently, the three equations become $q_1 = q_2$, $q_2 = q_3$, $q_1 = q_3$, or $q_1 = q_2 = q_3 = \frac{1}{3}$. Similarly the **p** values for $\mathcal{G}_a^N$ come from $-p_1 + p_2 = 0$, $p_3 = 0$, $p_1 - p_2 - p_3 = 0$, or $p_1 = p_2 = \frac{1}{2}$, $p_3 = 0$.

Caution is required. The $\mathcal{G}_c^N$ ambiguity is the loop among cells (2, 2), (2, 3), (3, 2), (3, 3), which is the subgame

$$\begin{array}{|cc|cc|}\hline 1 & 0 & -1 & 1 \\ 0 & 1 & 1 & -1 \\ \hline \end{array}.$$

This game is *not* in a $\mathcal{G}^N$ form, but the system of $q_2 - q_3 = \lambda_1$, $q_3 = \lambda_1$, or $q_2 - q_3 = q_3$, and $p_3 = p_2 - p_3$ is trivially solved for $q_2 = \frac{2}{3}$, $q_3 = \frac{1}{3}$, $p_2 = \frac{2}{3}$, $p_3 = \frac{1}{3}$.

Alternatively, the Nash component of this subgame is

$$\begin{array}{|cc|cc|}\hline 0.5 & -0.5 & -1 & 0.5 \\ -0.5 & 1 & 1 & -1 \\ \hline \end{array},$$

where, because it now must be that $\lambda = 0$, defines expressions such as $0.5q_2 - q_3 = 0$, or the same conclusion of $q_2 = \frac{2}{3}$, $q_3 = \frac{1}{3}$.

Here is a worry; this approach of adding an entry to a column ensures that a non-singular matrix and a solution always will emerge. This appears to be in conflict with the reality that an NME connecting the Nash cells may not exist. The answer is that in such settings, a component of an outcome will be negative.

This fact can simplify discovering whether such an NME exists. None exist with the full

$$\mathcal{G}^N = \begin{array}{|cc|cc|cc|}\hline 3 & 4 & -2 & -1 & -6 & -3 \\ 0 & -1 & 4 & 4 & 1 & -3 \\ -3 & 3 & -2 & -4 & 5 & 1 \\ \hline \end{array}.$$

To see why, as developed above, if one did exist, then Row's expected value for each row is zero, so

$$EV(T) = 3q_1 - 2q_2 - 6q_3 = 0, \quad EV(M) = 4q_2 + q_3 = 0, \quad EV(B) = -3q_1 - 2q_2 + 5q_3 = 0.$$

The $EV(M)$ coefficients are positive, so either $q_2 = q_3 = 0$, which forces $q_1 = 0$, or one of these terms is negative. Thus, if in a $\mathcal{G}^N$ row for Row, or a $\mathcal{G}^N$ column for Column, at least two entries are of the same sign and the rest are zero, a general NME does not exist.

This argument shows that the Eq. 4.8 game does not have an NME relating the three Nash cells with $v \geq 0$. To see this, notice that, for Row, the top row of the $\mathcal{G}^N$ is $(2, v, 2)$. So, for nonnegative v, the $EV(T) = 2q_1 + vq_2 + 2q_3 = 0$ expression, along with $q_1 + q_2 + q_3 = 1$, requires a q_j with a negative value.

Chapter 5
Extensive Form Games

There are other ways to represent games, and so we show how to adapt the coordinate system. (Readers more interested in normal form games with more players could jump ahead to the next chapter.) In a single-shot normal form game, it is standard to assume that each player makes their strategic choices simultaneously, or, at least, without knowledge of any other player's strategic choices. What would happen if a game consists of multiple rounds of decisions where players observe what others do before making their next choice; e.g., a game of chess?

This extensive form situation typically is modeled with a tree structure. What a rich environment; nirvana for designers of games providing a strong temptation to explore many heavenly avenues! But doing so would quickly overwhelm the themes of this book, so our more modest emphasis is to outline how to create coordinate systems for these different themes. Admittedly, some details are left to the reader and many topics are not covered, but the basic methodology is described. What makes the program feasible is that the decomposition principles remain the same.

After introducing the central concepts (Sect. 5.1), they are applied to the Centipede game in Sect. 5.2. Just the fact this game is known enough to have a name means that its coordinates will display a tension between individual and mutual benefits.

The chapter's real action starts in Sect. 5.5 with the arrival of Nature. Her role is to provide the players with one of two potentially dissimilar games. The decomposition turns out to be surprisingly simple, yet rewarding with an ability to unleash an amazing array of possible issues! As an advertisement for forthcoming events, Nature could select a game that dominates the general game's strategic Nash behavior even though Nature's other choice endows the general game with an inclination for cooperation and coordination.

What adds to the importance of Sect. 5.7 is that, beyond proving assertions, it outlines the basic procedure to decompose games with more strategies, players, and different information sets. The approach is not difficult; it is a natural, straightforward generalization of what is developed in Sect. 3.4.2.

© Springer Nature Switzerland AG 2019

D. T. Jessie and D. G. Saari, *Coordinate Systems for Games*, Static & Dynamic Game Theory: Foundations & Applications, https://doi.org/10.1007/978-3-030-35847-1_5

To adequately describe the technical approach of Sect. 5.7, some concept, such as subgame perfect, are only briefly mentioned. Readers familiar with these concepts could skim, or even ignore, what is said before moving to Sect. 5.4.

5.1   What Is with What?

A game tree is illustrated on the left side of Fig. 5.1. In this figure, Player 1 starts by playing either left or right (the two lines coming out of the top node), and Player 2 plays either left or right (the two lines emerging from two middle vertices). The payoffs are at the end of each of the four branches. The strategic decomposition of a normal form game exploits the fact that the difference in payoffs (rather than the actual values) is what matters. The same holds here; the only difficulty in computing the corresponding $\mathcal{G}^N$ and $\mathcal{G}^B$ is to identify the associated payoffs.

For a quick answer, associate Player 1 with Row, where L is treated as B and R with T. In this manner, where Player 2 is Column, the normal form and associated decomposition become

$$
\begin{array}{|cc|cc|}\hline -4 & -6 & 7 & -9 \\ -6 & 10 & 3 & 2 \\ \hline \end{array}
=
\begin{array}{|cc|cc|}\hline 1 & 3 & 2 & -3 \\ -1 & 4 & -2 & -4 \\ \hline \end{array}
+
\begin{array}{|cc|cc|}\hline -5 & -6 & 5 & -6 \\ -5 & 6 & 5 & 6 \\ \hline \end{array}
+ \mathcal{G}^K(0,0). \tag{5.1}
$$

With the bimatrix, Row's η values are determined by the column Column selects; e.g., $\eta_{1,1}$ is obtained by computing Player 1's values in the left column, which are her TL and BL entries. The approach is precisely the same with the tree. Should, for instance, Player 2 select L, Player 1's payoffs are determined by her L and R options. Thus, the two payoffs for the Nash analysis are her entries on the L legs coming out of both the Player 2's vertices, or RL and LL. (This pair is indicated by the dashed lines at the bottom left of Fig. 5.1.) The difference of Player 1's payoffs at these two nodes is 2 with an average of 1, so the Nash values replace the larger tree value (which is -4) with 1 and the smaller value (which is -6) with -1 as indicated with the dashed line below the middle tree. The average of these two values is -5

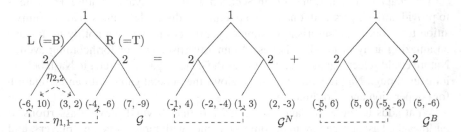

Fig. 5.1 Extensive form

and $\kappa_1 = \kappa_2 = 0$, so -5 is Player 1's β value in Player 2's L legs. (See the dashed lines below the third Fig. 5.1 tree.)

To further illustrate, Player 1's choice of L means that the $\mathcal{G}^N$ terms for Player 2 involve his payoffs on the LL and LR nodes. (This connection is indicated with the Fig. 5.1 dashed line above the payoff values.) The difference of these payoffs is 8, which is split with one $\mathcal{G}^N$ node receiving 4 and the other -4 allows games to be constructed with an uncountable number of $\mathcal{G}^B$ choices. This separation in constructing $\mathcal{G}^N$ and $\mathcal{G}^B$ allows inventing all sorts of extensive form games that exhibit different attributes.

5.2 Centipede Game

The η and β values for an extensive form game are computed as described in Chap. 2. This is illustrated with a version of the Centipede Game that Robert Rosenthal introduced in 1981 [22]. In this game, different $\mathcal{G}^N$ and $\mathcal{G}^B$ properties emerge.

To interpret Fig. 5.2, start at the first node on the left labeled 1: Player 1 has two strategies, down D and right R. For the D choice, the payoffs are 10 and 5 (for Players 1 and 2, respectively), and the game is over. For R, the game moves to the next node labeled 2, where Player 2 decides between D and R. Similar to the previous node, the D choice ends the game with payoffs 8 and 12, while R moves the game to the next node. Each node's label identifies the decision-maker; the arrows represent available choices. Play continues until a payoff is reached.

Clearly, each player's payoffs improve as the game progresses. The interesting hitch is that the values improve in an alternating fashion where if I agree to continue the game, my immediate payoff *decreases*; it will jump back up only if my partner continues. This statement, where a player's better outcome relies on the partner's actions, requires this game to have a crucial $\mathcal{G}^R$ component that precipitates a me-we tension. Answers require finding the Nash and Behavioral components.

The Nash component can be iteratively computed by starting at Player 2's final decision at the far right of the game tree. His choice is between D, with a personal payoff of 20, and R, with a payoff of 18. From a Nash best response perspective, Player 2 selects D, which yields his highest payoff. As in a normal form 2×2 game, it is the difference $20 - 18 = 2$ that determines the choice rather than the 20 and 18 values. Thus, in $\mathcal{G}^N$, his η value for R is -1 (far right of Fig. 5.3), and 1 for D (downward arrow from the last node).

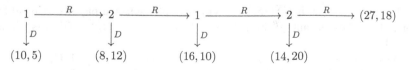

Fig. 5.2 A centipede game

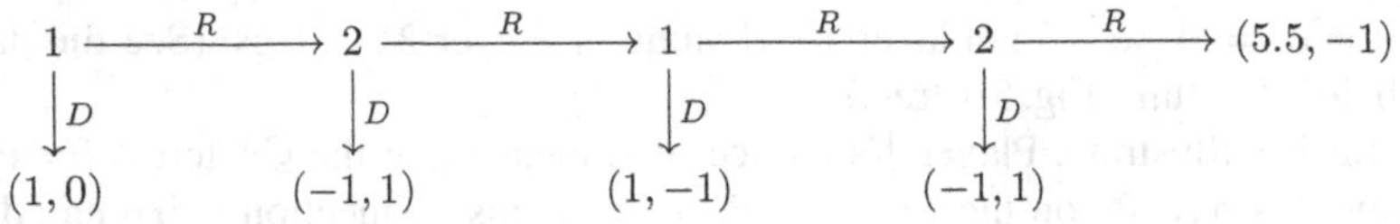

$$1 \xrightarrow{\ R\ } 2 \xrightarrow{\ R\ } 1 \xrightarrow{\ R\ } 2 \xrightarrow{\ R\ } (5.5, -1)$$

$$\downarrow D \qquad\qquad \downarrow D \qquad\qquad \downarrow D \qquad\qquad \downarrow D$$

$$(1, 0) \qquad\quad (-1, 1) \qquad\quad (1, -1) \qquad\quad (-1, 1)$$

Fig. 5.3 A $\mathcal{G}^N$ branch

The next step is to determine Player 1's actions at the penultimate node. By choosing D, she is guaranteed a payoff of 16. But her payoff with selecting R depends on what Player 2 would do in the next step. Following standard procedures, let q_1 and $q_2 = 1 - q_1$ represent the likelihood of Player 2 choosing, respectively, D and R. Thus, Player 1's expected value by selecting R is $EV(R) = 14q_1 + 27q_2$. When computing the $EV(D) - EV(R) = (16 - 14)q_1 + (16 - 27)q_2$ value, her best choice is D if $q_1 > \frac{11}{13}$. In determining $\mathcal{G}^N$, the q_1 and q_2 coefficients of $EV(D) - EV(R) = 2q_1 - 11q_2$ define two sets of η values. The yield from q_1 is 1 (for D, which is her payoff in Fig. 5.3 from the penultimate node's downward arrow) and -1 (which is her RD payoff with the first downward arrow from the right in Fig. 5.3). The η values from q_2 are -5.5 and 5.5, where 5.5 is her payoff with the RR node.

Here is a mystery; what happened to the η value of -5.5? A way to explain is to place this portion of the game in a 2×2 form of

	R		D	
R	27	18	14	20
D	16	10	16	10

$\rightarrow$ Nash terms

	R		D	
R	5.5	-1	-1	1
D	-5.5	0	1	0

As this expression makes clear, Player 1's η value of -5.5 belongs in the tree's DR branch. But neither the DR nor DD branch exists because, by Player 1 selecting D, the game is over. We will return to this missing payoff node later. (The -1 in the $(1, -1)$ payoff for the D payoff from the penultimate node is Player 2's payoff when analyzing what happens with the third node from the right.)

The next backward node has Player 2 selecting between D with a guaranteed payoff of 12, or R, which depends on Player 1's choice on the following node. This leads to the comparison of $EV(D) = 12p_1 + 12p_2$ and $EV(R) = 10p_1 + 20p_2$, which defines two sets of η values, $\{1, -1\}$ from p_1 and $\{4, -4\}$ from p_2. The $\{4, -4\}$ choices would introduce two new branches in the $\mathcal{G}^N$ tree of the DD and DR form. Again, these branches stretch the imagination: They cannot exist because the game is over. The rest of the tree is constructed as above leading to the single branch in Fig. 5.2.

This game's kernel is $\mathcal{G}^K (15, 13)$ (with carefully selected payoffs to avoid fractions). Thus the relevant $\mathcal{G}^B$ branch (which is $\mathcal{G}^B = \mathcal{G} - [\mathcal{G}^N + \mathcal{G}^K]$) is given in Fig. 5.4.

The decomposition explicitly displays the conflict in the Centipede Game. According to $\mathcal{G}^N$ (Fig. 5.3), the sole Nash cell is at the beginning where Player 1 ends the game at the first step by choosing D. Moreover, should the game proceed, at each

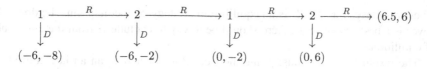

Fig. 5.4 $\mathcal{G}^B$

node, the relevant player's incentive is to play D. In contrast, $\mathcal{G}^B$'s Pareto superior cell (Fig. 5.4) is at the very end; it promotes the mutual benefit of continuing to play the game. This tension between individual opportunity and mutual benefit is captured (as always) by conflicting $\mathcal{G}^N$ and $\mathcal{G}^B$ entries.

The sum of the η values and the β values each should equal zero. They do not. The reason is that certain η and β values are missing when the tree is pruned to eliminate branches that cannot exist in practical terms. (Mathematics is not concerned with this feature, so should the missing branches be brought back, the summation results return.)

5.3 Equilibria and Subgame Equilibrium

When calculating expected values for the Nash analysis, the expected value need not be based just on what the other player might do at the *next* node. After all, an accurate assessment would include what the other player might do at subsequent nodes. How can this be handled?

A remedy is to require that a player's strategy specifies what the player would do for *all decisions* within the game. Do players actually play this manner? Of course not; consider a player's agony while deliberating over a next chess move. But agreeing upon what constitutes an acceptable strategy identifies how to handle expected value computations.

To illustrate with the Centipede game, should Player 1 want to stop the game whenever she can, her strategy would not be just D for the first node; it must specify her actions at each node whether the node can, or cannot, occur, so her strategy becomes (D, D). Similarly, Player 2's strategy must specify his action at any node, so it could be (D, D), or (R, D), or ....

While solving one problem, this approach creates others by spawning huge numbers of centipede game equilibria! With Player 1's choice of (D, D), Player 2's payoff is the same whether he chooses (D, D), (R, D), or (R, R); with Player 1's choice of (D, D), Player 2 never gets to make a decision. Consequently, all of his possible choices that are coupled with Player 1's (D, D) are Nash equilibria: In reality the game stops and neither player can improve upon the payoffs of $(10, 6)$, but those extraneous Nash equilibria remain. Similarly, if both players select (D, R), it is an equilibrium. If p and q are probabilities of choosing D for Players 1 and 2, respectively, (D, p) and (D, q) is an equilibrium. Playing R at any node is *not* a best response, yet (D, R) is an equilibrium strategy. Something must be done; R never is a best response, so there should be a way to exclude R from the elite club of equilibria.

The resolution is obvious: prune the tree. But rather than an ad hoc cutting, a systematic plan that applies to all settings is needed. An accepted method, called a subgame perfect equilibrium (SPE), is defined after introducing information sets. For now, consider a subgame to be a subtree that starts at a single node and contains all subsequent nodes in the original game. The subgames in the Fig. 5.2 centipede game are

$$2 \xrightarrow{\ \ R\ \ } (27, 18)$$
$$\downarrow D$$
$$(14, 20)$$

$$1 \xrightarrow{\ \ R\ \ } 2 \xrightarrow{\ \ R\ \ } (27, 18)$$
$$\downarrow D \qquad\qquad \downarrow D$$
$$(16, 10) \qquad\quad (14, 20)$$

$$2 \xrightarrow{\ \ R\ \ } 1 \xrightarrow{\ \ R\ \ } 2 \xrightarrow{\ \ R\ \ } (27, 18)$$
$$\downarrow D \qquad\qquad \downarrow D \qquad\qquad \downarrow D$$
$$(8, 12) \qquad\quad (16, 10) \qquad\quad (14, 20)$$

In contrast, the following is *not* a subgame because it does not contain all subsequent branches:

$$2 \xrightarrow{\ \ R\ \ } 1 \xrightarrow{\ \ R\ \ } 2 \xrightarrow{\ \ R\ \ } (27, 18)$$
$$\downarrow D \qquad\qquad\qquad\qquad\qquad \downarrow D$$
$$(8, 12) \qquad\qquad\qquad\qquad (14, 20)$$

The SPE equilibrium condition is that a strategy choice must be a best response in each subgame. As an immediate consequence, because R is *not a best response* in any subgame, it can never be part of an SPE. Thus the unique SPE of the centipede game is where both Player 1 and Player 2 choose (D, D).

5.4   Nature's Involvement in a 2 × 2 Game

Another scenario modeled by extensive form games is where an important player, Nature, determines the state of the world. As an example, Players 1 and 2 could be competing firms where Nature determines whether the future holds a high or low demand for their products. Or perhaps the players are two countries negotiating resource rights where Nature, perhaps aggravated by climate change, determines whether there will be a large or small supply of available water. While the players' interactions would differ from Nature's different whims, they must commit to their actions before Nature shows her hand.

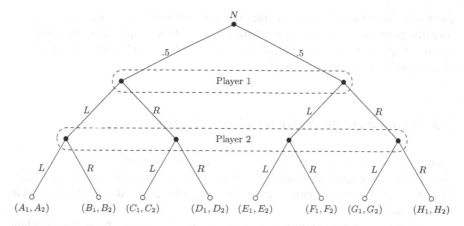

Fig. 5.5 Nature decides!

This means that each player makes a choice of L or R (e.g., Fig. 5.5) without knowing what will be Nature's or the other player's choice. But, each player does know that Nature chooses the states with equal probability. The payoffs of the eight possible Fig. 5.5 outcomes, labeled (A_1, A_2) through (H_1, H_2), consist of payoff pairs for each player.

The circles around nodes in Fig. 5.5 represent what information is available to each player: Of importance, the players cannot distinguish among nodes in the same circle. Thus, when Player 1 decides, she has no idea whether Nature has chosen state 1 or 2 because both are in the same top circle.

The set of possible states of the game, which is where decisions are made, is represented by a collection of nodes $C = \{n_{\alpha_1}, \ldots n_{\alpha_k}\}$. Figure 5.5, for instance, has seven nodes (bullets); label them as $n_0, n_1, \ldots, n_6$. Here n_0 is the root node of the tree where Nature determines the state; nodes n_1, n_2 represent the possible states for Player 1's decision, and nodes n_3, n_4, n_5, n_6 represent the possible states for Player 2's decision.

To model a player's ability to detect which nodes are relevant, let $\mathcal{H}$ be the information set; it is a partition of C where the different partition sets describe what nodes the players can distinguish. Player i's information set, $\mathcal{H}_i$, is a partition of the relevant subset of C; these are the nodes where this player makes a decision. With Fig. 5.5, Player 1's relevant nodes are n_1 and n_2, so her information set $\mathcal{H}_1$ is a partition of $\{n_1, n_2\}$. If she cannot distinguish between these nodes, her information set consists of the single element $\mathcal{H}_1 = \{\{n_1, n_2\}\}$, which means when deciding whether to select L or R, she has no idea whether she is at node n_1 or n_2. If she can distinguish between them, then $\mathcal{H}_1 = \{\{n_1\}, \{n_2\}\}$.

Similarly, if Player 2 cannot distinguish among his states of the game, his information set is $\mathcal{H}_2 = \{\{n_3, n_4, n_5, n_6\}\}$. This means that when making an L or R decision, he has no idea about which of the four nodes represent where he is located. But if he found some way to observe, for instance, what Player 1 does, but not the state of the world, his information set becomes the more refined $\mathcal{H}_2 = \{\{n_3, n_5\}, \{n_4, n_6\}\}$. The first partition set of $\{n_3, n_5\}$ is where he knows that she played L; the second is where she plays R. In mathematical terms and as illustrated below, a strategy is an action conditional on each element $h \in \mathcal{H}_i$.

5.5 From a Decomposition to Coordinates

Figure 5.5 identifies what information is available to which players; it will be described in more detail when justifying the decomposition, which is introduced next. For now, the special case given by Fig. 5.6 is used to motivate the decomposition.

First, a notational modification to facilitate the discussion. Nature is deciding between two games, L or R, which are expressed here as games 1 or 2. To compact the space-consuming trees, express these two games in the following bimatrix form where the "Join" symbol "⋈" reflects that the two games are connected through Nature's choice. In this manner, Fig. 5.5 has the representation

$$
\begin{array}{|cc|cc|}\hline A_1 & A_2 & B_1 & B_2 \\ C_1 & C_2 & D_1 & D_2 \\\hline\end{array} \bowtie \begin{array}{|cc|cc|}\hline E_1 & E_2 & F_1 & F_2 \\ G_1 & G_2 & H_1 & H_2 \\\hline\end{array} \tag{5.2}
$$

with Fig. 5.6 having

$$
\begin{array}{|cc|cc|}\hline -3 & -3 & 9 & 1 \\ 1 & 9 & 5 & 5 \\\hline\end{array} \bowtie \begin{array}{|cc|cc|}\hline 7 & 7 & -5 & 9 \\ 9 & -5 & -3 & -3 \\\hline\end{array} \tag{5.3}
$$

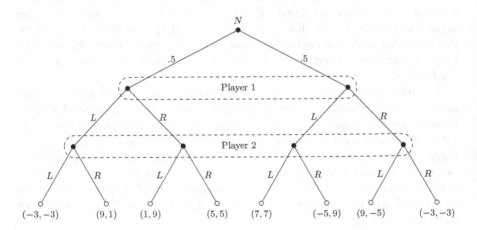

Fig. 5.6 What happens? Nature decides!

As Eq. 5.3 displays, Nature selects between the players experiencing a Hawk–Dove game (Nature goes to the Left) or a Prisoner's Dilemma game (Nature goes to the Right).

As it might be expected, building blocks for the decomposition of Fig. 5.6 are supplied by the coordinates of the two individual games. In representing these coordinates, let O represent the bimatrix with zero entries. In this way, the Nash structures for the two games are

$$
\mathcal{G}_1^N = \begin{array}{|cc|cc|} \hline -2 & -2 & 2 & 2 \\ \hline 2 & 2 & -2 & -2 \\ \hline \end{array} \bowtie O, \quad \mathcal{G}_2^N = O \bowtie \begin{array}{|cc|cc|} \hline -1 & -1 & -1 & 1 \\ \hline 1 & -1 & 1 & 1 \\ \hline \end{array} \qquad (5.4)
$$

The general coordinate representation, where the superscripts identify the game, is

$$
\mathcal{G}_1^N = \begin{array}{|cc|cc|} \hline \eta_{1,1}^1 & \eta_{1,2}^1 & \eta_{2,1}^1 & -\eta_{1,2}^1 \\ \hline -\eta_{1,1}^1 & \eta_{2,2}^1 & -\eta_{2,1}^1 & -\eta_{2,2}^1 \\ \hline \end{array} \bowtie O, \quad \mathcal{G}_2^N = O \bowtie \begin{array}{|cc|cc|} \hline \eta_{1,1}^2 & \eta_{1,2}^2 & \eta_{2,1}^2 & -\eta_{1,2}^2 \\ \hline -\eta_{1,1}^2 & \eta_{2,2}^2 & -\eta_{2,1}^2 & -\eta_{2,2}^2 \\ \hline \end{array}.
$$
$$(5.5)$$

The two behavioral structures are

$$
\mathcal{G}_1^B = \begin{array}{|cc|cc|} \hline -4 & -4 & 4 & -4 \\ \hline -4 & 4 & 4 & 4 \\ \hline \end{array} \bowtie O, \quad \mathcal{G}_2^B = O \bowtie \begin{array}{|cc|cc|} \hline 6 & 6 & -6 & 6 \\ \hline 6 & -6 & -6 & -6 \\ \hline \end{array} \qquad (5.6)
$$

and the kernel components are

$$
\mathcal{G}_1^K = \mathcal{G}^K(3,3) \bowtie O, \quad \mathcal{G}_2^K = O \bowtie \mathcal{G}^K(2,2). \qquad (5.7)
$$

The general representation of these components is

$$
\mathcal{G}_1^B = \begin{array}{|cc|cc|} \hline \beta_1^1 & \beta_2^1 & -\beta_1^1 & \beta_2^1 \\ \hline \beta_1^1 & -\beta_2^1 & -\beta_1^1 & -\beta_2^1 \\ \hline \end{array} \bowtie O, \quad \mathcal{G}_2^B = O \bowtie \begin{array}{|cc|cc|} \hline \beta_1^2 & \beta_2^2 & -\beta_1^2 & \beta_2^2 \\ \hline \beta_1^2 & -\beta_2^2 & -\beta_1^2 & -\beta_2^2 \\ \hline \end{array} \qquad (5.8)
$$

The kernel components are

$$
\mathcal{G}_1^K = \begin{array}{|cc|cc|} \hline \kappa_1^1 & \kappa_2^1 & \kappa_1^1 & \kappa_2^1 \\ \hline \kappa_1^1 & \kappa_2^1 & \kappa_1^1 & \kappa_2^1 \\ \hline \end{array} \bowtie O, \quad \mathcal{G}_2^K = O \bowtie \begin{array}{|cc|cc|} \hline \kappa_1^1 & \kappa_2^2 & \kappa_1^2 & \kappa_2^2 \\ \hline \kappa_1^2 & \kappa_2^2 & \kappa_1^2 & \kappa_2^2 \\ \hline \end{array}. \qquad (5.9)
$$

5.5.1 Nash Component for the General Game

It is reasonable to expect that a cell's Nash structure in the general Figs. 5.5 and 5.6 games combines the corresponding cells in the component games 1 and 2. After all, the players have no idea what Nature will do, so they cannot determine whether, say, the relevant cell BL comes from game 1 or game 2. The decomposition (developed below) proves this is correct: The Nash structure of the general game, $\mathcal{G}^N$, requires the corresponding cell in both bimatrices to be *identical!* This leads to the fascinating possibility that Nash structure of a $\mathcal{G}^N$ cell may differ from that of either of the two base games! After all, with Eq. 5.4 and game 1, BL is a Nash cell, while in game 2, BL is hyperbolic-column. It cannot be both types for $\mathcal{G}^N$.

To compute $\mathcal{G}^N$ for the general game, take the average of entries in the corresponding cells of both $\mathcal{G}_1^N$ and $\mathcal{G}_2^N$. So, with Eq. 5.4, the BL entries for both games in $\mathcal{G}^N$ are $\frac{1+2}{2} = 1.5$ for Player 1 and $\frac{2-1}{2} = 0.5$ for Player 2. In this manner, the Nash payoffs for the Fig. 5.6 full game are

$$\mathcal{G}^N = \begin{array}{|cc|cc|}\hline -1.5 & -1.5 & .5 & 1.5 \\\hline 1.5 & .5 & -.5 & -.5 \\\hline\end{array} \bowtie \begin{array}{|cc|cc|}\hline -1.5 & -1.5 & .5 & 1.5 \\\hline 1.5 & .5 & -.5 & -.5 \\\hline\end{array} \qquad (5.10)$$

defining the Nash dynamic that consists of two attracting Nash cells (BL and TR) and two repelling cells (BR and TL) for each component game. In general, the $\eta_{i,j}$ value for both bimatrices of $\mathcal{G}^N$ is

$$\eta_{i,j} = \frac{\eta_{i,j}^1 + \eta_{i,j}^2}{2}, \quad i, j = 1, 2. \qquad (5.11)$$

Terms left behind after this subtraction define the *Nash context* component

$$\mathcal{G}^{NC} = \begin{array}{|cc|cc|}\hline -.5 & -.5 & 1.5 & .5 \\\hline .5 & 1.5 & -1.5 & -1.5 \\\hline\end{array} \bowtie \begin{array}{|cc|cc|}\hline .5 & .5 & -1.5 & -.5 \\\hline -.5 & -1.5 & 1.5 & 1.5 \\\hline\end{array} \qquad (5.12)$$

Before discussing these components, the general form of $\mathcal{G}^N$ (as in Chap. 2) has four η variables expressed in the familiar Eq. 2.21 form

$$\mathcal{G}^N = \begin{array}{|cc|cc|}\hline \eta_{1,1} & \eta_{1,2} & \eta_{2,1} & -\eta_{1,2} \\\hline -\eta_{1,1} & \eta_{2,2} & -\eta_{2,1} & -\eta_{2,2} \\\hline\end{array} \bowtie \begin{array}{|cc|cc|}\hline \eta_{1,1} & \eta_{1,2} & \eta_{2,1} & -\eta_{1,2} \\\hline -\eta_{1,1} & \eta_{2,2} & -\eta_{2,1} & -\eta_{2,2} \\\hline\end{array} \qquad (5.13)$$

As required of Fig. 5.5 games, the two bimatrix components of $\mathcal{G}^N$ are identical.

The Nash context component also has four γ variables that define

$$\mathcal{G}^{NC} = \begin{array}{|cc|cc|}\hline -\gamma_{1,1} & -\gamma_{1,2} & -\gamma_{2,1} & \gamma_{1,2} \\\hline \gamma_{1,1} & -\gamma_{2,2} & \gamma_{2,1} & \gamma_{2,2} \\\hline\end{array} \bowtie \begin{array}{|cc|cc|}\hline \gamma_{1,1} & \gamma_{1,2} & \gamma_{2,1} & -\gamma_{1,2} \\\hline -\gamma_{1,1} & \gamma_{2,2} & -\gamma_{2,1} & -\gamma_{2,2} \\\hline\end{array}. \qquad (5.14)$$

While both bimatrices of $\mathcal{G}^N$ are identical, one $\mathcal{G}^{NC}$ bimatrix is the negative of the other. The γ values are given by

$$\gamma_{i,j} = \eta_{i,j}^2 - \eta_{i,j} = \frac{\eta_{i,j}^2 - \eta_{i,j}^1}{2}. \qquad (5.15)$$

Here is a concern; each bimatrix of $\mathcal{G}^{NC}$ has the Eq. 2.21 Nash form: Does $\mathcal{G}^{NC}$ smuggle in Nash information? Of course it does; the $\mathcal{G}^{NC}$ bimatrices capture what needs to be added to each $\mathcal{G}^N$ bimatrix to recover $\mathcal{G}_1^N$ and $\mathcal{G}_2^N$. Of particular importance (proved later), each $\mathcal{G}^{NC}$ bimatrix component contains Nash information for the relevant *component* game, *but $\mathcal{G}^{NC}$ contains no Nash information of the general game!*

5.5.2 Changing Probabilities

For a brief discussion about ramifications of this decomposition, Eq. 5.10 shows that the general game has a Nash structure of two Nash cells (BL and TR of both games) and two repellers (TL and BR). Because of the form of Eq. 5.11, the structure of a $\mathcal{G}^N$ cell for the general game is dictated by the larger η values from the corresponding cell in $\mathcal{G}_1^N$. Had, for instance, the Nash structure of the second game been

$$\mathcal{G}_2^N = \mathbf{O} \bowtie \begin{array}{|cc|cc|} \hline -3 & -3 & -3 & 3 \\ \hline 3 & -3 & 3 & 3 \\ \hline \end{array} \tag{5.16}$$

then the common bimatrix for $\mathcal{G}^N$ would have been

$$\begin{array}{|cc|cc|} \hline -2.5 & -2.5 & -.5 & 2.5 \\ \hline 2.5 & -.5 & .5 & .5 \\ \hline \end{array}$$

where each $\mathcal{G}^N$ bimatrix now has a single Nash cell and a diametrically opposite repeller.

This comment indicates how the general game's Nash structure changes as Nature's probabilities of selecting L or R vary. The reason is that Nature's changing likelihoods can be handled by treating the payoffs as expected values.[1] Suppose, for instance, information shows that rather than a $\frac{1}{2}$–$\frac{1}{2}$ split in Fig. 5.6, Nature has a $\frac{3}{4}$ chance of selecting R. Thus, game 1's payoffs are multiplied by $\frac{1}{4}$ while those of game 2 by $\frac{3}{4}$. Dropping the denominator, Nature's probability change enhances game 2's payoffs by a multiple of 3, which leads to Eq. 5.16. So, this change in probability causes the game's Nash structure to move from having two Nash cells and two repellers to having one Nash cell and one repeller. Expanding on this comment, it follows that as the likelihood of Nature moves continuously from selecting game 1 with certainty to selecting game 2 with certainty, the form of $\mathcal{G}^N$ moves continuously from the form of $\mathcal{G}_1^N$ to that of $\mathcal{G}_2^N$.

---

[1]Do so with care because of a possible discontinuity should some game occur with certainty.

5.5.3 The Behavioral Terms for the General Game

The behavioral components also have a structure parallel to that of $\mathcal{G}^N$; a form that respects the information sets. Doing so captures the reality that the players have no knowledge about whether game 1 or 2 will prevail and this lack of information should be imbedded in the behavioral term.

The construct is similar to how $\mathcal{G}^N$ was defined.[2] Namely, replace $\mathcal{G}_1^B$ and $\mathcal{G}_2^B$ with $\mathcal{G}^B$, which has identical bimatrices. To do so, in each $\mathcal{G}^B$ cell place the average of the cell entries from the two games. The differences from the values and average define the *separated behavioral component*, $\mathcal{G}^{SB}$. Here each bimatrix has the Chap. 2 form of a behavioral component, but one bimatrix is the negative of the other.

Illustrating with the Eq. 5.6 values,

$$\mathcal{G}^B = \begin{array}{|c|c|}\hline 1 & 1 \\\hline 1 & -1 \\\hline\end{array}\begin{array}{|c|c|}\hline -1 & 1 \\\hline -1 & -1 \\\hline\end{array} \bowtie \begin{array}{|c|c|}\hline 1 & 1 \\\hline 1 & -1 \\\hline\end{array}\begin{array}{|c|c|}\hline -1 & 1 \\\hline -1 & -1 \\\hline\end{array} \tag{5.17}$$

and

$$\mathcal{G}^{SB} = \begin{array}{|c|c|}\hline -5 & -5 \\\hline -5 & 5 \\\hline\end{array}\begin{array}{|c|c|}\hline 5 & -5 \\\hline 5 & 5 \\\hline\end{array} \bowtie \begin{array}{|c|c|}\hline 5 & 5 \\\hline 5 & -5 \\\hline\end{array}\begin{array}{|c|c|}\hline -5 & 5 \\\hline -5 & -5 \\\hline\end{array} \tag{5.18}$$

The general representations of these behavioral components are

$$\mathcal{G}^B = \begin{array}{|c|c|}\hline \beta_1 & \beta_2 \\\hline \beta_1 & -\beta_2 \\\hline\end{array}\begin{array}{|c|c|}\hline -\beta_1 & \beta_2 \\\hline -\beta_1 & -\beta_2 \\\hline\end{array} \bowtie \begin{array}{|c|c|}\hline \beta_1 & \beta_2 \\\hline \beta_1 & -\beta_2 \\\hline\end{array}\begin{array}{|c|c|}\hline -\beta_1 & \beta_2 \\\hline -\beta_1 & -\beta_2 \\\hline\end{array} \tag{5.19}$$

and

$$\mathcal{G}^{SB} = \begin{array}{|c|c|}\hline \xi_1 & \xi_2 \\\hline \xi_1 & -\xi_2 \\\hline\end{array}\begin{array}{|c|c|}\hline -\xi_1 & \xi_2 \\\hline -\xi_1 & -\xi_2 \\\hline\end{array} \bowtie \begin{array}{|c|c|}\hline -\xi_1 & -\xi_2 \\\hline -\xi_1 & \xi_2 \\\hline\end{array}\begin{array}{|c|c|}\hline \xi_1 & -\xi_2 \\\hline \xi_1 & \xi_2 \\\hline\end{array} \tag{5.20}$$

Namely, in both bimatrices of $\mathcal{G}^B$, let β_i be player i's β value. The value is

$$\beta_i = \frac{\beta_i^1 + \beta_i^2}{2}, \quad i = 1, 2, \tag{5.21}$$

while

$$\xi_i = \beta_i^1 - \beta_i = \frac{\beta_i^1 - \beta_i^2}{2}, \quad i = 1, 2. \tag{5.22}$$

This construction completes the decomposition.

Theorem 5.1 *A Fig. 5.5 game can be uniquely decomposed into a Nash component with the Eq. 5.13 structure, a Nash context component $\mathcal{G}^{NC}$ with the Eq. 5.14 form, the behavioral term $\mathcal{G}^B$ (Eq. 5.19) and a separated behavioral term $\mathcal{G}^{SB}$ (Eq. 5.20),*

---

[2]The same approach holds for the kernel terms, which are important when constructing games for experiments or when resources can be transferred, but they are essentially ignored here.

and the kernel components (Eq. 5.9). The relevant coefficients for these components, as defined by Eqs. 5.11, 5.15, 5.21, 5.22, come from the decomposition of the two games.

As the probability of Nature varies from certainty of selecting L to certainty of selecting R, the common structure of the bimatrices of $\mathcal{G}^N$ and of $\mathcal{G}^B$ shifts, respectively, from that of $\mathcal{G}_1^N$ to $\mathcal{G}_2^N$ and $\mathcal{G}_1^B$ to $\mathcal{G}_2^B$.

5.6 Paradise for Designers!

With so many variables, all sorts of intriguing games can be created. For instance, Fig. 5.6 game inherits the Hawk–Dove Nash structure but the Prisoner's Dilemma's $\mathcal{G}^B$ behavioral structure. The resulting general game roughly resembles a Battle of the Sexes where each player exhibits some concern, or coordination, for the choice of the other player.

This phenomenon, where full game's behavior need not reflect either of the two-component games, manifests what can occur with the averaging of cell entries (e.g., Eqs. 5.11 and 5.21) of the component games that is required by Nature's indecision. In this averaging with Fig. 5.6, one game had a more dominant Nash structure (the first game) while the other had a more dominant cooperative, behavioral structure. These comments describe what happens with the theoretical analysis. It would be of empirical interest to determine whether this split would be picked up by participants in a lab experiment.

The averaging procedure permits the design of all sorts of structures. Suppose that TL is the sole Nash cell for game 1 while TL is a repeller for game 2. Depending on the η values, the general game's Nash structure of this TL cell could be almost anything. If the $\mathcal{G}^N$ choices for TL are $(2, 4)$ and $(-1, -1)$, then the TL cell remains Nash in the general game. But if the choices are $(2, 4)$ and $(-3, -3)$, then the TL cell for both $\mathcal{G}^N$ bimatrices is hyperbolic-row; size matters. The same kind of behavior arises for $\mathcal{G}^B$, which leads to bountiful buffet of delicious structures admitted by the general game that differs from the component games.

Still other variables, such as the structure of the information sets, produce other possibilities. For instance, following the procedure described in Sect. 5.7, the decomposition structure can quickly be found for different Fig. 5.5 information settings. As an example, suppose Player 1 knows that Nature will select R, but Player 2 is stuck with the described indecision. As one might guess, Player 1's Nash and Behavioral structures of the general game are that of her game 2'a Nash and Behavioral structures; Player 2, however, has the Nash and Behavioral structure of the original Fig. 5.5 game.

Illustrating this new information set with Fig. 5.6, the decomposition structure is

$$\mathcal{G}^N = \begin{array}{|c c|c c|} \hline 0 & -1.5 & 0 & 1.5 \\ \hline 0 & .5 & 0 & -.5 \\ \hline \end{array} \bowtie \begin{array}{|c c|c c|} \hline -1 & -1.5 & -1 & 1.5 \\ \hline 1 & .5 & 1 & -.5 \\ \hline \end{array},$$

where in the second bimatrix the Nash cell jumped from BR to BL, which requires an adjustment in the players' strategic approaches.

The new behavioral component is

$$\mathcal{G}^B = \begin{array}{|cc|cc|} \hline 0 & 1 & 0 & 1 \\ 0 & -1 & 0 & -1 \\ \hline \end{array} \bowtie \begin{array}{|cc|cc|} \hline 6 & 1 & -6 & 1 \\ 6 & -1 & -6 & -1 \\ \hline \end{array}$$

causing the second bimatrix of the general game to move from a Prisoner's Dilemma to where TL becomes more attractive.

This general manner makes it possible to construct all sorts of games, where, say, even should Player 1 has private information about Nature's action, Player 2 could have an advantage. Or, it can identify games where it is to Player 1's advantage to let Player 2 know what Nature will do. Suppose the players have different estimates on the probabilities of what Nature will do; what can happen? The possibilities are without limit.

5.7 Details of the Decomposition

What adds interest about the proof of Theorem 5.1 is that it serves as a template to find decompositions of other extensive form games that have different information sets. As a brief sketch of what is presented, the first subsection (Sect. 5.7.1) outlines difference in the analysis caused by lack of information. Then a standard best response analysis is carried out where, in the analysis, each player is hindered by not knowing what Nature will do.

The decomposition of the general game mimics the Sect. 3.4.2 procedure. First, find vectors (game components) that capture the barest payoff information needed for a Nash analysis. These components reflect the minimal details required to compute the usual best response in terms of the expected values, where these values now cover information sets. Vectors orthogonal to these special choices contain no information about the Nash analysis; they are redundant aspects of the payoffs relative to the Nash terms, so they become the behavioral and kernel contributions.

5.7.1 Subgames and Decisions

Information sets require reformulating the concept of a subgame, which is done in a natural manner. In Sect. 5.3, a subgame started at a node and contained all subsequent nodes in the original tree. That is, a subgame identifies all consequences and actions that follow from an original node. In Fig. 5.1, if Player 1 selects L, the subgame from Player 2's node identifies the relevant payoffs needed to compute certain $\mathcal{G}^N$ entries.

With information sets, the goal remains to start at a node. What differs is that a subsequent step may not be a node; it may be an information set such as $\{n_i, n_j\}$. But the player does not know which of these two nodes is the relevant one, so the subgame should include all subsequent nodes for *both* n_i and n_j. This is the third of the following requirements for a subgame.

1. It has a single initial node n_i where $\{n_i\} \in \mathcal{H}$
2. If a node is in the subgame, then so are all of its descendants in the game tree
3. If a subgame node $n_j \in h$ for $h \in \mathcal{H}$, then every other node in h is also in the subgame

To illustrate with Fig. 5.5, consider its subtree that begins at node n_1; this is where Player 1 makes decision after Nature chooses state L. A subgame must contain all of its descendants, so it includes the nodes n_3 and n_4 where Player 2 decides whether to play L or R, and it includes the outcomes A, B, C, D.

In Fig. 5.5, Player 1 does *not* know whether Nature selected state L or R; a condition specified by the information set $\mathcal{H}_1 = \{\{n_1, n_2\}\}$. This means (the above third condition) that all nodes subsequent to n_2 must also be included, which includes all nodes below n_0. *However,* this choice starts at a two-member partition set, which violates the first requirement that a subgame starts at a node. Consequently, Fig. 5.5 has only the subgame beginning at the initial node n_0; i.e., it has no proper subgame.

5.7.2 Player 2

The information sets slightly complicate the decomposition analysis. Fortunately, guidance comes from the vector analysis approach developed in Sect. 3.4.2.

Start with Player 2's payoffs found at the bottom of the Fig. 5.5 tree, where he must decide between L and R. All he knows are the payoffs; he knows nothing about Player 1's choice or the state of Nature. As in Chap. 2, should p_1 be the probability that Player 1 selects L, so $p_2 = 1 - p_1$, Player 2's preference is given by

$$L \succ R \Leftrightarrow \frac{1}{2}(p_1 A_2 + p_2 C_2 + p_1 E_2 + p_2 G_2) > \frac{1}{2}(p_1 B_2 + p_2 D_2 + p_1 F_2 + p_2 H_2)$$
$$\Leftrightarrow p_1(A_2 - B_2 + E_2 - F_2) + p_2(C_2 - D_2 + G_2 - H_2) > 0$$

$$(5.23)$$

In the Nash analysis (as in Chap. 2), rather than the individual payoffs, of importance are the values of the Eq. 5.23 coefficient sums $A_2 - B_2 + E_2 - F_2$ and $C_2 - D_2 + G_2 - H_2$. In examining these expressions, the goal is to determine what payoff information is relevant, and what is redundant, for the Nash analysis. The approach follows the lead of Sect. 3.4.2; that is

- find vectors whereby a dot product with the game produces the $A_2 - B_2 + E_2 - F_2$ and $C_2 - D_2 + G_2 - H_2$ values,
- then complete the vector space by finding orthogonal vectors.

Player 2's eight different payoffs reside in an eight-dimensional space. To keep the arguments understandable, the two four-dimensional subspaces consisting, respectively, of the A_2, B_2, E_2, F_2 and C_2, D_2, G_2, H_2 variables, are analyzed first; that is, only the variables actually being used are considered. In the first subspace (the A_2, B_2, E_2, F_2 variables), Player 2's portion of the game is given by the vector $\hat{\mathbf{g}}_{1,2} = (A_2, B_2, E_2, F_2)$, so the vector leading to the desired $A_2 - B_2 + E_2 - F_2$ value with a dot product is $\hat{\mathbf{n}}_{1,2} = (1, -1, 1, -1)$. Similarly, in the second subspace, Player 2's game portion is the vector $\hat{\mathbf{g}}_{2,2} = (C_2, D_2, G_2, H_2)$, where a dot product with the vector $\hat{\mathbf{n}}_{2,2} = (1, -1, 1, -1)$ yields $C_2 - D_2 + G_2 - H_2$.

Following Sect. 3.4.2, completing each subspace requires adding three more linearly independent vectors. Finding such vectors that are orthogonal to $\hat{\mathbf{n}}_{j,2}$ is immediate; just match up in pairs one of the 1's with one of the -1's in $\hat{\mathbf{n}}_{j,2}$, $j = 1, 2$. One set of choices is

$$\hat{\mathbf{a}}_{j,2}^1 = (1, 1, 0, 0), \ \hat{\mathbf{a}}_{j,2}^2 = (0, 0, 1, 1), \ \hat{\mathbf{k}}_{j,2} = (1, 1, 1, 1), \ \text{and} \ \hat{\mathbf{c}}_{j,2} = (-1, 1, 1, -1), \qquad (5.24)$$

(where j identifies the first or second subspace and the superscript identifies which bimatrix). For instance, the dot product of $\hat{\mathbf{n}}_{1,2}$ with $\hat{\mathbf{c}}_{1,2}$ is $(1)(-1) + (-1)(1) + (1)(1) + (-1)(-1) = 0$. The $\mathbf{a}_{j,2}^k$ terms play the role of $\mathcal{G}^{Average}$ from Chap. 2.

The vectors $\{\hat{\mathbf{a}}_{j,2}^1, \hat{\mathbf{a}}_{j,2}^2, \hat{\mathbf{c}}_{j,2}, \hat{\mathbf{n}}_{j,2}\}$ are pairwise orthogonal, so they form a basis. Thus, Player 2's payoffs can be uniquely expressed as

$$\mathbf{g}_{j,2} = \eta_{j,2}\hat{\mathbf{n}}_{j,2} + a_{j,2}^1\hat{\mathbf{a}}_{j,2}^1 + a_{j,2}^2\hat{\mathbf{a}}_{j,2}^2 + \gamma_{j,2}\hat{\mathbf{c}}_{j,2}, \quad j = 1, 2. \qquad (5.25)$$

The only difference between a vector with a hat and the general vector it represents is that the general vector includes the ignored coordinates, which are all zeros. In this manner

$$\eta_{1,2}\mathbf{n}_{12} + \eta_{2,2}\mathbf{n}_{2,2} = \begin{array}{|c|c|c|c|} \hline 0 & \eta_{1,2} & 0 & -\eta_{1,2} \\ \hline 0 & \eta_{2,2} & 0 & -\eta_{2,2} \\ \hline \end{array} \bowtie \begin{array}{|c|c|c|c|} \hline 0 & \eta_{1,2} & 0 & -\eta_{1,2} \\ \hline 0 & \eta_{2,2} & 0 & -\eta_{2,2} \\ \hline \end{array} \qquad (5.26)$$

while

$$\gamma_{1,2}\mathbf{c}_{1,2} + \gamma_{2,2}\mathbf{c}_{2,2} = \begin{array}{|c|c|c|c|} \hline 0 & -\gamma_{1,2} & 0 & \gamma_{1,2} \\ \hline 0 & -\gamma_{2,2} & 0 & \gamma_{2,2} \\ \hline \end{array} \bowtie \begin{array}{|c|c|c|c|} \hline 0 & \gamma_{1,2} & 0 & -\gamma_{1,2} \\ \hline 0 & \gamma_{2,2} & 0 & -\gamma_{2,2} \\ \hline \end{array} \qquad (5.27)$$

These expressions are, respectively, the second agent's contributions to Eqs. 5.13 and 5.14.

Finding the coefficient values (e.g., the η and γ values) follows the lead of Sect. 3.4.2. Illustrating by finding $\eta_{1,2}$, the dot product of $\mathbf{g}_{1,2}$ with $\mathbf{n}_{1,2}$ is $[\mathbf{g}_{1,2}, \mathbf{n}_{i,2}] = (A_2 - B_2 + E_2 - F_2)$. Taking the same dot product with the Eq. 5.25 representation of $\mathbf{g}_{1,2}$ leads to $[\mathbf{g}_{1,2}, \mathbf{n}_{i,2}] = \eta_{1,2}[\mathbf{n}_{1,2}, \mathbf{n}_{1,2}] = 4\eta_{1,2}$. These two equations represent the same dot product, so $\eta_{1,2} = \frac{A_2 - B_2 + E_2 - F_2}{4} = \frac{1}{2}\left(\frac{A_2 - B_2}{2} + \frac{E_2 - F_2}{2}\right) = \frac{\eta_{1,2}^1 + \eta_{1,2}^2}{2}$, which is Eq. 5.11.

Similarly, to find the $\gamma_{1,2}$ value, take the dot product of $\mathbf{c}_{1,2}$ with both representations of $\mathbf{g}_{1,2}$. The first yields $[\mathbf{g}_{1,2}, \mathbf{c}_{1,2}] = (-A_2 + B_2 + E_2 - F_2)$, while the second (using Eq. 5.25) has $[\mathbf{g}_{1,2}, \mathbf{c}_{1,2}] = \gamma_{1,2}[\mathbf{c}_{1,2}, \mathbf{c}_{1,2}] = 4\gamma_{1,2}$. Consequently $\gamma_{1,2} = \frac{-A_2+B_2+E_2-F_2}{4} = \frac{1}{2}\left(-\frac{A_2-B_2}{2} + \frac{E_2-F_2}{2}\right) = \frac{\eta_{1,2}^2-\eta_{1,2}^1}{2}$, which agrees with Eq. 5.15. All coefficients are found in the same manner.

The next step is to convert the "average payoff" terms $\mathbf{a}_{i,j}^k$ into behavioral and kernel contributions. The approach mimics much of what was done in Chap. 2. Namely, define $\mathbf{b}_2 = [\mathbf{a}_{1,2}^1 - \mathbf{a}_{2,2}^1] + [\mathbf{a}_{1,2}^2 - \mathbf{a}_{2,2}^2]$, $\mathbf{e}_2 = [\mathbf{a}_{1,2}^1 - \mathbf{a}_{2,2}^1] - [\mathbf{a}_{1,2}^2 - \mathbf{a}_{2,2}^2]$, $\mathbf{k}_2^1 = [\mathbf{a}_{1,2}^1 + \mathbf{a}_{2,2}^1]$, and $\mathbf{k}_2^2 = [\mathbf{a}_{1,2}^2 + \mathbf{a}_{2,2}^2]$. A standard computation shows that these vectors are orthogonal. By construction, Eq. 5.25 can be replaced with

$$\mathbf{g}_{1,2} + \mathbf{g}_{2,2} = \eta_{1,2}\mathbf{n}_{1,2} + \eta_{2,2}\mathbf{n}_{2,2} + \gamma_{1,2}\mathbf{c}_{1,2} + \gamma_{2,2}\mathbf{c}_{2,2} + \beta_2\mathbf{b}_2 + \xi_2\mathbf{e}_2 + \kappa_1\mathbf{k}_2^1 + \kappa_2^2\mathbf{k}_2^2, \quad (5.28)$$

where the $\mathbf{k}_2^j$ terms capture the second agent's kernel terms for the two-component games. It remains to interpret $\beta_2\mathbf{b}_2$ and $\xi_2\mathbf{e}_2$, which will be, respectively, Player 2's contribution to $\mathcal{G}^B$ (Eq. 5.19) and $\mathcal{G}^{SB}$ (Eq. 5.20).
Indeed, by definition

$$\beta_2\mathbf{b}_2 = \begin{array}{|c|c|c|c|}\hline 0 & \beta_2 & 0 & \beta_2 \\\hline 0 & -\beta_2 & 0 & -\beta_2 \\\hline\end{array} \boxtimes \begin{array}{|c|c|c|c|}\hline 0 & \beta_2 & 0 & \beta_2 \\\hline 0 & -\beta_2 & 0 & -\beta_2 \\\hline\end{array}$$

and

$$\gamma_2\mathbf{e}_2 = \begin{array}{|c|c|c|c|}\hline 0 & \gamma_2 & 0 & \gamma_2 \\\hline 0 & -\gamma_2 & 0 & -\gamma_2 \\\hline\end{array} \boxtimes \begin{array}{|c|c|c|c|}\hline 0 & -\gamma_2 & 0 & -\gamma_2 \\\hline 0 & \gamma_2 & 0 & \gamma_2 \\\hline\end{array}.$$

To compute, say, β_2 take the dot product of $\mathbf{b}_2$ with both representations of $\mathbf{g}_{1,2} + \mathbf{g}_{2,2}$. The first is $A_2 + B_2 - C_2 - D_2 + E_2 + F_2 - G_2 - H_2$ while the second (with Eq. 5.28) is $\beta_2[\mathbf{b}_2, \mathbf{b}_2] = 8\beta_2$. Consequently

$$\beta_2 = \tfrac{1}{8}[A_2 + B_2 - C_2 - D_2 + E_2 + F_2 - G_2 - H_2] = $$
$$\tfrac{1}{2}\left[\tfrac{A_2+B_2-C_2-D_2}{4} + \tfrac{E_2+F_2-G_2-H_2}{4}\right] = \tfrac{1}{2}[\beta_2^1 + \beta_2^2],$$

where the algebra in the penultimate step uses the footnote comment in Sect. 3.3.2.

5.7.3 Player 1

An identical analysis applies to Player 1, where, if Player 2 selects L with probability q_1 and R with probability q_2 (so $q_2 = 1 - q_1$), her best response expression is

$$L \succ R \Leftrightarrow \frac{1}{2}(q_1 A_1 + q_2 B_1 + q_1 E_1 + q_2 F_1) > \frac{1}{2}(q_1 C_1 + q_2 D_1 + q_1 G_1 + q_2 H_1)$$
$$\Leftrightarrow q_1(A_1 - C_1 + E_1 - G_1) + q_2(B_1 - D_1 + F_1 - H_1) > 0.$$

The main difference is that the emphasis now is on Player 1's column entries for the different components. This arises with the first step of finding a vector so that the dot product with the game yields the crucial $A_1 - C_1 + E_1 - G_1$ and $B_1 - D_1 + F_1 - H_1$ values. They are, respectively, $\mathbf{n}_{1,1}$ and $\mathbf{n}_{2,1}$ with the representation

$$\eta_{1,1}\mathbf{n}_{1,1} + \eta_{2,1}\mathbf{n}_{2,1} = \begin{array}{|cc|cc|} \hline \eta_{1,1} & 0 & \eta_{2,1} & 0 \\ \hline -\eta_{1,1} & 0 & -\eta_{2,1} & 0 \\ \hline \end{array} \bowtie \begin{array}{|cc|cc|} \hline \eta_{1,1} & 0 & \eta_{2,1} & 0 \\ \hline -\eta_{1,1} & 0 & -\eta_{2,1} & 0 \\ \hline \end{array},$$

which is Player 1's contribution to $\mathcal{G}^N$.

Remaining details are essentially the same, which completes the decomposition.

5.7.4 Alternative Basis for Behavioral Coordinates

A vector space admits multiple bases, where some are more appropriate for different needs. In this spirit, alternative choices for the behavioral coordinates are offered.

A first choice is to replace the four-dimensional $\mathcal{G}^B$ and $\mathcal{G}^{SB}$ with $\mathcal{G}_1^B$ and $\mathcal{G}_2^B$ as given in Eq. 5.6. These choices have the advantage of being familiar, which makes them much easier to use particularly when designing a game, but they have a dis-advantage of ignoring the structure of information sets. Implicitly, they provide a sense that there is an understanding what Nature might do. But $\mathcal{G}^B$ does not have this feature; with the two identical component games, it mimics what happens with the $\mathcal{G}^N$ structure by reflecting the spirit that the players have no true idea what Nature might do.

Another choice, albeit with the same criticism, is to replace the two $\mathbf{a}_{i,j}^k$ vectors in Eq. 5.24 with $\mathbf{d}_{i,j} = (1, 1, -1, -1)$ and $\mathbf{k}_j^i$. By doing so, the $\mathcal{G}^B$ and $\mathcal{G}^{SB}$ terms are replaced by

$$\delta_{1,1}\mathbf{d}_{1,1} + \delta_{2,1}\mathbf{d}_{2,1} + \delta_{1,2}\mathbf{d}_{1,2} + \delta_{2,2}\mathbf{d}_{2,2} = \begin{array}{|cc|cc|} \hline \delta_{1,1} & \delta_{1,2} & \delta_{2,1} & \delta_{1,2} \\ \hline \delta_{1,1} & \delta_{2,2} & \delta_{2,1} & \delta_{2,2} \\ \hline \end{array} \bowtie \begin{array}{|cc|cc|} \hline -\delta_{1,1} & -\delta_{1,2} & -\delta_{2,1} & -\delta_{1,2} \\ \hline -\delta_{1,1} & -\delta_{2,2} & -\delta_{2,1} & -\delta_{2,2} \\ \hline \end{array}.$$

Here the behavioral component structure of combining a β_i with a $-\beta_i$ in the same game is replaced with having the negative value of $\delta_{i,j}$ in same cell in the companion game.

Computing $\delta_{i,j}$ values is as above. Illustrating with $\delta_{1,1}$, the program is to carry out the dot product of $\mathbf{d}_{1,1}$ with the two representations of $\mathbf{g}_1 + \mathbf{g}_2$. One dot product has the outcome $A_1 + B_1 - E_1 - F_1$, while the other is $\delta_{1,1}[\mathbf{d}_{1,1}, \mathbf{d}_{1,1}] = 4\delta_{1,1}$ Consequently

$$\delta_{1,1} = \frac{A_1 + B_1 - E_1 - F_1}{4} = \frac{1}{2}[\frac{A_1 + B_1}{2} - \frac{E_1 + F_2}{2}],$$

or the difference between Player 1's average payoff in the first row of game one and the first row of game two.

Chapter 6
Multiplayer Games

Echoing the "Two is company, but three is a crowd" expression, adding players to a game can create interesting, unexpected, and perhaps unwanted differences. On the other hand, modeling truly multiplayer settings with two-player games can be counterproductive.

To assist in understanding what can happen, we identify some subtleties that accompany games with three or more players. To offer a preview, in a two-person game, should Row decide not to cooperate, Column can adopt an appropriate response. But with five players where three want to follow a particular agenda, what can the other two do?

This question reflects a societal behavior of majority coalitions. Expressing this reality in terms of how it affects the structure of the games, adding agents introduces new symmetries that allow outcomes to differ in unexpected ways from what we have learned to embrace with two-person situations. Something stronger than Nash cells, for instance, may be needed. Stated differently, this connection with majority voting accurately suggests that the analysis will involve a wedding of convenience between voting and game theory.

As a starting point, certain features—such as the game-theoretic dynamics out of equilibrium—are introduced by expanding the Hawk–Dove game to three players. Motivated by actual situations where a flock of smaller birds chase off a larger one, it is reasonable to extend the game to where several cooperating doves can chase off a hawk. Moving from birds that fly overhead to birds-of-an-illicit-feather, suppose the authorities pick up three alleged criminals; what is the resulting structure of the associated Prisoner's Dilemma game? Can standard approaches, such as tit-for-tat, have new features? (Yes.)

© Springer Nature Switzerland AG 2019
D. T. Jessie and D. G. Saari, *Coordinate Systems for Games*, Static & Dynamic Game Theory: Foundations & Applications, https://doi.org/10.1007/978-3-030-35847-1_6

6.1 Hawk–Dove–Dove

This section adds a player to the earlier Hawk–Dove games in order to explore three player $2 \times 2 \times 2$ settings. The decomposition identifies features that differ from, or are similar to, two-person situations. Then, using the associated coordinate system as a designing tool, desired features of a game are used to construct its representation.

6.1.1 Analyzing Hawk–Dove–Dove

Start with the three-player Hawk–Dove game (the two matrices correspond to Player 3's choice of Hawk or Dove)

$$
\mathcal{G}_{Hawk} = \begin{array}{c} \\ \text{Hawk} \\ \text{Dove} \end{array}
\begin{array}{c} \text{Hawk} \quad\;\; \text{Dove} \\ \begin{array}{|ccc|ccc|} \hline 0 & 0 & 0 & 0 & 4 & 0 \\ 4 & 0 & 0 & 4 & 4 & 17 \\ \hline \end{array} \end{array}, \;
\mathcal{G}_{Dove} = \begin{array}{c} \\ \text{Hawk} \\ \text{Dove} \end{array}
\begin{array}{c} \text{Hawk} \quad\;\; \text{Dove} \\ \begin{array}{|ccc|ccc|} \hline 0 & 0 & 4 & 17 & 4 & 4 \\ 4 & 17 & 4 & 11 & 11 & 11 \\ \hline \end{array} \end{array}
$$

$$(6.1)$$

where, should two hawks engage in combat, the lonely dove standing off to the side can sneak off with some of the pickings. Should there be a single Hawk, it will dominate, but it leaves some leftovers for the hungry doves. If all three are doves, the outcome is shared.

The kernel term is $\mathcal{G}^K(5, 5, 5)$, and the important Nash component is

$$
\mathcal{G}^N_{Hawk} = \begin{array}{c} \\ \text{Hawk} \\ \text{Dove} \end{array}
\begin{array}{c} \text{Hawk} \quad\;\; \text{Dove} \\ \begin{array}{|ccc|ccc|} \hline -2 & -2 & -2 & -2 & 2 & -2 \\ 2 & -2 & -2 & 2 & 2 & 3 \\ \hline \end{array} \end{array}, \;
\mathcal{G}^N_{Dove} = \begin{array}{c} \\ \text{Hawk} \\ \text{Dove} \end{array}
\begin{array}{c} \text{Hawk} \quad\;\; \text{Dove} \\ \begin{array}{|ccc|ccc|} \hline -2 & -2 & 2 & 3 & 2 & 2 \\ 2 & 3 & 2 & -3 & -3 & -3 \\ \hline \end{array} \end{array}
$$

which clearly identifies the competitive advantage of being a lone hawk: These are the three Nash cells (with positive entries). The game has two repellers (hawk-hawk-hawk and dove–dove–dove) and three hyperbolic cells of order $(1, 2)$. Recall (Definition 1.2), being hyperbolic of order $(1, 2)$ means that one agent cannot unilaterally move to a better outcome (the one with a positive cell entry), but the other two can.

The game's largest joint payoff (Eq. 6.1) is in the all-dove setting. But this feature cannot reflect the me-$\mathcal{G}^N$ component because this cell is a repeller. Consequently, what makes this cell of interest is the we-$\mathcal{G}^B$ structure.

$$
\mathcal{G}^B_{Hawk} = \begin{array}{c} \\ \text{Hawk} \\ \text{Dove} \end{array}
\begin{array}{c} \text{Hawk} \qquad\quad \text{Dove} \\ \begin{array}{|ccc|ccc|} \hline -3 & -3 & -3 & -3 & -3 & -3 \\ -3 & -3 & -3 & -3 & -3 & 9 \\ \hline \end{array} \end{array}, \;
\mathcal{G}^B_{Dove} = \begin{array}{c} \\ \text{Hawk} \\ \text{Dove} \end{array}
\begin{array}{c} \text{Hawk} \qquad\quad \text{Dove} \\ \begin{array}{|ccc|ccc|} \hline -3 & -3 & -3 & 9 & -3 & -3 \\ -3 & 9 & -3 & 9 & 9 & 9 \\ \hline \end{array} \end{array}
$$

Indeed, dove–dove–dove is the $\mathcal{G}^B$ Pareto superior cell; a Pareto inferior cell is diametrically opposite at hawk-hawk-hawk. But rather than a single $\mathcal{G}^B$ Pareto inferior cell (as true with 2×2 games), with more players it becomes possible for several cells (here, four of them) to vie for the dubious honor of being Pareto inferior. (Each

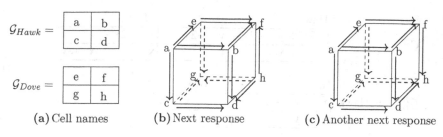

$$\mathcal{G}_{Hawk} = \begin{array}{|c|c|} \hline a & b \\ \hline c & d \\ \hline \end{array}$$

$$\mathcal{G}_{Dove} = \begin{array}{|c|c|} \hline e & f \\ \hline g & h \\ \hline \end{array}$$

(a) Cell names   (b) Next response   (c) Another next response

Fig. 6.1 Hawk–Dove–Dove structures

of these cells has at least two hawks.) With respect to creating games, this $\mathcal{G}^B$ feature has the power to dampen the appeal of several cells.

To identify the Nash dynamics, the next response dynamic is plotted in Fig. 6.1b. Using Fig. 6.1a cell names, the fact cells a and h are repellers that require the three arrows from each of these cells to point outward. Similarly, d, f, and g are Nash cells, so all three arrows point toward each of these cells. The structures for the remaining cells, b, c, and e, become uniquely determined by the arrows assigned to the repelling and attracting cells. All of this is indicated in Fig. 6.1b cube.

Three of the cube's faces have a twisting action on opposing edges, which identifies the existence of mixed equilibria (Sect. 1.4.4). They are the right face (cells b, d, h, f), back face (cells e, f, h, g), and bottom face (cells c, d, h, g). The NME in each face serves as guideposts toward selecting between the face's two Nash cells. A fourth NME in the interior directs traffic toward an appropriate face.

6.1.2 Designing a Hawk–Dove–Dove Game

Our recurring theme is to emphasize classes of games. As developed next, game coordinates continue to be a useful tool to design classes of multiplayer games.

Suppose the goal is to create a class of games that captures the observed behavior where several peace-loving doves get stiff enough feathers to chase off the tormenting hawk. This is a cooperative, rather than an individual action, so the modeling starts with $\mathcal{G}^B$. In any $\mathcal{G}^B$ cell with two doves and one hawk, let $x > 0$ be the incentive to unite to push out the bully, while $-y < 0$ is the penalty suffered by the escaping hawk; these are the bold entries in Eq. 6.2. What may be unexpected is that these entries completely determine the $\mathcal{G}^B$ component! (This is a consequence of Theorem 4.1.) A surprise is that the doves uniting to chase off the intrusive hawk makes the dove–dove–dove cell a candidate to be the $\mathcal{G}^B$ Pareto *inferior* cell! (The other candidate is hawk-hawk-hawk.) Namely, the negative $-y$ values ensure that dove–dove–dove is not a lovey-dovey cooperative choice.

$$\mathcal{G}^B_{Hawk} = \begin{array}{c} \\ \text{Hawk} \\ \text{Dove} \end{array} \begin{array}{c} \text{Hawk} \end{array}$$

	Hawk			Dove		
Hawk	$-2x+y$	$-2x+y$	$-2x+y$	x	$-2x+y$	x
Dove	$-2x+y$	x	x	$\mathbf{x}$	x	$-y$

$$\mathcal{G}^B_{Hawk} = \qquad\qquad\qquad\qquad\qquad\qquad\qquad , \tag{6.2}$$

	Hawk			Dove		
Hawk	x	x	$-2x+y$	$-\mathbf{y}$	x	x
Dove	$\mathbf{x}$	$-\mathbf{y}$	x	$-y$	$-y$	$-y$

$$\mathcal{G}^B_{Dove} =$$

Because hawk-hawk-hawk is not a cooperative option, it must be that $-2x + y < 0$ or $y < 2x$. Whether the three hawks, or the three doves, constitute the $\mathcal{G}^B$ Pareto inferior cell depends, respectively, on whether $-2x + y < -y$ (or $y < x$) or $-y < -2x + y$ (or $x < y$). Stated in words, a strong penalty for Hawk being chased off makes hawk-hawk-hawk a $\mathcal{G}^B$ Pareto inferior choice.

With surprisingly innocuous assumptions, $\mathcal{G}^B$ is completely determined. Turning to the me-Nash component, basic assumptions again determine $\mathcal{G}^N$, which are captured by the bold faced $\mathbf{u}$, $\mathbf{v}$, $\mathbf{w}$ entries.

	Hawk			Dove		
Hawk	$-\mathbf{u}$	$-\mathbf{u}$	$-\mathbf{u}$	$-\mathbf{w}$	u	$-\mathbf{w}$
Dove	u	$-\mathbf{w}$	$-\mathbf{w}$	w	w	$-v$

$$\mathcal{G}^N_{Hawk} = \qquad\qquad\qquad\qquad\qquad\qquad\qquad ,$$

	Hawk			Dove		
Hawk	$-\mathbf{w}$	$-\mathbf{w}$	u	$-v$	w	w
Dove	w	$-v$	w	$\mathbf{v}$	$\mathbf{v}$	$\mathbf{v}$

$$\mathcal{G}^N_{Dove} =$$

A cell with two or three hawks is a combat zone that the hawks would try to avoid; this explains the $-u < 0$ and $-w < 0$ entries. The remaining $v > 0$ choice means that a player is better off in a three-dove cell than being cast into a setting where two doves will chase the player away. These simple assumptions about personal behavior complete the $\mathcal{G}^N$ form; they endow dove–dove–dove as the sole Nash cell! In contrast, the $\mathcal{G}^B$ structure of this cell *discourages* cooperation.

Illustrating with $x = 2$, $y = 3$, an example of this class of games is

	Hawk			Dove		
Hawk	$-1-\mathbf{u}$	$-1-\mathbf{u}$	$-1-\mathbf{u}$	$2-\mathbf{w}$	$-1+u$	$2-\mathbf{w}$
Dove	$-1+u$	$2-\mathbf{w}$	$2-\mathbf{w}$	$2+w$	$2+w$	$-3-v$

$$\mathcal{G}_{Hawk} = \qquad\qquad\qquad\qquad\qquad\qquad\qquad ,$$

	Hawk			Dove		
Hawk	$2-\mathbf{w}$	$2-\mathbf{w}$	$-1+u$	$-3-v$	$2+w$	$2+w$
Dove	$2+w$	$-3-v$	$2+w$	$-3+\mathbf{v}$	$-3+\mathbf{v}$	$-3+\mathbf{v}$

$$\mathcal{G}_{Dove} =$$

where certain u, v, w positive values make some cells more attractive than others: These differences reflect Nash, rather than behavioral terms. With $u = v = w = 1$, for instance, no cell is of particular interest to all three players.

	Hawk			Dove		
$\mathcal{G}_{Hawk} =$ Hawk	-2	-2	-2	1	0	1
Dove	0	1	1	3	3	-4

	Hawk			Dove		
$\mathcal{G}_{Dove} =$ Hawk	1	1	0	-4	3	3
Dove	3	-4	3	-2	-2	-2

Figure 6.1c displays the next response dynamic, where after using the arrows required for the Nash and repelling cells, not much is left to complete the dynamic. There are no twisting actions on the faces or edges (for the cube), so there are no NME; the Nash cell dominates even though any {two hawks, one dove} cell is a Pareto improvement.

Theorem 6.1 *A three-player Hawk–Dove game, where two doves can chase off the hawk, has a single Nash cell at dove–dove–dove. Contrary to what happens in a two-person Hawk–Dove game, this is not a $\mathcal{G}^B$ Pareto superior cell; instead, either this cell, or hawk-hawk-hawk is the $\mathcal{G}^B$ Pareto inferior cell.*

Construction of other classes of games follows the same approach. The Battle of the Sexes game can evolve into the Battle of the Family game after a child is born. Now the list of activities expands from {Culture, Sports} to include {Amusement parks}. Designing a class of associated games with different features follows the above template to create a $3 \times 3 \times 3$ game. Similarly, with three alleged criminals, where at least two must defect for a conviction, leads to an extended $2 \times 2 \times 2$ Prisoner's Dilemma game.

6.2 Nash Dynamic Structure

When designing games, it helps to know all possible Nash dynamic structures. Returning to Definition 4.1 and $2 \times 2 \times 2$ games, each player's $\mathcal{G}^{N,signs}$ entry is either $+1$ or -1, so the cell values are $+3$ for a Nash cell, 1 for (2, 1) hyperbolic cells (only one agent can unilaterally move to a preferred cell), -1 for (1, 2) hyperbolic cells, and -3 for a repeller. In this manner, the dynamic structure of all such games follows immediately from the result that the sum of a game's cell values equals zero. (Everything extends to n-person $2 \times 2 \times \cdots \times 2$ games, or even $k_1 \times \cdots \times k_n$ games.)

Theorem 6.2 *A $2 \times 2 \times 2$ game (with no $\mathcal{G}^{N,signs}$ cells with a zero entry) can have*

1. *Four Nash cells accompanied by four repelling cells,*
2. *Three Nash cells accompanied by*

   (a) *three repelling cells, one (2, 1) hyperbolic cell, and one (1, 2) hyperbolic cell, or*
   (b) *two repelling, and three (1, 2) hyperbolic cells,*

3. *Two Nash cells accompanied by*

   (a) *three repelling cells and three (1, 2) hyperbolic cells, or*
   (b) *two repelling cells, two (1, 2) hyperbolic cells, and two (2, 1) hyperbolic cells, or*
   (c) *one repelling cell, four (1, 2) hyperbolic cells, and one (2, 1) hyperbolic cell, or*
   (d) *zero repelling cells, and six (1, 2) hyperbolic cells,*

4. *One Nash cell accompanied by*

   (a) *two repelling cells, four (2, 1) hyperbolic cells, and one (1, 2) hyperbolic cell, or*
   (b) *one repelling cell, three (2, 1) and three (1, 2) hyperbolic cells, or*
   (c) *zero repelling cells, five (1, 2) hyperbolic cells, and two (2, 1) hyperbolic cells*

5. *Zero Nash cells accompanied by*

   (a) *two repelling cells and six (2, 1) hyperbolic cells, or*
   (b) *one repelling cell, five (2, 1) hyperbolic, and two (1, 2) hyperbolic cells, or*
   (c) *zero repelling cells, four (2, 1), and four (1, 2) hyperbolic cells.*

According to (3), for example, a $2 \times 2 \times 2$ game can be fashioned with two Nash cells accompanied with a surprising array of possible defecting behavior. The theorem provides guidance in creating games by specifying what structures (e.g., kinds of defecting or stability choices) can occur and where they are located.

The $\mathcal{G}^N$ properties simplify designing examples and positioning the cells. The six cell entries required to construct a $\mathcal{G}^N$ with BRBa as the sole Nash cell and TLF as a repeller, for instance, can determine half of the $\mathcal{G}^N$ choices (if they are not adjacent). This is illustrated in Eq. 6.3 with the twelve bold entries.

$$
\mathcal{G}^N_{Front} = \begin{array}{|cc c|cc c|} \hline \mathbf{-2} & \mathbf{-2} & \mathbf{-2} & -2 & 2 & -2 \\ \hline \mathbf{2} & \mathbf{-2} & \mathbf{-2} & 2 & 2 & \mathbf{-2} \\ \hline \end{array}, \quad \mathcal{G}^N_{Back} = \begin{array}{|cc c|cc c|} \hline -2 & -2 & 2 & \mathbf{-2} & \mathbf{2} & \mathbf{2} \\ \hline 2 & -2 & 2 & \mathbf{2} & \mathbf{2} & \mathbf{2} \\ \hline \end{array} \quad (6.3)
$$

There remains flexibility in selecting which of the remaining cells have what properties. Cell TRF, for instance, is hyperbolic (1, 2). To convert it to hyperbolic (2, 1), change one of the player's entries from -2 to 2; e.g., doing this for Row makes the cell hyperbolic (2, 1), but BRF now must be hyperbolic (1, 2).

Other structures come from the geometry of three-player games; e.g., a $k \times k \times k$ game can be represented with a cube and its six faces. Using the simpler $2 \times 2 \times 2$ setting for descriptive purposes, each of the six Eq. 6.3 faces has one of the 2×2 game structures illustrated in Fig. 1.2—one player determines which face is being considered while the other two define the dynamic. Thus the front and back faces in Eq. 6.3 have identical 2×2 structures, which can be seen by ignoring Face's payoffs; her role is to determine whether $\mathcal{G}^N_{Front}$ or $\mathcal{G}^N_{Back}$ is the relevant one. In fact, for Eq. 6.3, all six 2×2 games have a Nash and diametrically opposite repelling cell with two hyperbolic cells. (In contrast, the full game has one Nash and one repelling cell.)

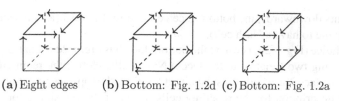

(a) Eight edges   (b) Bottom: Fig. 1.2d (c) Bottom: Fig. 1.2a

Fig. 6.2 Constructing the Nash dynamic

The Nash behavior on these six faces closely determines that of the general game: This leads to a statement similar to Theorem 6.2. As an example, a Nash or a repelling cell in the full game endows a cell in each of three faces with that Nash status. In 2×2 games, two Nash cells or two repelling cells cannot be adjacent, which imposes a constraint on the locations of Nash and repelling cells in a $2 \times 2 \times 2$ game. A similar behavior arises with hyperbolic $(2, 1)$ cells: This cell is on three faces, and it is a Nash cell for one of them. This fact, alone, provides information about whether the Nash structure of a 2×2 game (one face) reasonably represents what happens in a $2 \times 2 \times 2$ game.

A reason to examine face structures is that each face captures the relevant information for a particular coalition; information that can affect their actions. Demonstrating with Eq. 6.3, should Face select Front, the {Row, Column} coalition sees the game as

$$
\begin{array}{|cc|cc|}
\hline
-2 & -2 & - & -2 & 2 & - \\
\hline
2 & -2 & - & 2 & 2 & - \\
\hline
\end{array}, \tag{6.4}
$$

which, for this coalition's purpose, has a Nash cell at BR. This BRF cell is *not* Nash, but should the {Row, Column} coalition be on the front face, BR is of strategic interest.

Understanding how the cells can be positioned receives a strong assist from geometry. For instance, suppose a $2 \times 2 \times 2$ game has only hyperbolic cells. According to Theorem 6.2, there must be four hyperbolic $(2, 1)$ cells and four hyperbolic $(1, 2)$ cells. As a cell is on three faces, on two of the faces a hyperbolic $(2, 1)$ cell is hyperbolic and on one it is Nash. Although the actual game has *no* Nash *nor* repelling cells, on *six* of the cube faces there is a total of four Nash cells and four repelling cells—the remaining 16 face cells are hyperbolic. These face structures capture potential strategic interests of the appropriate coalitions, so of interest for modeling is to appreciate how they can be positioned.

To create a "trickle-down game theory," assume that the cube's top face has two Nash cells (Fig. 6.2a). Thus the top face has Fig. 1.2d structure with two repelling cells. In the full cube, each cell has three directions; each top-face "Nash cell" has its remaining arrow pointing away from the top face to trickle down to the bottom face. In contrast, each repelling cell's remaining arrow is uplifting. Thus, this assumption about the top face already identifies two-thirds (eight of the twelve edges) of the full game structure; all that remains are the four edge directions for the bottom face. Each Fig. 1.2 choice is available, but with some restrictions; e.g., because the front-right edge points downwards, the bottom-face cell cannot be Nash (as this would require the full game to have a Nash cell).

One choice (Fig. 6.2b) is where the bottom face has Fig. 1.2d format, so it contains the remaining two (relative to the faces) Nash cells. Notice the required structure: The bottom-face Nash cells must be directly below the top-face repelling cells, with a similar description for the repelling cells. Each of the remaining four faces (the vertical ones) has a Fig. 1.2a construct with no Nash cell and an NME.

A second possibility is for the bottom face to have a Fig. 1.2b or c appearance, with one Nash and one repelling cell. Again, the bottom-face Nash cell must be below a top-face repeller and the bottom-face repeller must be below a top-face Nash. To obtain Fig. 1.2b appearance of the bottom face's Nash and repeller being adjacent, reverse one of the bottom-face arrows from Fig. 6.2b. For instance, consider the arrow on the bottom-face, right edge, which now is pointing upward. Changing it to point downward creates a Fig. 1.2b situation of the bottom face's repeller and Nash cells being adjacent. This construction places the remaining face-Nash cell on the lower right corner of the right face. To force the bottom face to assume a Fig. 1.2c structure, where the bottom-face's Nash and repeller are opposite, reverse the dashed and the solid arrows coming out of the bottom-face north-east corner. Here the remaining face-Nash cell is on the back face.

It remains for the bottom to have a Fig. 1.2a setting of no face-Nash cells, which is in Fig. 6.2c. Here, the front, bottom, and back cube faces have Fig. 1.2a form of no Nash cells on the faces, while the right and left faces have Nash cells underneath (but separated) from the Nash cells on the top face. In this manner, where certain assumptions dictate what happens on other faces, all possibilities admitted by Theorem 6.2 can be represented along with restrictions. (Readers comfortable with algebraic topology can craft stronger assertions.)

As a bonus, this Fig. 6.2 construction underscores what should be obvious: Trying to use a 2×2 game to capture what happens with three agents in a $2 \times 2 \times 2$ game can lead to misleading conclusions. Figure 6.2 has no Nash cells, yet four of the 2×2 associated games could each have such a cell. This possibility strongly suggests that game-theoretic conclusions reached from 2×2 approximations can incorrectly indicate $2 \times 2 \times 2$ behavior. As also indicated below, these face-Nash structures can introduce complexity issues.

6.2.1   Adding Complexity

Beyond the Nash structure, the $\mathcal{G}^B$ component has a strong impact. A $\mathcal{G}^B$ feature that differs from two-player games is that with more players, $\mathcal{G}^B$ *need not have a Pareto superior and/or inferior cell.* The following three games,[1] which add a $\mathcal{G}^B$ term to the $\mathcal{G}^N$ of Eq. 6.3, show how this difference makes a difference. (The kernel for all three games is $\mathcal{G}^K (6, 6, 6)$.)

---

[1] These games and some of the section's discussion, come from [13].

The first choice has a $\mathcal{G}^B$ that positions a Pareto superior and inferior cell, respectively, over Eq. 6.3 Nash and repelling cell. Consequently this behavioral component

$$\mathcal{G}^B_{1,front} = \begin{array}{|ccc|ccc|} \hline -4 & -4 & -4 & 0 & -4 & 0 \\ \hline -4 & 0 & 1 & 0 & 0 & 3 \\ \hline \end{array} \quad \mathcal{G}^B_{1,back} = \begin{array}{|ccc|ccc|} \hline 1 & 1 & -4 & 3 & 1 & 0 \\ \hline 1 & 3 & 1 & 3 & 3 & 3 \\ \hline \end{array}$$

reinforces the Nash structure, which ensures that the simplicity of the resulting game

$$\mathcal{G}_{1,\,front} = \begin{array}{|ccc|ccc|} \hline 0 & 0 & 0 & 4 & 4 & 4 \\ \hline 4 & 4 & 5 & 8 & 8 & 7 \\ \hline \end{array} \quad \mathcal{G}_{1,\,back} = \begin{array}{|ccc|ccc|} \hline 5 & 5 & 4 & 7 & 9 & 8 \\ \hline 9 & 7 & 9 & 11 & 11 & 11 \\ \hline \end{array}$$

holds no surprises: BRBa is a desired destination for all players without any of them needing to have even a passing thought about what the others might do.

The simplicity of the game changes should the $\mathcal{G}^B$ Pareto superior and inferior points be located, respectively, over the $\mathcal{G}^N$ repeller and Nash cell, as with

$$\mathcal{G}^B_{2,\,front} = \begin{array}{|ccc|ccc|} \hline 3 & 3 & 3 & 1 & 3 & 1 \\ \hline 3 & 1 & 0 & 1 & 1 & -4 \\ \hline \end{array} \quad \mathcal{G}^B_{2,\,back} = \begin{array}{|ccc|ccc|} \hline 0 & 0 & 3 & -4 & 0 & 1 \\ \hline 0 & -4 & 0 & -4 & -4 & -4 \\ \hline \end{array}, \quad (6.5)$$

This choice serves as a recipe for the Prisoner's Dilemma game

$$\mathcal{G}_{2,\,front} = \begin{array}{|ccc|ccc|} \hline 7 & 7 & 7 & 5 & 11 & 5 \\ \hline 11 & 5 & 4 & 9 & 9 & 0 \\ \hline \end{array} \quad \mathcal{G}_{2,\,back} = \begin{array}{|ccc|ccc|} \hline 4 & 4 & 11 & 0 & 8 & 9 \\ \hline 8 & 0 & 8 & 4 & 4 & 4 \\ \hline \end{array} \quad (6.6)$$

This $\mathcal{G}^B$ component converts the game's repelling point into a choice more appealing than the Nash cell. Both $\mathcal{G}_1$, $\mathcal{G}_2$ have the same Nash structure, but changes in the $\mathcal{G}^B$ Pareto locations generate very different games. As developed in the next section, $\mathcal{G}_2$'s more appealing TLF cell, which is Pareto superior to the Nash cell, can be sustained with grim trigger or tit-for-tat in a repeated game.

Suppose $\mathcal{G}^B$ does *not* have a Pareto superior cell (in the sense that it contains each player's maximal $\mathcal{G}^B$ entry), as with

$$\mathcal{G}^B_{3,\,front} = \begin{array}{|ccc|ccc|} \hline 3 & 0 & 1 & 1 & 0 & 3 \\ \hline 3 & 2 & 0 & 1 & 2 & -4 \\ \hline \end{array} \quad \mathcal{G}^B_{3,\,back} = \begin{array}{|ccc|ccc|} \hline 0 & 2 & 1 & -4 & 2 & 3 \\ \hline 0 & -4 & 0 & -4 & -4 & -4 \\ \hline \end{array} \quad (6.7)$$

With the $\mathcal{G}^N$ of Eq. 6.3, this behavioral component creates the puzzling game

$$\mathcal{G}_{3,\,front} = \begin{array}{|ccc|ccc|} \hline 7 & 4 & 5 & 5 & 8 & 7 \\ \hline 11 & 7 & 4 & 9 & 11 & 0 \\ \hline \end{array} \quad \mathcal{G}_{3,\,back} = \begin{array}{|ccc|ccc|} \hline 4 & 7 & 9 & 0 & 11 & 11 \\ \hline 8 & 0 & 8 & 4 & 4 & 4 \\ \hline \end{array} \quad (6.8)$$

that fails to have a clear optimal state. Yes, the game's Nash cell (BRBa) is Pareto inferior to TLF, BLF, TRF, TLBa. But, as shown with the material developed next, none of these preferred choices can be sustained with grim trigger or tit-for-tat.

6.2.2   Those Retaliatory Strategies

The above $\mathcal{G}_2$ and $\mathcal{G}_3$ have cells that are Pareto superior to the game's Nash cell, which, for the players, make them more desirable. To sustain such an outcome in a repeated setting, cooperation is needed. But these desirable cells are not Nash, so at least one player, say Player 1, can select something personally better. Knowing that Player 1 has a more enticing option means that cooperation requires Player 1 to accept the poorer of two choices.

Why would Player 1 do this? Forget altruism; the reason Player 1 would incur the personal expense of adopting a poorer choice is that, as with Theorem 2.4, the game can be changed so that the $\mathcal{G}^B$ portion is equipped with a Nash structure: With the altered repeated game, cooperation becomes a best response. Everything that follows could be carried out with $k_1 \times k_2 \times \cdots \times k_n$ games, but with the expense of miserable notation. While only $n \geq 2$ player $2 \times 2 \times \cdots \times 2$ games are considered, it should be obvious from the description how to handle more general games.

To describe the punishment clause, suppose the poorer payoff that Player 1 must accept to cooperate is *better than any payoff* Player 1 would face in some other array. If so, and if Player 1 does not cooperate, the other players can force Player 1 to select from that lousy list of bad choices: They impose a huge negative externality on Player 1. With Eq. 6.6, for instance, if the players seek the cooperative TLF outcome of $(7, 7, 7)$, and Player 1 reneges by playing B to grab the 11, in subsequent plays of the game, the other players play RBa to force Player 1 to select between the poorer choices of 0 and 4.

The notational advantage of considering games where each player has two strategies is that Player 1's two entries in the jth array are $-|\eta_{j,1}| + \beta_{j,1}$ and $|\eta_{j,1}| + \beta_{j,1}$; cooperation requires Player 1 to take the poorer $-|\eta_{j,1}| + \beta_{j,1}$ choice. To enforce this selection, there must be another array (i.e., choice of the other players' pure strategies), say the ith, where all of Player 1's payoffs are smaller than $-|\eta_{j,1}| + \beta_{j,1}$. For this property to hold, the payoffs must satisfy

$$|\eta_{j,1}| + \beta_{j,1} > -|\eta_{j,1}| + \beta_{j,1} > |\eta_{i,1}| + \beta_{i,1}. \tag{6.9}$$

This leads to the following result establishing a connection between cooperation and $\mathcal{G}^B$ values.

Theorem 6.3 *In an $n \geq 2$ player $2 \times \cdots \times 2$ game, suppose with a targeted, non-Nash cooperative outcome, Player 1, with discount rate $\delta \in (0, 1)$, must select a strategy yielding the smaller of $\{-|\eta_{j,1}| + \beta_{j,1}, |\eta_{j,1}| + \beta_{j,1}\}$ as identified in Eq. 6.9. To enforce this cooperative action, the other $(n-1)$ players use the grim trigger option that would force Player 1 to select between $|\eta_{i,1}| + \beta_{i,1}$ and $-|\eta_{i,1}| + \beta_{i,1}$. It is in Player 1's interest to cooperate if*

$$\delta > \frac{2|\eta_{j,1}|}{[\beta_{j,1} - \beta_{i,1}] + [|\eta_{j,1}| - |\eta_{i,1}|]} \tag{6.10}$$

If the other players adopt a tit-for-tat strategy, it is in Player 1's interest to cooperate in the special case after a first defection if

$$\delta > \frac{2|\eta_{j,1}|}{[\beta_{j,1} - \beta_{i,1}] + [|\eta_{i,1}| - |\eta_{j,1}|]} \tag{6.11}$$

To compare the above assertion with Theorem 2.4, in a 2×2 game, $\beta_{1,1} = -\beta_{2,1}$ such that $[\beta_{j,1} - \beta_{i,1}]$ entry becomes $2\beta_{1,1}$. Illustrating with the Prisoner's Dilemma of Eq. 6.6 where the goal is to sustain TLF, it follows from Eq. 6.5 that this cell's β value for all three players is $\beta = 3$ while (Eq. 6.3) $|\eta_{i,j}| = 2$ for all players. For a player to satisfy Eq. 6.9, the penalty must force the player into an array with the BRBa cell, where each player's β value is $\beta = -4$. Thus, TLF can be sustained with $\delta > \frac{2|2|}{[3-(-4)]+[2-2]} = \frac{4}{7}$.

There is more: Imagine Column salivating over the personally attractive payoff of 11 in the neighboring TRF cell (Eq. 6.6). Could Column secure this outcome at the expense of the other two players? (Moving from TLF to TRF would reduce Row's and Face's payoff from 7 to 5, and increase Column's payoff from 7 to 11.) Using Eqs. 6.10 and 6.11, it is in Players 1 and 3 interest to support this personally poorer TRF choice for $\delta > \frac{2|2|}{1-(-4))} = \frac{4}{5}$, while Column's interest is aroused already at $\delta > \frac{2|2|}{3+4} = \frac{4}{7}$. Thus an interesting behavior emerges: TLF can be sustained for $\frac{4}{7} < \delta \le \frac{4}{5}$, but Column can shift the attention to a personally preferred TRF for $\delta > \frac{4}{5}$. What a lopsided morality lesson! Overly conscientious players (who respect the future as manifested by having a larger δ value) can be punished! With reflection, this behavior should be expected because a larger δ value means that the player is willing to accept a lower payoff. This example captures one of several possible $\mathcal{G}^B$ consequences that differs from two-player games.

Now turn to game $\mathcal{G}_3$ (Eq. 6.8). Because of the $|\eta_{i,j}| = 2$ terms, Eqs. 6.10, 6.11 reduce to $\delta > \frac{4}{\beta_{j,1}-\beta_{i,1}}$. Of the four cells that are Pareto preferred to the Nash BRBa (using the β values from Eq. 6.7), TLF and TRF *cannot* be supported because, for Column, both cells require $\delta > \frac{4}{0-(-4)} = 1$. (By definition, $0 \le \delta \le 1$.) Similarly, BRF cannot be supported by Face, and Row cannot support the last TLBa (as both would require $\delta > \frac{4}{0-(-4)} = 1$ for the appropriate player). The complexity of this game is captured by it not having a natural solution concept, which remains true even in a repeated setting. The source of games of this type is addressed below.

Proof of Theorem 6.3 Suppose Player 1 is facing opponents who have implemented a grim trigger in which he can play Top (Cooperate) for a payoff of $-|\eta_{j,1}| + \beta_{j,1}$ ad infinitum, or play Bottom (Defect) for a one-time payoff of $|\eta_{j,1}| + \beta_{j,1}$. If our player does not cooperate, the grim trigger ensures that the subsequent payoffs are $|\eta_{i,1}| + \beta_{i,1}$ ad infinitum. So, Player 1 will cooperate if

$$\sum_{t=1}^{\infty} \delta^{t-1}(-|\eta_{j,1}| + \beta_{j,1}) > (|\eta_{j,1}| + \beta_{j,1}) + \sum_{t=2}^{\infty} \delta^{t-1}(|\eta_{i,1}| + \beta_{i,1}),$$

$$\frac{-|\eta_{j,1}| + \beta_{j,1}}{1 - \delta} > |\eta_{j,1}| + \beta_{j,1} + \delta\left(\frac{|\eta_{i,1}| + \beta_{i,1}}{1 - \delta}\right),$$

$$-|\eta_{j,1}| + \beta_{j,1} - (1 - \delta)(|\eta_{j,1}| + \beta_{j,1}) > \delta(|\eta_{i,1}| + \beta_{i,1}),$$

$$(\delta - 2)|\eta_{j,1}| + \delta\beta_{j,1} > \delta(|\eta_{i,1}| + \beta_{i,1})$$

$$(6.12)$$

After some algebraic computations, Eq. 6.10 is obtained.

If Player 1 is faced with opponents who are playing tit-for-tat, then the grim trigger in Eq. 6.10 must hold or Player 1 will never cooperate, as he retains the ability to defect ad infinitum against. However, there is also the possibility of the first cooperation, then defection, then cooperation, etc. In order for cooperation to hold in this case, it is also needed that

$$\sum_{t=1}^{\infty} \delta^{t-1}(-|\eta_{j,1}| + \beta_{j,1}) > |\eta_{j,1}| + \beta_{j,1} + \delta\left(-|\eta_{i,1}| + \beta_{i,1}\right)$$

$$+ \delta^2(|\eta_{j,1}| + \beta_{j,1}) + \delta^3\left(-|\eta_{i,1}| + \beta_{i,1}\right) + \cdots$$

$$\sum_{t=1}^{\infty} \delta^{t-1}(-|\eta_{j,1}| + \beta_{j,1}) > \sum_{t=1}^{\infty} \delta^{2(t-1)}(|\eta_{j,1}| + \beta_{j,1}) + \delta \sum_{t=1}^{\infty} \delta^{2(t-1)}(-|\eta_{i,1}| + \beta_{i,1})$$

$$\frac{-|\eta_{j,1}| + \beta_{j,1}}{1 - \delta} > \frac{|\eta_{j,1}| + \beta_{j,1}}{1 - \delta^2} + \delta\left(\frac{-|\eta_{i,1}| + \beta_{i,1}}{1 - \delta^2}\right)$$

$$\left(\frac{1 + \delta}{1 + \delta}\right)\frac{-|\eta_{j,1}| + \beta_{j,1}}{1 - \delta} > \frac{|\eta_{j,1}| + \beta_{j,1}}{1 - \delta^2} + \delta\left(\frac{-|\eta_{i,1}| + \beta_{i,1}}{1 - \delta^2}\right)$$

$$(1 + \delta)(-|\eta_{j,1}| + \beta_{j,1}) > |\eta_{j,1}| + \beta_{j,1} + \delta\left(-|\eta_{i,1}| + \beta_{i,1}\right)$$

$$(-\delta - 2)|\eta_{j,1}| + \delta\beta_{j,1} > \delta\left(-|\eta_{i,1}| + \beta_{i,1}\right)$$

This inequality is equivalent to Eq. 6.11.   □

6.3   New Symmetries

Attributes of multiplayer games that differ from two-player games started with Eq. 6.7 where $\mathcal{G}^B$ does not have a Pareto superior cell. Although $\mathcal{G}^B$ need not have a Pareto superior cell, whatever strategy any player selects, the $\mathcal{G}^B$ portion that is left for the *coalition* of the two remaining agents *does have* a Pareto superior and inferior cell.

To illustrate, if Player 3 selects Front, then Eq. 6.7 portion of $\mathcal{G}^B$ for Row and Column is

$$\mathcal{G}_{3,\,front}^{B,\,pair} = \begin{array}{|cc|c|cc|c|} \hline 3 & 0 & - & 1 & 0 & - \\ \hline 3 & 2 & - & 1 & 2 & - \\ \hline \end{array} \qquad (6.13)$$

with a Pareto superior cell at BL and a Pareto inferior cell diametrically opposite at TR. Indeed, should Player 3 stay with the Front choice, for Row and Column the game becomes

7 4	–	5 8	–
11 7	–	9 11	–

where they could treat BR as a Nash cell with payoff entries enhanced by $\mathcal{G}_{3,\,front}^{B,\,pair}$. Stated in another manner, the absence of a Pareto superior cell in $\mathcal{G}^B$ can generate opportunities that motivate coalitions to explore advantages.

Similarly, should Column selects L, the $\mathcal{G}_3^B$ portion considered by Row and Face is

$$\mathcal{G}_{3,L}^{B,\,pair} = \begin{array}{|ccc|ccc|} \hline 3 & - & 1 & - & - & - \\ \hline 3 & - & 0 & - & - & - \\ \hline \end{array} , \quad \begin{array}{|ccc|ccc|} \hline 0 & - & 1 & - & - & - \\ \hline 0 & - & 0 & - & - & - \\ \hline \end{array}$$

with TF as the Pareto superior cell and BBa as the Pareto inferior cell.

The cube representing the $2 \times 2 \times 2$ game has six faces, so six $\mathcal{G}_{3,\,face\,name}^{B,\,pair}$ components can be defined; each represents the choices offered by a particular pair. (Similarly, a $3 \times 3 \times 3$ game will have nine components.) Because each $\mathcal{G}^B$ entry will be in two of these components, the following follows immediately.

Proposition 6.1 *For a $2 \times 2 \times 2$ game $\mathcal{G}$, the $\mathcal{G}^B$ component can be represented as*

$$\mathcal{G}^B = \frac{1}{2} \sum_{face\ names} \mathcal{G}_{face\ name}^{B,\,pair}. \tag{6.14}$$

If the entries in a $\mathcal{G}_{face\ name}^{B,\,pair}$ are not all equal, then $\mathcal{G}_{face\ name}^{B,\,pair}$ has a Pareto superior cell with a Pareto inferior cell diametrically opposite.

Illustrating the last comment with $\mathcal{G}_3^B$, the Pareto superior and inferior cells of each of the six faces are listed next.

Face Name	P sup	P inf	Face Name	P sup	P inf
Front	BL	TR	Back	TL	BR
Left	TF	BBa	Right	TF	BBa
Top	RBa	LF	Bottom	LF	RBa

These $\mathcal{G}_{face\ name}^{B,\,pair}$ structures can convert a game into a setting where a majority of players prefer a particular cell on a particular face. As discussed next, this coalitional advantage can launch instabilities within the game structure.

6.3.1 A Wedding Between Voting and Game Theory

To see how coalitions, which are of deep interest to game theory, arise in a natural manner, start with the definition of a Nash equilibrium (Definition 1.1); this is where *each player's* strategy is a personal maximum. With three or more players, an obvious generalization is to explore what a *coalition of players*, rather than individuals, can attain. These concerns were raised in several places, such as by Aumann [1] and by Bernheim, Peleg, and Whinston [3] in their analysis of coalition-proof Nash equilibria.

To appreciate what can go wrong, the first step is to understand what structural factors (symmetries) generate unexpected coalitional behavior with multiple players. As the following description shows, increasing the number of players and strategies allows the complexity to soar. While our attention is on pairs of cells (primarily with three players), everything extends. A word to the wise; these behaviors crop up with a positive probability.

An advantage in comparing what can happen between pairs of cells is that the source of all paired comparison difficulties, whether from statistics, majority voting, and so forth, now is completely understood ([28, 30]). Expressed in terms of multi-player games, rather than enjoying a Pareto structure, it can be that a majority of the players prefer cell A to B, cell B to C, cell C to D, and cell D to A to cause a cycle. Resembling the behavior described in Chap. 3, where a game can be represented by the sum of a zero-sum and an identical play game, or the sum of a symmetric and an antisymmetric game, or ..., it is not that difficult to represent all $\mathcal{G}^B$ terms as the sum of a $\mathcal{G}^B$ with a Pareto structure and a $\mathcal{G}^B$ with a cyclic behavior: Expect this feature.

The source of cyclic behavior

This cyclic attribute is due to a natural symmetry that emerges with three or more players, so expect it to plague $\mathcal{G}^B$ *and* $\mathcal{G}^N$ terms. As the first example created below shows, majority coalitions of players could prefer Nash cell A to C, Nash cell C to B, and Nash cell B to A to generate a cycle. After explaining this behavior, consequences are briefly explored.

Paired comparison problems are *completely caused* by what are called *ranking wheel configurations*[2] [27, 28, 30]. In the current context, consider a freely rotating wheel (Fig. 6.3a) attached to a wall. Equally spaced on the wheel's edge are the names of the three players: Row, Column, and Face. Should there be more players, equally space their names; e.g., Fig. 6.3c has the four players Row, Column, Face, and Set.

On the wall, write down different positive values next to the names. In Fig. 6.3a, the choices are 1, 2, 3 while Fig. 6.3b admits any different positive values. Positive values are specified to generate Nash cells for a $\mathcal{G}^N$, but this constraint can be relaxed when creating a $\mathcal{G}^B$ that has a cyclic structure.

The wheel's position defines a cell's entries. With Fig. 6.3a, the number 3 is by Row, 1 is by Column, and 2 is by Face, so cell A is (3, 1, 2). (The cell defined by

---

[2]In an appropriate space, they define the coordinate directions of payoffs, or inputs, that lead to cycles.

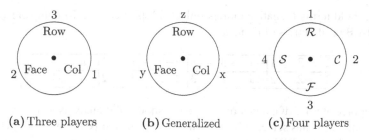

Fig. 6.3 Ranking wheel configurations

Fig. 6.3b is (z, x, y).) Next, rotate the wheel so that the name "Row" is positioned by the following value. Doing so with Fig. 6.3a defines cell B with entries (1, 2, 3). The final rotation (that is, each name has been by each value once) is cell C with entries (2, 3, 1). This construction is summarized with

Cell	Entries
A	(3, 1, 2)
B	(1, 2, 3)
C	(2, 3, 1)

With these cells,

1. A coalition of {Column, Face} prefers B to A,
2. A coalition of {Row, Column} prefers C to B, and
3. A coalition of {Row, Face} prefers A to C.

Reading from the bottom line up, these choices create the $A \succ C$, $C \succ B$, $B \succ A$ cycle. It remains to construct a $\mathcal{G}^N$ with these cells.

- From line 1, for the {Column, Face} coalition to change the outcome without interference from Row, cells A and B must be on a face defined by a choice of Row.
- From line 2, cells B and C must be on a face defined by a choice of Face.
- From line 3, cells A and C must be on a face defined by a choice of Column.

The construction of a $\mathcal{G}^N$ follows immediately. This is because, by selecting a cell location for A, the rest of the structure is dictated by the constraint that two Nash cells on a face cannot be adjacent. So, position cell A at, say, BLBa; this is on the bottom face. The only way for B to be a second Nash cell on the bottom face is to be at BRF. Similarly, to be at BRF, Face selected Front; the only way Nash cells B and C can be on the front face is for C to be at TLF.

The cycle is complete; Column's selection of L means that the coalition of Row and Face can jump the cell choice to the starting Nash cell. These values[3] define

---

[3]The values can differ as long as obvious inequalities are respected.

Eq. 6.15 bold terms (negative entries reflect $\mathcal{G}^N$ properties). To complete $\mathcal{G}^N$, just select BLBa cell entries of u, v, w.

$$\mathcal{G}^N = \begin{array}{|ccc|ccc|}\hline \mathbf{2} & \mathbf{3} & \mathbf{1} & -1 & -3 & -w \\ -2 & -2 & -2 & 1 & 2 & 3 \\\hline\end{array} \quad \begin{array}{|ccc|ccc|}\hline -3 & -v & -1 & u & v & w \\ 3 & 1 & 2 & -u & -1 & -3 \\\hline\end{array} \tag{6.15}$$

To appreciate the dynamics of this game, should the analysis start at cell A (at BLBa), then Row has selected Bottom, which permits the {Column, Face} coalition to select cell B (at BRF). But at cell B, Face selected Front, so the new {Row, Column} coalition can move to their preferred cell C (at TLF). At cell C, Column commits to Left, which allows the {Row, Face} coalition to select cell A. And the cycle goes on, and on, and on.

This ranking wheel mechanism also can define a cyclic $\mathcal{G}^B$: The entries (because of $\mathcal{G}^B$ properties) completely determine $\mathcal{G}^B$. Using Eq. 6.15 bold entries, where $\mathbf{1}$ is replaced with $-\mathbf{1}$ as a reminder that $\mathcal{G}^B$ ranking wheel entries can be negative, the resulting component (with *no* Pareto superior cell) is

$$\mathcal{G}^B = \begin{array}{|ccc|ccc|}\hline \mathbf{2} & \mathbf{3} & \mathbf{-1} & -1 & 3 & -4 \\ 2 & 2 & 2 & -1 & 2 & 3 \\\hline\end{array} \quad \begin{array}{|ccc|ccc|}\hline 3 & -4 & -1 & -4 & -4 & -4 \\ 3 & -1 & 2 & -4 & -1 & 3 \\\hline\end{array}$$

As Eq. 6.15 proves, this ranking wheel methodology simplifies the design of examples where three Nash cells are locked in a majority vote cycle. More generally, it can be used to create a $\mathcal{G}^N$ for an $n \geq 3$ player game (each player has at least two strategies) where n of the Nash cells are caught in a majority vote cycle. (See [29] to discover what can happen if a coalition must have more than a majority to change the outcome.)

A $2 \times 2 \times 2$ game can have four Nash cells, so it is reasonable to wonder whether a cycle can involve all four of them. To be specific, can u, v, w values (Eq. 6.15) be selected so that BLBa is a Nash cell creating the cycle

$$(1, 2, 3) \to (2, 3, 1) \to (3, 1, 2) \to (u, v, w) \to (1, 2, 3). \tag{6.16}$$

The answer, as the following asserts, is no.

Theorem 6.4 *A $2 \times 2 \times 2$ game with three Nash cells, A, B, C can be created that has a cycle in that a majority coalition can change the outcome from A to B, a majority coalition can change the outcome from B to C, and a majority coalition can change the outcome from C to A. Although a $2 \times 2 \times 2$ game can have four Nash cells, it is impossible to have a majority vote cycle among all four.*

Proof With a $2 \times 2 \times 2$ games and four Nash cells, by selecting where to position one of them, the other three locations are determined. For instance,

$$\begin{array}{|ccc|ccc|}\hline \mathbf{u_1} & \mathbf{v_1} & \mathbf{w_1} & & & \\ & & & \mathbf{u_2} & \mathbf{v_2} & \mathbf{w_2} \\\hline\end{array} \quad \begin{array}{|ccc|ccc|}\hline & & & \mathbf{u_3} & \mathbf{v_3} & \mathbf{w_3} \\ \mathbf{u_4} & \mathbf{v_4} & \mathbf{w_4} & & & \\\hline\end{array} \tag{6.17}$$

If there is a cycle among all four, the order can be written down. At each step, list the coalition causing the change. For instance, if the order is

$$(u_1, v_1, w_1) \rightarrow (u_2, v_2, w_2) \rightarrow (u_3, v_3, w_3) \rightarrow (u_4, v_4, w_4) \rightarrow (u_1, v_1, w_1),$$

it follows from Eq. 6.17 that the first change is made by {Row, Column}, the second by {Row, Face}, the third by {Row, Column}, and the last by {Row, Face}. Row is involved in *all of four changes,* which requires Row's payoff at each stage to increase. Consequently, $u_1 < u_2 < u_3 < u_4 < u_1$. This $u_1 < u_1$ contradiction proves that this cycle is impossible.

More generally, as a cycle returns to the starting cell, in any listing of the coalitions, each player must appear an even number of times. (With an odd number of times, the cell is on the wrong face.) The only possible list of three players where each appears twice is {Row, Column}, {Row, Face}, and {Column, Face}. A four cycle requires one more transition, so at least one player appears three times. But, to have a four-cycle, a player appearing three times requires the player to appear in all four transitions, which leads to the above contradiction.   □

Although it is impossible to have a four-cycle among Nash cells for $2 \times 2 \times 2$ games, all four Nash cells can be involved in some sense. One possibility is where a coalition of players wishes to abandon one Nash cell only to become immersed in a cycle. According to the theorem, the cycle never returns to the original Nash cell.

As an example with Eq. 6.15, let $u = w = 0.5$ and $v = 2$. Here, Row and Face find a move to BRF beneficial, while Column is indifferent. (Anyway, by selecting R, Column is powerless.) This move unleashes a cycle that never returns to the original Nash cell. For a different behavior, let $u = v = w = 4$ where the Nash cell is "above" the cycle. The cycle could continue with myopic players who fail to seek full payoff advantage, or it could end after any Nash cell where the appropriate coalition selects this optimal TRBa Nash cell. After moving to (4, 4, 4), there are no incentives for any coalition to consider any other cell. Thus (4, 4, 4) is *coalition proof* in that a coalition cannot attain a preferred outcome.

6.3.2 Another Designer's Delight! Coalition-Free Nash Points?

So many competing tensions! The me-factors can dominate when interactions among individuals are limited (so $\mathcal{G}^B$ has limited influence); expect we-cooperative opportunities in interacting communities with externalities (manifested by a stronger $\mathcal{G}^B$). Coalitional stresses, which permit cyclic exchanges of Nash equilibria, must be anticipated with larger groups. What a delight for anyone wishing to explore intriguing games!

Knowing what can happen uncovers fascinating behavior. Even more, if cycles can arise among Nash points, expect similar actions to accompany hyperbolic cells. (The discussion associated with Fig. 6.2 proves that this is the case.) Adding interest to this observation is that (as with Fig. 6.2) an appropriate coalition can treat a hyperbolic point as a Nash cell. After all, on a $\mathcal{G}^N$ face, a cell with positive entries for both players in the coalition is, for their purposes, Nash. Consequently, hyperbolic cells can introduce coalitional complexities that affect a game.

As an example, on the front face, position what the {Row, Column} coalition might view as Nash cells at TL and BR. In the actual $2 \times 2 \times 2$ game, that TL cell could be hyperbolic $(2, 1)$ where only Face has an incentive to move from Front to Back. The following entries generate an interesting dynamic:

$$
\begin{array}{|ccc|ccc|}
\hline
\mathbf{4} & \mathbf{4} & \mathbf{-5} & -3 & -4 & 1 \\
\hline
-4 & -3 & -1 & 3 & 3 & 3 \\
\hline
\end{array}
\quad
\begin{array}{|ccc|ccc|}
\hline
\mathbf{1} & \mathbf{1} & \mathbf{5} & -2 & -1 & -1 \\
\hline
-1 & 2 & 1 & 2 & -2 & -3 \\
\hline
\end{array}
\qquad (6.18)
$$

For an accompanying story, follow the bold-entry road starting with BRBa. All players would welcome a move from BRBa to the BRF; thanks for this change belonging to Face. Although BRF is a true Nash cell, should Face commit to Front, the ungrateful, greedy {Row, Column} coalition can force the outcome to their preferred TRF; a choice giving each a personal high: From their coalitional perspective, TRF is a face-Nash cell.

Face would suffer at TRF, but TRF is hyperbolic $(2, 1)$, so she could reap revenge: After Row and Column make their TL choice, Face could move to Back (the TLBa cell) where she reaches a personal maximal payoff. As TLBa is a legitimate Nash cell located on the back face, there is nothing the {Row, Column} coalition can do to change the outcome. Had they stayed at BRF and resisted the temptation of moving to TLF, they would have enjoyed more comfort!

Indeed, TLBa is a *coalition proof Nash point* for Eq. 6.18 game.[4] To use this example to demonstrate the difference between Nash and coalition proof Nash, with Column's and Face's selecting L and Ba, Row is confronted with selecting from the game

$$
\textbf{back face, left column} = \begin{array}{|ccc|ccc|}
\hline
1 & - & - & - & - & - \\
\hline
-1 & - & - & - & - & - \\
\hline
\end{array}
$$

where the TLBa choice of 1 is optimal. That is, with Nash, each player's search for an improved outcome is over an appropriate one-dimensional array.

For coalition proof Nash cell, each *coalition's* search is over a *two-dimensional face*. To illustrate with Face's choice of Ba, the {Row, Column} coalition seeks a Pareto improvement over $(1, 1)$ in

$$
\textbf{back face} = \begin{array}{|ccc|ccc|}
\hline
\mathbf{1} & \mathbf{1} & - & -2 & -1 & - \\
\hline
-1 & 2 & - & 2 & -2 & - \\
\hline
\end{array}
$$

No improvement over the bold entries exists, so this coalition search is stilted.

---

[4]But, BRF is the risk-dominant cell with the $3 \times 3 \times 3 = 27$ product of $\mathcal{G}^N$ entries compared to the $1 \times 1 \times 5 = 5$ product for TLBa.

The {Row, Face} coalition searches over the options (Column selecting L) of

$$
\textbf{left face} = \begin{array}{|ccc|ccc|} \hline 4 & - & -5 & - & - & - \\ \hline -4 & - & -1 & - & - & - \\ \hline \end{array}, \quad \begin{array}{|ccc|ccc|} \hline 1 & - & 5 & - & - & - \\ \hline -1 & - & 1 & - & - & - \\ \hline \end{array}
$$

Again, no cell is a Pareto improvement over the $(1, 5)$ option of TLBa. Similarly, the remaining coalition of {Column, Face} explores possibilities in

$$
\textbf{top face} = \begin{array}{|ccc|ccc|} \hline - & 4 & -5 & - & -4 & 1 \\ \hline - & - & - & - & - & - \\ \hline \end{array}, \quad \begin{array}{|ccc|ccc|} \hline - & 1 & 5 & - & -1 & -1 \\ \hline - & - & - & - & - & - \\ \hline \end{array}
$$

where there is no Pareto improvement over the TLBa choice. Thus, not only is TLBa a Nash cell for this game, it also is a coalition proof Nash cell.

Here is an interesting feature: The behavioral component $\mathcal{G}^B$ plays no role in analyzing whether a cell is Nash. But, $\mathcal{G}^B$ can be a major actor in coalitional considerations. To support this comment, enhance Eq. 6.18 morality story by adding

$$
\mathcal{G}^B = \begin{array}{|ccc|ccc|} \hline 2 & 2 & 2 & 1 & 2 & -1 \\ \hline 2 & 1 & -2 & 1 & 1 & 1 \\ \hline \end{array}, \quad \begin{array}{|ccc|ccc|} \hline -5 & -5 & 2 & 2 & -5 & -1 \\ \hline -5 & 2 & -2 & 2 & 2 & 1 \\ \hline \end{array} \tag{6.19}
$$

to obtain

$$
\mathcal{G} = \begin{array}{|ccc|ccc|} \hline 6 & 6 & -3 & -2 & 4 & 0 \\ \hline -2 & -2 & -3 & 4 & 4 & 4 \\ \hline \end{array}, \quad \begin{array}{|ccc|ccc|} \hline -4 & -4 & 7 & 0 & -6 & -2 \\ \hline -6 & 4 & -1 & 4 & 0 & -2 \\ \hline \end{array} \tag{6.20}
$$

Instead of enjoying the cozy 4's from one of the original cells of this story (cell BRF), each of Row and Column is stuck with a miserable -4; this lousy outcome is a consequence of the unleashed dynamic.

But, is {Row, Column}, or any other coalition, stuck at this horrible choice? That is, in this enhanced setting, is TLBa still a *coalition proof Nash point*? Answers require examining the three faces passing through TLBa to discover whether any coalition can do better. Skip the two coalitions involving Face because there is no way to entice Face away from the personal maximum value of 7. It remains to examine what happens with {Row, Column} over the back face, which is

$$
\textbf{back face} = \begin{array}{|ccc|ccc|} \hline -4 & -4 & - & 0 & -6 & - \\ \hline -6 & 4 & - & 4 & 0 & - \\ \hline \end{array}
$$

Here, {Row, Column} *does* have a Pareto improved choice over $(-4, -4)$ at BRBa, which offers them $(4, 0)$!

This improved outcome comes from Eq. 6.19; the behavioral $\mathcal{G}^{B, parts}_{backface}$ places a severe Pareto inferior cell over TLBa and a Pareto superior cell over BRBa. Actually, the fact that $\mathcal{G}^B$ can influence whether a Nash cell is, or is not, coalition proof must be expected. After all, $\mathcal{G}^B$ identifies options that a *group* of players, not individual players, might find attractive. The search revolves about a Nash cell, so information in any of the one- dimensional arrays coming out of the cell are not affected by $\mathcal{G}^B$ entries: These directions are part of the Nash analysis. (This $\mathcal{G}^B$ feature is captured by Fig. 2.2.) And so, as with this example, if there is a Pareto preferred term, it *must* be in the region *not* covered by Nash considerations. Either this diametrically opposite cell is Nash, or its attractive payoff values are introduced (as in this example) by $\mathcal{G}^B$. Thus, critical to a coalition proof analysis is the $\mathcal{G}^B$ component, where the modifier "critical" is appropriate.

The story continues; BRBa is *not* a Nash cell, so at least one player can change to a preferred outcome. Indeed, BRBa is hyperbolic (1, 2), so two players (Column, Face) have such an incentive. But Face's move to the front unleashes a cycle. So, Eq. 6.18 game enjoys a coalition proof Nash cell at TLBa, but the associated Eq. 6.20 game does *not*.

This difference is manifested by a cyclic behavior caused by Nash and hyperbolic cells along with $\mathcal{G}^B$ features. Lessons learned are immediate: With $2 \times 2 \times 2$ games, $\mathcal{G}^B$ entries can convert a hyperbolic cell into an alluring choice for some coalition. As for coalitional choices, for each possible coalition, only one cell must be examined; on each face, it is the Nash cell diametrically opposite the cell of interest. (In a natural way, this generalizes to games with more strategies.)

6.3.3 More Strategies, More Cycles

To conclude, two examples are offered that show how more players and/or strategies can admit more cycles. The first example is the $\mathcal{G}^N$ component of a $2 \times 2 \times 2 \times 2$ game.

Such a game can have up to eight Nash cells; all eight can be involved in cycles. The entries for one cycle are the permutations of (2, 4, 6, 8) created by Fig. 6.3c ranking wheel. Those for the second cycle come from (1, 3, 5, 7). Once the cells are defined, the game's construction follows the pattern used to develop Eq. 6.15.

Only the Nash cells are specified, but they uniquely complete the game (Corollary 4.1). As an illustration, the TLF cell of the first set is $(-5, -8, -3, -2)$ coming, respectively, from Row's entry in BLF set 1, Column's entry in TRF set 1, Face' entry in TLBa set 1, and Set's entry in TLF in set 2.

$$
\text{Set 1} = \begin{array}{|cccc|cccc|cccc|cccc|}
\hline
- & - & - & - & 6 & 8 & 2 & 4 & 7 & 5 & 3 & 1 & - & - & - & - \\
\hline
5 & 3 & 1 & 7 & - & - & - & - & - & - & - & - & 8 & 2 & 4 & 6 \\
\hline
\end{array}
$$

$$
\text{Set 2} = \begin{array}{|cccc|cccc|cccc|cccc|}
\hline
4 & 6 & 8 & 2 & - & - & - & - & - & - & - & - & 1 & 7 & 5 & 3 \\
\hline
- & - & - & - & 3 & 1 & 7 & 5 & 2 & 4 & 6 & 8 & - & - & - & - \\
\hline
\end{array}
$$

(6.21)

To follow the cycles, the player with a cell's largest payoff has no interest in moving. But that player's position determines which options the remaining three players can explore. In the TRF cell of set 1, Column is content with the largest payoff. But

Column's choice of R allows the {Row, Face, Set} coalition to explore possibilities in *any* R column in the four matrices. The unique Pareto improvement for these three players is BRBa in set 1; this choice hurts Column, but the new payoffs are better for each of the coalition players.

In this new cell, Row is delighted with the strong payoff; Row's choice of Bottom leaves the new coalition of {Column, Face, Set} free to explore whether there are Pareto improving opportunities in the bottom row of the four matrices. The only choice is at BLBa in set 2. As this cell contains Set's largest payoff, Set commits to set 2, where {Row, Column, Face} finds that TLF improves all coalition members. At this cell, Face is happy, which allows {Row, Column, Set} to explore Front face choices, where the only Pareto improvement for all players is TRF in set 1—this choice completes the cycle. A similar argument follows the cycle with odd entries.

Here is a question that is left to the reader. Can there be a cycle uniting all eight Nash cells? Eq. 6.21 shows that the question must be more carefully posed. This is because with the "odd integer" cycle at the BLF cell of set 1, Set has committed to set 1, so the {Row, Column, Face} coalition can explore options only in this set. Here the coalition has *two choices*: TLBa *or* TLF both of set 1. The first continues along the odd-integer cycle; the second smooths out the odd-integer journey to embark on the even-integer cycle. This construct does include all eight Nash cells.

However, with choices along the even-integer cycle, never does a coalition give a thought about joining its odd counterpart. Instead, each coalition in the even group has a unique choice, and it belongs to the fraternity. Rather than this cycle uniting all eight Nash cells, one cycle is above the other; entry is admitted through a single entrance.

The question remains: Can there be a cycle that cannot be reduced to smaller cycles and that includes eight Nash cells? (For a hint, see the proof of Theorem 6.4.)

Now turn to $3 \times 3 \times 3$ games, where the $\mathcal{G}^N$ component can have up to nine Nash cells. The following $\mathcal{G}^N$ exhibits two cycles, where each involves three Nash cells. The cell entries are constructed with Fig. 6.3b ranking wheel mechanism where one cycle used entries from $(1, 3, 5)$ and the second used $(2, 4, 6)$ values. The matrices are listed in the Front, Middle, Back face order. Only the six Nash cells are specified.

1 3 5	– – –	– – –
– – –	3 5 1	– – –
– – –	– – –	– – –

– – –	– – –	– – –
– – –	– – –	6 2 4
– – –	– – –	– – –

– – –	5 1 3	– – –
2 4 6	– – –	– – –
– – –	– – –	4 6 2

$$(6.22)$$

Notice the number of cells that are without an assignment! One must believe it is possible to have three separate cycles where each involves three cells. In this way, all nine cells are involved in some cyclic scenario. While the possibility of whether this can occur is left to the reader, hints are supplied.

First, Eq. 6.22 does *not have* the suggested flexibility. The reason is the geometry of Nash cells; no other Nash cells can be on an array emanating from a given Nash cell. Indicating by x the cells precluded by the Nash cell at TLF, y the cells from MMF, and so forth, the marked-up matrices

1	3	5	x	y	t	x	–	–
x	y	u	3	5	1	–	y	z
x	–	–	–	y	–	–	–	v

,

x	–	–	–	–	t	–	–	z
–	u	z	–	y	z	6	2	4
						–	v	z

,

x	u	t	5	1	3	–	v	t
2	4	6	u	y	t	u	v	z
–	u	v	–	v	t	4	6	2

prove that the only cells eligible for Nash status would be one of the two blank choices on the bottom row of the middle face: Eq. 6.22 *cannot* be extended to have three cycles.

For a general proof, the $3 \times 3 \times 3$ structures with nine Nash cells are special; the cells tend to be along diagonals and super-diagonals. But, as the above arguments demonstrate, a three-cycle construction has a specialized twisting structure. The proof is to show that the two are not compatible.

6.4 Comments

These various structures provide so many opportunities that a complete exposition is impossible. In the above, for instance, creating wild examples with the combination of $\mathcal{G}^B$ components with a cyclic nature were not explored.

And so, bowing to realism, the goal has been to select various themes to identify different directions that can arise with changes in the underlying game structure. Of interest is how symmetries play a key role in everything that has been described. Admittedly, the associated symmetries permitting the decomposition, which is central to all that has been discussed, is not obvious. This gap is corrected in the next chapter.

Chapter 7
The Underlying Mathematics

This chapter, which essentially is the concluding one (it is followed by a brief summary), develops the mathematical foundation for our decomposition. It is written in a manner to encourage readers to apply these notions to other concerns. Admittedly, the approach involves a higher level of mathematics than typically found in the social sciences, but these new ways to resolve long-standing puzzles[1] make it worth learning these abstract algebraic concepts, which have roots tracing back to those elementary algebra exercises with more variables than equations.

Many readers will not be familiar with these mathematical concepts, so our presentation has an expository flavor where, in places, liberties are taken with the mathematics in order to suggest what other topics can be explored. Details are left to references; e.g., see Dummit and Foote [6] for an introduction to abstract algebra and Sagan [32] to learn about representations of the symmetric group.

Game theoretic modeling typically begins by specifying a game's payoffs, which become central for the analysis. But as argued in Sect. 1.2, the usual objective is not a specific game; the true intent is to model a phenomenon. For this to hold, lessons learned from an analysis must not strictly depend upon $\mathcal{G}$'s specified payoffs.

To illustrate with the Eq. 2.1 Prisoner's Dilemma

$$\mathcal{G}_2 = \begin{array}{|cc|cc|} \hline 4 & 4 & -2 & 6 \\ 6 & -2 & 0 & 0 \\ \hline \end{array}, \tag{7.1}$$

the game's BR Nash equilibrium is Pareto inferior to the TL option. But TL is a repeller, which permits either player to defect with profit from any cooperative TL intent. To determine whether this defection is a relevant behavior, rather than an

---

[1] Earlier, these algebraic approaches were applied by one of us (Saari) to uncover hidden structures of topics such as voting theory (e.g., [24, 25]), mathematical psychology (e.g., [26]), and decision theory [28]. The results were discovered by using these symmetry considerations, but details of the abstract analysis were purposely omitted to make the papers accessible to a wider audience.

© Springer Nature Switzerland AG 2019
D. T. Jessie and D. G. Saari, *Coordinate Systems for Games*, Static & Dynamic Game
Theory: Foundations & Applications, https://doi.org/10.1007/978-3-030-35847-1_7

Eq. 7.1 oddity, conclusions from the analysis must *not* depend on the precise payoff choices. Adding small ε values to each payoff in $\mathcal{G}_2$, for instance, should not change the conclusions.

Yet, a typical analysis centers on the game's specific entries, which can obscure whether conclusions reflect $\mathcal{G}$'s structure. Does the general behavior of the Battle of the Sexes (Eq. 1.7, Sect. 2.5.5), for instance, depend on those zeros in the game's bimatrix, which manifest a distaste of opposing a partner's wishes, or would basic assertions remain after replacing the zeros with small ϵ values of one sign or the other (i.e., representing opposition, or stronger support).

To determine whether consequences are captive to $\mathcal{G}$'s specific entries, or reflect a general phenomenon, we must be able to identify which conclusions are caused by which structures of a game. These comments, which argue for a nonlocal analysis, are relevant across the social sciences. After all, precise information about a game—as true with many social science models—can be difficult, if not impossible, to obtain. The entries of the introductory $\mathcal{G}_{Traffic}$ (Eq. 1.2), for instance, most surely *fail* to precisely represent each driver's intent; this means that answers that strictly depend on these particular values are of minimal to zero interest. Rather than $\mathcal{G}_{Traffic}$'s specific terms, the real goal should be to identify the structural aspects of $\mathcal{G}_{Traffic}$ that encourage defection from cooperation.

One way to move beyond a local analysis is to use probability and statistics. Instead of specifying a particular game, say the above $\mathcal{G}_2$, a level of uncertainty in the payoffs can be incorporated by specifying $\mathcal{G}_2 + \varepsilon$ leading to

$$\mathcal{G}_2 + \boldsymbol{\varepsilon} = \frac{\begin{array}{|cc|cc|} 4 + \epsilon_1 & 4 + \epsilon_2 & -2 + \epsilon_3 & 6 + \epsilon_4 \\ 6 + \epsilon_5 & -2 + \epsilon_6 & 0 + \epsilon_7 & 0 + \epsilon_8 \end{array}}{}$$

where each ϵ_j is a random variable with small positive or negative values. This approach merely replaces the point $\mathcal{G}_2$ (in the space of games) with a small neighborhood of $\mathcal{G}_2$; the analysis and conclusions remain essentially local with the cost of a more difficult analysis.

7.1 Decompositions

A different global approach is to decompose $\mathbb{G}$, the space of games, into components that emphasize properties of interest. The goal might be to stress Nash equilibria, maybe Pareto information, or perhaps a potential-game component. Each choice requires discovering the suitable decomposition where features of interest are separated from everything else. The intent is to represent each game $\mathcal{G} \in \mathbb{G}$ as a combination of separate, independent components: The separation allows the contributing factors to be individually examined.

Compare this separation of parts with buying a car; in deciding, features of reliability and engine power are separated from those of amenities. Similarly, in the decomposition of $\mathbb{G}$, the goal is to separate the information of a game that is of particular interest (e.g., Nash structure, Pareto structure, etc.) from accompanying factors. Then the roles played by competing parts are identified. Knowing how terms from different components influence a model's behavior makes it possible to understand the kinds of modeling changes that alter general outcomes. This decomposition approach is a form of a global analysis.

The choice of decomposition depends upon what is selected as important. The natural decomposition for the strategic structure of the Nash equilibria is developed in this book (also see Jessie and Saari [12–14]). A study of cooperative games and bargaining solutions can be approached with the Kalai and Kalai [15] framework. Then there is the class of potential games for which Candogan et al. [5] developed a decomposition.

As it should be expected, different themes are typically analyzed in dissimilar ways. Candogan et al., view games as a flow on a graph; they use tools from algebraic topology (i.e., the Helmholtz decomposition) to identify potential games. Kalai and Kalai begin with relevant properties of a cooperative solution concept to show that their CoCo value, which uses an intuitive decomposition of $\mathbb{G}$, is the unique solution concept that satisfies these properties. We (Jessie and Saari) present our decomposition of $\mathbb{G}$ in terms of linear algebra and prove that it is the unique decomposition that isolates the Nash equilibria information.

It would be useful, as done here, to unify the analysis; that is, create a general method whereby all three decompositions are found in the same way. For each class, our approach characterizes the relevant structures of $\mathbb{G}$ by using symmetry tools from abstract algebra. A delightful bonus is that the emerging foundation can be described in terms of linear algebra; this makes it easier to carry out further research. (See Sects. 3.3.1 and 3.4.5.)

To offer insight, notice that $F(x, y, x)$ depends upon two rather than three variables because of a $x = z$ link. So, if $F(x, y, z) = F(z, y, x)$ is always true, it is reasonable to expect that F only uses portions of the variables. determined by some $x = z$ connection. Similarly, if function f always satisfies $f(x_1, x_2, x_3, x_4) = -f(x_3, x_4, x_1, x_2)$, then f must possess special properties that relate the variables in some remarkable manner. But this functional connection holds for a Nash analysis when computing $f = EV(Top) - EV(Bottom)$ by interchanging rows. The decomposition extracts how the variables are related.

Notice the message: Typically the symmetries where rows or columns can be interchanged without changing the game are noted and then ignored. These seemingly trivial symmetries, however, restrict the admissible functions $f(\mathbf{x})$. In turn, the properties of these functions are full of hidden information that lead to the appropriate decompositions.

For an outline, the following starts with definitions and a brief discussion of "symmetry." After that, the algebraic structure of games is examined, from which we show how the space of games, $\mathbb{G}$, can be decomposed in different ways.

7.2 Definitions

Everything holds for any finite number of players where each has an arbitrary but finite number of strategies. To keep the exposition manageable, only 2×2 games are considered. (Games with more players/strategies have more symmetries—and still other solutions concepts.) A game $\mathcal{G}$ consists of a set of players $I = \{1, 2\}$, where each player, $i = 1, 2$, has a set of strategies $S_i = \{s_{1,i}, s_{2,i}\}$ and a payoff function $\pi_i : S_1 \times S_2 \to \mathbb{R}$. To play the game, each player simultaneously chooses a strategy, which is a probability distribution over S_i for Player i, given by $\mathbf{p} = (p_1, p_2)$ for Player 1 and $\mathbf{q} = (q_1, q_2)$ for Player 2. Denote the space of all 2×2 games by $\mathbb{G}$.

As for algebraic structures, start with a *group*. Think of this as abstracting how to solve $3x = 9$ in order to address similar concerns from other disciplines; e.g., a crystal might be in configuration a where the goal is to figure how to rotate it, x, into a desired symmetry b, or the objective may be to solve the Rubik cube puzzle. A group specifies the properties that a set of elements and an associated operation "$\circ$" must have to solve $a \circ x = b$ and $y \circ a = b$ problems. Group properties describe, for instance, whether $\mathcal{A}\mathcal{X} = \mathcal{B}$ matrix problems can be solved where $\mathcal{A}$ and $\mathcal{B}$ are given matrices and $\mathcal{X}$ is unknown.

To motivate key conditions for a group, solving $3x = 9$ requires an *identity* $\mathbf{1}$ (i.e., $\mathbf{1} \circ a = a \circ \mathbf{1} = a$ for any a) and an *inverse* a^{-1} for any element a, where $a \circ a^{-1} = a^{-1} \circ a = \mathbf{1}$. For $3x = 9$, the inverse is 3^{-1}. With these properties, multiply both sides by 3^{-1} where

$$3^{-1}(3x) = (3^{-1}3)x = (1)(x) = x = (3^{-1})9 = 3. \tag{7.2}$$

Similarly, a group's critical components are an *identity* (i.e., $\mathbf{1} = \begin{pmatrix} 1 & 0 \\ 0 & 1 \end{pmatrix}$ for 2×2 matrices) and, for any a in the group, an *inverse* a^{-1} (so that $a^{-1} \circ a = a \circ a^{-1} = \mathbf{1}$). Only two more properties are needed: mimicking the requirement that $(3^{-1})9$ is defined, a *closure* property is imposed asserting that if d and e are in the set, then so is $d \circ e$. Finally, to solve a problem, those parenthesis must be moved, which is *associativity* where $a \circ (b \circ c) = (a \circ b) \circ c$. With these properties, the Eq. 7.2 solution approach always applies.

A second algebraic structure is a *ring:* this generalizes standard arithmetic of addition and multiplication that are connected with distribution laws $a(b + c) = ab + ac$. Call the ring's two operations addition and multiplication. Basic conditions (see references for all requirements) include where addition defines an abelian (i.e., the order, $a + b = b + a$, does not matter) group, which, when combined with multiplication, satisfies the distributive rule. As an example, the space of $n \times n$ matrices satisfies the distributive laws

$$\mathcal{A}(\mathcal{B} + \mathcal{C}) = \mathcal{A}\mathcal{B} + \mathcal{A}\mathcal{C} \text{ and } (\mathcal{B} + \mathcal{C})\mathcal{A} = \mathcal{B}\mathcal{A} + \mathcal{C}\mathcal{A}$$

along with other conditions to define a ring.

Of importance is the familiar *vector space,* which includes the space of 2×2 games, $\mathbb{G}$. To identify $\mathbb{G}$ with the eight-dimensional vector space $\mathbb{R}^8$, order the bimatrix's entries in the TL, TR, BL, and BR cell order for the first player and then the same for the second player. In this way the $\mathcal{G}_2$ payoffs (Eq. 7.1) can be written as the vector $(4, -2, 6, 0; 4, 6, -2, 0)$.

The addition and scalar multiplication of games are illustrated by

$$
\begin{array}{|cc|cc|}
\hline
4 & 4 & -2 & 6 \\
6 & -2 & 0 & 0 \\
\hline
\end{array}
+
\begin{array}{|cc|cc|}
\hline
-2 & -3 & 4 & -2 \\
-3 & 5 & 2 & 1 \\
\hline
\end{array}
=
\begin{array}{|cc|cc|}
\hline
2 & 1 & 2 & 4 \\
3 & 3 & 2 & 1 \\
\hline
\end{array}
\tag{7.3}
$$

$$
2 \cdot
\begin{array}{|cc|cc|}
\hline
4 & 4 & -2 & 6 \\
6 & -2 & 0 & 0 \\
\hline
\end{array}
=
\begin{array}{|cc|cc|}
\hline
8 & 8 & -4 & 12 \\
12 & -4 & 0 & 0 \\
\hline
\end{array}
\tag{7.4}
$$

With the identification $\mathbb{G} \cong \mathbb{R}^8$, Eq. 7.3 becomes the vector addition

$$
(4, -2, 6, 0; 4, 6, -2, 0) + (-2, 4, -3, 2; -3, -2, 5, 1) = (2, 2, 3, 2; 1, 4, 3, 1)
\tag{7.5}
$$

and the Eq. 7.4 multiplication is identified with a scalar multiple of a vector as

$$
2(4, -2, 6, 0; 4, 6, -2, 0) = (8, -4, 12, 0; 8, 12, -4, 0).
$$

Not only is $\mathbb{G}$ a vector space, but it also is an abelian group where the additive identity $\mathbf{1}$ is the game with zero entries.

Treat the space $\mathbb{G} \cong \mathbb{R}^8$ as the direct sum $\mathbb{G} \cong \mathbb{R}^4 \oplus \mathbb{R}^4$, where Player 1's and Player 2's payoffs can be handled separately: This disconnect, which is key in the Nash decomposition, reflects the reality that Player 2's *payoffs* are not needed to determine Player 1's Nash best response; only Player 2's action (given by $\mathbf{q}$) is required. The transition from each player's best response to the equilibrium is what involves both players' payoffs.

7.3 Symmetry

Our main tool is to determine the symmetry structures of the space of games $\mathbb{G}$, where "symmetry" is identified with "sameness." Indeed, a valued use of groups is to characterize an object's symmetries, which are the transformations that keep the object the same.

For instance, rotating a square (with horizontal and vertical edges) in a clockwise direction by 90°, or reflecting it about a diagonal, keeps the same shape. To say the square is symmetric (i.e., sameness) means that the shape remains unchanged after the operation. So if S denotes the square and $\sigma = $ "rotate S by 90°," the symmetry condition is $\sigma \cdot S = S$.

The relevant symmetries are those that reflect properties of interest. If the square's top two corners are colored red, then, although a rotation preserves the square's shape, it fails to respect this coloring. An operation that does not preserve the color of the top corners generates a *path dependency* problem whereby coloring the corners and performing the operation differs from first performing the operation and then coloring the corners. Mathematically, if $f =$ "color the top corners red," this lack of symmetry is expressed as

$$f(\sigma \cdot S) \neq \sigma f(S). \tag{7.6}$$

This example captures a widely used mathematical technique: Analyze an object by determining which functions acting on an object avoid the Eq. 7.6 dependency concern. A goal is to understand how a group's actions can characterize an object X. Suppose, for instance, an unknown object X is found to be symmetric with respect to any rotation and any reflection about any line through its center. These symmetry actions immediately identify X as a circle, or a disc.

As this "circle or disc" comment demonstrates many objects can satisfy given conditions. The symmetries of the square with colored corners are shared by points placed at -1 and 1 on the real line, or the set of positive and the set of negative integers. Rather than a liability, this commonality provides a mathematical opportunity by allowing a familiar object to be analyzed to discover properties of a desired but, perhaps, unfamiliar structure. This is a feature of representation theory where consequences of symmetries are extracted from properties of matrices.

Our goal is to discover which symmetries characterize the space $\mathbb{G}$. Doing so uncovers the structure of $\mathbb{G}$ and properties of desired functions on $\mathbb{G}$, such as the Nash best response function or a bargaining solution. An advantage of these tools (symmetry groups and their representations) is that results can be expressed in the concrete language of linear algebra. This permits using familiar tools of linear algebra to address specific problems, which underscores the power of this approach.

7.4   Algebraic Structure of Games

To motivate what follows, describe the symmetries of a square by how the corners are permuted. Starting at the upper left corner and labeling them in a clockwise direction as $\{1, 2, 3, 4\}$, reflecting the square about the 1–3 diagonal creates the $\{1, 4, 3, 2\}$ ordering, where 2 and 4 are interchanged and 1 and 3 remain fixed; denote this permutation by $(2, 4)$. Reflecting about the 2–4 diagonal converts the original $\{1, 2, 3, 4\}$ into $\{3, 2, 1, 4\}$ where (because 2 and 4 remain unchanged but 1 and 3 are interchanged) the permutation is $(1, 3)$. Rotating the square 90° in a clockwise direction leads to $(1, 2, 3, 4)$ meaning that corner 1 is rotated to corner 2, corner 2 to corner 3, corner 3 to corner 4, and corner 4 to corner 1. The various powers (e.g, apply the rotation several times) and combinations of these operations define a group; e.g., the inverse of $(1, 3)$ is $(1, 3)$.

Similarly, our analysis of the transformations of $\mathcal{G}$ is based on the general permutation groups $\mathfrak{S}_n$ that describes all ways to permute the terms $\{1, 2, \ldots, n\}$.[2] This requires examining how a permutation $\sigma \in \mathfrak{S}_n$ affects $\mathcal{G}$. As a first choice, let $\sigma \in \mathfrak{S}_2$ on the set of Player 1's strategies S_1 be $\sigma(S_1) = \sigma\{s_{1,1}, s_{2,1}\} = \{s_{\sigma(1),1}, s_{\sigma(2),1}\}$. Using $\sigma = (1, 2)$ (remember, this permutation interchanges 1 and 2) yields

$$\sigma \cdot \begin{pmatrix} s_{1,1} & \boxed{\begin{array}{cc|cc} 4 & 4 & -2 & 6 \end{array}} \\ s_{2,1} & \boxed{\begin{array}{cc|cc} 6 & -2 & 0 & 0 \end{array}} \end{pmatrix} = \begin{array}{c} s_{2,1} \\ s_{1,1} \end{array} \boxed{\begin{array}{cc|cc} 6 & -2 & 0 & 0 \\ 4 & 4 & -2 & 6 \end{array}} \tag{7.7}$$

which interchanges the rows of the game.[3] A similar action $\tau \in \mathfrak{S}_2$ can be defined on S_2 that interchanges the columns.

$$\tau \cdot \left(\begin{array}{cc} s_{1,2} & s_{2,2} \\ \boxed{\begin{array}{cc|cc} 4 & 4 & -2 & 6 \\ 6 & -2 & 0 & 0 \end{array}} \end{array} \right) = \begin{array}{cc} s_{2,2} & s_{1,2} \\ \boxed{\begin{array}{cc|cc} -2 & 6 & 4 & 4 \\ 0 & 0 & 6 & -2 \end{array}} \end{array} \tag{7.8}$$

7.4.1 Symmetries of the Space of Games

These operations provide three views of $\mathbb{G}$: For the first two, $\mathfrak{S}_2$ is acting on either S_1 or S_2; the third is where $\mathfrak{S}_2 \otimes \mathfrak{S}_2 \cong V_4$ (the Klein four-group) acts on $S_1 \times S_2$ (which allows changes in rows and columns). Distinctions among these changes are important because different actions lead to distinctly different decompositions of $\mathbb{G}$. An additional action of $\mathfrak{S}_2$ on $\mathbb{G}$ leads to a third decomposition.

The meaning of $\mathfrak{S}_2$, which interchanges rows or columns, is depicted with the first two sets of squares of Fig. 7.1a, b[4]; the numbers identify the square's quadrants (or a game's bimatrix). Rotating the knob in Fig. 7.1a interchanges columns, where the change is indicated by the arrow: This is a $\mathfrak{S}_2$ symmetry on S_2. (The group's symmetry property of having an inverse is captured by the fact that doing this rotation twice returns to the original setting.) Similarly, Fig. 7.1b represents the $\mathfrak{S}_2$ action on S_1; it interchanges rows.

A difference arises with $\mathfrak{S}_2 \otimes \mathfrak{S}_2$ because it allows combinations of actions from both symmetries. The geometry is represented in Fig. 7.1c where first the columns are interchanged and then the rows; the resulting scrambling of the square's quadrants is given by the arrow. (Doing all of this again leads to the original setting.)

The Fig. 7.1 geometry reflects the $\mathfrak{S}_2 \otimes \mathfrak{S}_2$ symmetries of a bimatrix. Of importance, although the form of the game changes, the game remains unchanged for purposes of game theory. It is this invariance that identifies these symmetries as being

---

[2]Recall, a game $\mathcal{G} = \langle I, S_1, S_2, \pi_1, \pi_2 \rangle$ consists of a set of players I, a set of strategies S_i and a payoff function π_i for each player.

[3]Note that Player 2's payoffs are interchanged, as the action changes the labeling of outcomes, which is the domain of the payoff functions $\pi_i : S_1 \times S_2 \to \mathbb{R}$.

[4]This is related to Fig. 4.5 in [30, Chap. 4]. Indeed, voting and game theory share similar symmetries.

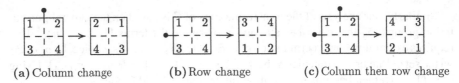

(a) Column change   (b) Row change   (c) Column then row change

Fig. 7.1 Ranking changes

basic for 2×2 games; the goal is to extract consequences. Also, Fig. 7.1 indicates what kinds of related symmetries arise in a $k_1 \times k_2 \times \cdots \times k_n$ game.

$$\begin{array}{|c c|} \hline a & -b & - \\ c & -d & - \\ \hline \end{array} \rightarrow \begin{array}{|c c|} \hline c & -d & - \\ a & -b & - \\ \hline \end{array}, \quad \text{Rotate } 90° = \begin{array}{|c c|} \hline c & -a & - \\ d & -b & - \\ \hline \end{array} \qquad (7.9)$$

Stated in a slightly different manner, Row's entries for the original game are in the first part of Eq. 7.9; interchanging the rows (a Fig. 7.1b rotation) creates the second setting. Of importance, whatever Column chooses, in either situation Row selects between {a and c} or {b and d}. This consistency holds for all Fig. 7.1 orientations, with a similar statement applying to what Column selects based on Row's choice. But now consider the earlier symmetry of the square, which rotates its vertices 90° in a clockwise direction[5]; this yields the third matrix of Eq. 7.9. Depending on what Column chooses, Row now selects between {c and d} or {a and b}, which defines a *different game*. Thus, this rotation is *not* a "sameness" symmetry; it does *not* preserve the game!

7.4.2   Homomorphisms

Central for our analysis is the concept of an homomorphism. Start with an "isomorphism," which equates operations on different spaces. The above "sameness" of the square with colored top corners is equated with points placed at -1 and $+1$; the sameness is captured by an isomorphism that identifies the right and left square corners with, respectively, the point at $+1$ and -1. Flipping the square about the middle vertical axis is the same as reversing the signs of the points. More generally, an isomorphism is a one-to-one mapping where the actions on the domain transfer to the actions on the range. The inverse mapping identifies actions on the range with corresponding actions on the domain.

A standard example is the exponential mapping, $f(x) = e^x$, that maps the real line to positive real values. Addition on the line, $a + b$, transfers to multiplication on the range; i.e., addition of $a + b = c$ transfers to multiplication of $e^c = e^{a+b} = e^a e^b$. The inverse mapping is the natural logarithm $\ln(y)$. Another example is the identification

---

[5] In voting theory, this rotation symmetry (which captures all possible paired comparison difficulties) is given by a "ranking wheel" as described in [30, Chap. 4].

of $\mathbb{G}$ with $\mathbb{R}^8$; part of the isomorphism is illustrated with Eq. 7.5. If two spaces are isomorphic, they are essentially (from the perspective of the operations) the same.

Central for what follows (and many social science applications) is the associated concept of an *homomorphism*; while it shares the objective of an isomorphism to identify operations on the domain and range, it includes mappings that are not one-to-one. As a typical example, consider the mapping $f(x)$ from the integers on the real line to the integers $\{0, 1, \ldots, 10\}$ given by the remainder after dividing x by 11. This mapping, denoted by "x mod 11," is many-to-one; e.g., $9 = 20 \bmod 11 = 31 \bmod 11$ (so $f(20) = f(31)$) while $4 = 37 \bmod 11 = 48 \bmod 11$. The goal is to determine whether addition on the integers transfers to addition on remainders with mod 11; namely, is $f(x) = x \bmod 11$ a homomorphism? This requires showing whether $f(x + y) = f(x) + f(y)$, or that

$$(x + y) \bmod 11 = [(x \bmod 11) + (y \bmod 11)] \bmod 11,$$

which it is. As an illustration, which shows that $f(20 + 37) = f(20) + f(37)$,

$$2 = (20 + 37) \bmod 11 = (9 + 4) \bmod 11 = [(20 \bmod 11) + (37 \bmod 11)] \bmod 11.$$

A commonly encountered homomorphism is a linear mapping $\mathcal{L} : \mathbb{R}^3 \to \mathbb{R}^2$, such as

$$\mathcal{L} = \begin{pmatrix} 3 & 1 & 2 \\ 1 & -3 & 4 \end{pmatrix}. \tag{7.10}$$

Key to the definition of an homomorphism is that it avoids the Eq. 7.6 path dependency. This means that carrying out the vector operation $x\mathbf{v}_1 + \mathbf{v}_2$ and mapping by $\mathcal{L}$ is the same as first mapping $\mathbf{v}_1$ and $\mathbf{v}_2$, and then carrying out the vector operation as expressed by

$$\mathcal{L}(x\mathbf{v}_1 + \mathbf{v}_2) = x\mathcal{L}(\mathbf{v}_1) + \mathcal{L}(\mathbf{v}_2).$$

This $\mathcal{L}$ example resembles the earlier mentioned "more variables than equations" algebra lesson. An isomorphism's inverse is easy to define, but what happens with a homomorphism? That algebra lesson showed how to partition the domain into sets of solutions. Namely, $\mathcal{L}$ divides $\mathbb{R}^3$ according to the kernel of $\mathcal{L}$ whereby the system of linear equations $3x + y + 2z = u$, $x - 3y + 4z = v$ has the solution expressing x and y in terms of z

$$\begin{pmatrix} x \\ y \end{pmatrix} = \begin{pmatrix} 0.3 & 0.1 \\ 0.1 & -0.3 \end{pmatrix} \begin{pmatrix} u \\ v \end{pmatrix} + \begin{pmatrix} -z \\ z \end{pmatrix} = \begin{pmatrix} 0.3u + 0.1v \\ 0.1u - 0.3v \end{pmatrix} + \begin{pmatrix} -z \\ z \end{pmatrix}, \tag{7.11}$$

where the lines of solutions decompose $\mathbb{R}^3$ into translates of the $y = -x = z$ line that passes through the point in the x-y plane defined by $z = 0$. The image of $\mathcal{L}$ for any point on such a line is (u, v); so, any two points on the line are solutions.

Of importance for our interests, a homomorphism *decomposes* its domain *relative to the structure of the given mapping*. The mod 11 example separates integers into 11 classes of the $w + 11n$ form where the w value is an integer between 0 and 10, and n is any integer; the image of f for any entry from the class is the same w.

This phenomenon is central to our approach: describe a solution concept as homomorphisms from the space of games, $\mathbb{G}$, to an appropriate space of outputs. The choice of the homomorphism (determined by the selected solution concept) decomposes $\mathbb{G}$ into the relevant component parts. As above, any two inputs from the same component share the same solution outcome. Notice, many other social science concerns can be described in terms of homomorphisms, which means that these spaces also can be decomposed into its relevant parts. As special cases, consider any setting expressed with linear equations.

7.4.3 Module

To apply this homomorphism methodology to $\mathbb{G}$, the concept of a vector space must be extended in order to incorporate the earlier described symmetry properties of games; this is done in terms of a *module*. In a vector space, the scalars act on the vectors where scalar multiplication is subject to the distributive law. What is needed is to replace the scalars with appropriate symmetry structures.

One choice is to replace the role of scalars with a ring (which has been done to analyze the Rubik cube). The choice used here captures the inherent group symmetries of $\mathbb{G}$. Doing so defines a group, or G-module, where the actions of a group must be compatible with the algebraic structures of the space of games $\mathbb{G}$. More precisely, for all $\mathcal{G}_i \in \mathbb{G}$, $\sigma, \tau \in \mathfrak{S}_2$, scalars $x, y \in \mathbb{R}$, and $\mathbf{1}$ is the permutation where nothing is changed, it must be that the action of the group on the vector space $\mathbb{G}$ satisfies

1. $\sigma\mathcal{G} \in \mathbb{G}$
2. $\sigma(x\mathcal{G}_1 + y\mathcal{G}_2) = x\sigma\mathcal{G}_1 + y\sigma\mathcal{G}_2$
3. $(\sigma\tau)\mathcal{G} = \sigma(\tau\mathcal{G})$
4. $\mathbf{1}\mathcal{G} = \mathcal{G}$

That these actions are satisfied is an exercise by using $\mathbb{G} \cong \mathbb{R}^8$ where the relevant algebraic structures are addition of vectors and scalar multiplication.

7.5 Best Response Algebraic Structures

As $\mathbb{G}$ admits different algebraic structures, it is necessary to determine which ones are compatible with desired solution concepts. As a metaphor, a student's correct choice of a university is one that reflects her goals. Similarly, an appropriate choice from the different $\mathbb{G}$ structures is one that is compatible with an intended solution concept.

Nash equilibria are determined by best response functions, so start with Player i's Nash best response function φ_i. For Row (Player 1) and Column's given strategy $\mathbf{q}$, the best response is a probability distribution $\mathbf{p} = (p_1, p_2)$ over the set of strategies S_1.[6] This distribution is determined from the expected values of playing $s_{1,1}$ and $s_{2,1}$ with Column's given strategy. So, let $\varphi_1 : \mathbb{G} \times \mathbb{R}^2 \to \mathbb{R}^2$ map a game $\mathcal{G} \in \mathbb{G}$ and the opponent's choice of $\mathbf{q} \in \mathbb{R}^2$ to Player 1's expected values for the two pure-strategy choices,

$$\varphi_1(\mathcal{G}|\mathbf{q}) = \big(EV(s_{1,1}), EV(s_{2,1})\big) \in \mathbb{R}^2. \tag{7.12}$$

Illustrating with $\mathcal{G}_2$ (Eq. 7.1), Eq. 7.12 is $\varphi_1(\mathcal{G}_2|\mathbf{q}) = (4q_1 - 2q_2, 6q_1 + 0q_2) \in \mathbb{R}^2$: Notice the similarity with $\mathcal{L}$ (Eq. 7.10); the general setting replaces specific coefficients $(4, -2, 6, 0)$ with Player 1's payoffs $(g_{1,1}, g_{2,1}, g_{3,1}, g_{4,1})$ to make Eq. 7.12 a system of two equations with four unknowns. If φ_1 is a homomorphism, perhaps something similar to the structure of "more variables than equations" type solutions applies. This requires discovering whether φ_1 *is* a homomorphism with appropriate group operations. (Stated technically, is φ_i is a module homomorphism with respect to some group action?)

First consider the $\sigma \in \mathfrak{S}_2$ action of switching rows. If φ_1 is an $\mathfrak{S}_2$-module homomorphism, then, for any two games $\mathcal{G}_1$ and $\mathcal{G}_2$ along with permutation σ, it must be that

$$\varphi_1(\sigma\mathcal{G}_1 + \mathcal{G}_2|\mathbf{q}) = \sigma\varphi_1(\mathcal{G}_1|\mathbf{q}) + \varphi_1(\mathcal{G}_1|\mathbf{q}), \tag{7.13}$$

to avoid the Eq. 7.6 path dependency issue. On the right-hand side of Eq. 7.13, σ acting on $\mathbb{R}^2$ is the standard permutation $\sigma (x_1, x_2) = \big(x_{\sigma(1)}, x_{\sigma(2)}\big)$.

As an example, let $\mathcal{G}_2$ be from Eq. 7.1, $\mathcal{G}_1 = \mathcal{G}_{Stag}$ (Eq. 1.9) from the stag hunt,

$$\mathcal{G}_1 = \begin{array}{|cc|cc|} \hline 7 & 7 & -1 & 3 \\ \hline 3 & -1 & 3 & 3 \\ \hline \end{array},$$

and σ be where the two rows are interchanged (Eq. 7.7). Then

$$\sigma\mathcal{G}_1 + \mathcal{G}_2 = \begin{array}{|cc|cc|} \hline 3 & -1 & 3 & 3 \\ \hline 7 & 7 & -1 & 3 \\ \hline \end{array} + \begin{array}{|cc|cc|} \hline 4 & 4 & -2 & 6 \\ \hline 6 & -2 & 0 & 0 \\ \hline \end{array} = \begin{array}{|cc|cc|} \hline 7 & 3 & 1 & 9 \\ \hline 13 & 5 & -1 & 3 \\ \hline \end{array}. \tag{7.14}$$

Therefore,

$$\varphi_1(\sigma\mathcal{G}_1 + \mathcal{G}_2|\mathbf{q}) = (7q_1 + q_2, 13q_1 - q_2). \tag{7.15}$$

Computing the right-hand side of Eq. 7.13, $\varphi_1(\mathcal{G}_1|\mathbf{q}) = (7q_1 - q_2, 3q_1 + 3q_2)$, so $\sigma\varphi_1(\mathcal{G}_1|\mathbf{q}) = (3q_1 + 3q_2, 7q_1 - q_2)$. Similarly, $\varphi_1(\mathcal{G}_2|\mathbf{q}) = (4q_1 - 2q_2, 6q_1 + 0q_2)$. Thus the sum is

$$\sigma\varphi_1(\mathcal{G}_1|\mathbf{q}) + \varphi_1(\mathcal{G}_2|\mathbf{q}) = (3q_1 + 3q_2, 7q_1 - q_2) + (4q_1 - 2q_2, 6q_1 + 0q_2),$$

---

[6]From this point on, calculations will be done with respect to Player 1. Those for Player 2 are analogous.

which equals the right-hand side of Eq. 7.15. At least this example satisfies Eq. 7.13.

To show that Eq. 7.13 holds in general, represent game $\mathcal{G}_i$ with payoffs in vector form with the $\Pi^i = (\pi^i_{1,1}, \pi^i_{2,1}, \pi^i_{3,1}, \pi^i_{4,1}; \pi^i_{1,2}, \pi^i_{2,2}, \pi^i_{3,2}, \pi^i_{4,2})$ ordering. (Because φ_1 calculates the expected values of Player 1's pure strategies, Player 2's payoffs are dropped in the following calculations.) Checking the homomorphism condition for $\sigma = (1, 2)$ gives

$$
\begin{aligned}
\varphi_1(\sigma\mathcal{G}_1 + \mathcal{G}_2|\mathbf{q}) &= \varphi_1\left(\left(\begin{bmatrix} \pi^1_{3,1} \\ \pi^1_{4,1} \\ \pi^1_{1,1} \\ \pi^1_{2,1} \end{bmatrix} + \begin{bmatrix} \pi^2_{1,1} \\ \pi^2_{2,1} \\ \pi^2_{3,1} \\ \pi^2_{4,1} \end{bmatrix}\right)\bigg|\mathbf{q}\right) \\
&= \varphi_1\left(\begin{bmatrix} \pi^1_{3,1} + \pi^2_{1,1} \\ \pi^1_{4,1} + \pi^2_{2,1} \\ \pi^1_{1,1} + \pi^2_{3,1} \\ \pi^1_{2,1} + \pi^2_{4,1} \end{bmatrix}\bigg|\mathbf{q}\right) \\
&= \begin{bmatrix} q(\pi^1_{3,1} + \pi^2_{1,1}) + (1-q)(\pi^1_{4,1} + \pi^2_{2,1}) \\ q(\pi^1_{1,1} + \pi^2_{3,1}) + (1-q)(\pi^1_{2,1} + \pi^2_{4,1}) \end{bmatrix} \\
&= \begin{bmatrix} q\pi^1_{3,1} + (1-q)\pi^1_{4,1} \\ q\pi^1_{1,1} + (1-q)\pi^1_{2,1} \end{bmatrix} + \begin{bmatrix} q\pi^2_{1,1} + (1-q)\pi^2_{2,1} \\ q\pi^2_{3,1} + (1-q)\pi^2_{4,1} \end{bmatrix} \\
&= \sigma\begin{bmatrix} q\pi^1_{1,1} + (1-q)\pi^1_{2,1} \\ q\pi^1_{3,1} + (1-q)\pi^1_{4,1} \end{bmatrix} + \begin{bmatrix} q\pi^2_{1,1} + (1-q)\pi^2_{2,1} \\ q\pi^2_{3,1} + (1-q)\pi^2_{4,1} \end{bmatrix} \\
&= \sigma\varphi_1(\mathcal{G}_1|\mathbf{q}) + \varphi_1(\mathcal{G}_2|\mathbf{q})
\end{aligned}
$$

(7.16)

The agreement proves the compatibility of φ_1 with the symmetry property of interchanging rows. Similarly, φ_2 is compatible with the symmetry of interchanging columns. Consequently, these best response choices, φ_1 and φ_2, are $\mathfrak{S}_2$-module homomorphisms.

It remains to check what happens with the Klein four-group. An allowable Klein action (e.g., Fig. 7.1) is to interchange the two columns; denote this permutation by σ'. The path dependency issue is whether $\varphi_1(\sigma'\mathcal{G}_1 + \mathcal{G}_2|\mathbf{q}) = \sigma'\varphi_1(\mathcal{G}_1|\mathbf{q}) + \varphi_1(\mathcal{G}_2|\mathbf{q})$. This replacement of σ with σ' converts Eq. 7.14 into

$$
\sigma'\mathcal{G}_1 + \mathcal{G}_2 = \begin{array}{|cc|cc|} \hline 3 & 3 & 3 & -1 \\ \hline -1 & 3 & 7 & 7 \\ \hline \end{array} + \begin{array}{|cc|cc|} \hline 4 & 4 & -2 & 6 \\ \hline 6 & -2 & 0 & 0 \\ \hline \end{array} = \begin{array}{|cc|cc|} \hline 7 & 7 & 1 & 5 \\ \hline 5 & 1 & 7 & 7 \\ \hline \end{array}
$$

where

$$
\varphi_1(\sigma'\mathcal{G}_1 + \mathcal{G}_2|\mathbf{q}) = (7q_1 + q_2, 5q_1 + 7q_2).
$$

(7.17)

As computed following Eq. 7.14, $\varphi_1(\mathcal{G}_2|\mathbf{q}) = (4q_1 - 2q_2, 6q_1 + 0q_2)$ and $\varphi_1(\mathcal{G}_1|\mathbf{q}) = (7q_1 - q_2, 3q_1 + 3q_2)$. There is the problem of interpreting what permutation σ' means when acting on a vector (x_1, x_2). A permutation cannot act on portions of a component, so the only choices are $\sigma'(x_1, x_2) = (x_1, x_2)$ or

$\sigma'(x_1, x_2) = (x_2, x_1)$. It does not matter which interpretation prevails because the first yields $\sigma'\varphi_1(\mathcal{G}_1|\mathbf{q}) + \varphi(\mathcal{G}_2|\mathbf{q}) = (7q_1 - q_2, 3q_1 + 3q_2) + (4q_1 - 2q_2, 6q_1 + 0q_2) = (11q_1 - 3q_2, 9q_1 + 3q_2)$, while the second would be $\sigma'\varphi_1(\mathcal{G}_1|\mathbf{q}) + \varphi(\mathcal{G}_2|\mathbf{q}) = (3q_1 + 3q_2, 7q_1 - q_2) + (4q_1 - 2q_2, 6q_1 + 0q_2) = (7q_1 + q_2, 13q_1 - q_2)$: neither equals Eq. 7.17. Consequently, φ_1 is not a $\mathfrak{S}_2 \times \mathfrak{S}_2$-module homomorphism.

Theorem 7.1 *The Nash best response function for Player i, $\varphi_i : \mathbb{G}^i \times \mathbb{R}^2 \to \mathbb{R}^2$ is an $\mathfrak{S}_2$-module homomorphism, where $\mathfrak{S}_2$ acts on Player i's strategies. But φ_i is not an $\mathfrak{S}_2$-module homomorphism when $\mathfrak{S}_2$ acts on the opponent's strategies, which means that φ_i is not an $\mathfrak{S}_2 \times \mathfrak{S}_2$-module homomorphism.*

Theorem 7.1 underscores the important distinction between viewing $\mathbb{G}$ as an $\mathfrak{S}_2$-module versus an $\mathfrak{S}_2 \times \mathfrak{S}_2$-module. With the Nash equilibrium structure, the natural view of $\mathbb{G}$ is that it affords two distinct $\mathfrak{S}_2$ module structures, but not as a single $\mathfrak{S}_2 \times \mathfrak{S}_2$ module.

7.6 Decompositions of Games

As shown, $\mathbb{G}$ is a module under actions of $\mathfrak{S}_2$ and φ_i is an $\mathfrak{S}_2$-module homomorphism (Theorem 7.1). Motivated by the homomorphisms of Sect. 7.4.2, which decomposed the appropriate domains into relevant component parts, the hope is for a homomorphism to decompose $\mathbb{G}$ into components reflecting Nash properties. Restated technically, does this module structure split $\mathbb{G}$ into smaller submodules? The next theorem asserts that this happens: Because φ_i, is an $\mathfrak{S}_2$-module homomorphism, $\mathbb{G}$ can be decomposed into irreducible submodules, which is the decomposition analyzed in this book.

Theorem 7.2 (Maschke's Theorem) *Let G be a finite group and let V be a nonzero G-module. Then*

$$V = W^{(1)} \oplus \cdots \oplus W^{(k)}$$

where each $W^{(i)}$ is an irreducible G-submodule of V.

Are there other decompositions? Is this decomposition unique, or, with cleverness, could a different one be designed? For instance, Eq. 7.11 is not the only representation of the $3x + y + 2z = u$, $x - 3y + 4z = v$ problem; another solution of y and z in terms of x is

$$\begin{pmatrix} y \\ z \end{pmatrix} = \begin{pmatrix} 0.4 & -0.1 \\ 0.3 & 0.1 \end{pmatrix} \begin{pmatrix} u \\ v \end{pmatrix} - \begin{pmatrix} x \\ x \end{pmatrix} = \begin{pmatrix} 0.4u + 0.1v \\ 0.3u - 0.3v \end{pmatrix} - \begin{pmatrix} x \\ x \end{pmatrix},$$

Fortunately, these solutions are equivalent; algebraic manipulations can convert each into the other.[7] As asserted next, this "uniqueness" property holds in general.

---

[7] The division of $\mathbb{R}^3$ involves parallel translates of the $y = -x = z$ line: two $\mathbb{R}^3$ points are equivalent solutions if their difference is on the line.

Theorem 7.3 (Schur's Lemma) *Let V and W be two irreducible G-modules. If* $\theta : V \to W$ *is a G-homomorphism, then either*

1. θ is a G-isomorphism, or
2. θ is the zero map.

Schur's lemma tells us that φ_i is either an isomorphism or the zero map[8] on each of these submodules; consequently, this decomposition is unique up to isomorphism.[9] This decomposition reveals precisely what information is relevant, and what is irrelevant, for determining which $\mathbb{G}$ structures affect φ_i outcomes—the Nash outcomes.

7.7 Finding the Structure

These theorems provide the powerful information that $\mathbb{G}$ has a unique Nash decomposition. Excellent! But, how do we find it?

An important tool in doing so is the "character table," which is a mathematical development adopted in areas such as chemistry, crystallography, and physics to analyze symmetries. By being widely used, expository articles and worked out examples are on the web, which can be useful for readers interested in applying these tools to other topics. So, at this stage, some liberties are taken, while the elegant mathematical results that we used for our decomposition are described, the discussion is accompanied with a more familiar eigenvalue–eigenvector analysis.

Indeed, what makes it possible to describe the decomposition of $\mathbb{G}$ in terms of familiar linear algebra tools is that the group operations can be represented as the matrices of linear transformations. To illustrate, rewriting the Eq. 7.7 bimatrix as a vector shows that σ maps vector $(4, -2, 6, 0; 4, 6, -2, 0)$ to vector $(6, 0, 4, -2; -2, 0, 4, 6)$. Only components are interchanged (e.g., (x_1, x_2) with (x_3, x_4) and (y_1, y_2) with (y_3, y_4)); as this permutation is a linear transformation, it can be described as matrix from $\mathbb{R}^8$ to $\mathbb{R}^8$. Thus, $\mathfrak{S}_2$ group actions on $\mathbb{G}$ can be expressed as linear transformations of $\mathbb{G}$, which is written as a homomorphism

$$\rho : \mathfrak{S}_2 \to GL(\mathbb{G}) \tag{7.18}$$

(where $GL(\mathbb{G})$ is the *general linear group* consisting of invertible matrices acting on $\mathbb{G}$). Function ρ is called a *representation* of $\mathfrak{S}_2$.

The fact ρ maps $\sigma \in \mathfrak{S}_2$ to a linear transformation of $\mathbb{G}$ means that ρ assigns a matrix to each $\sigma \in \mathfrak{S}_2$; in this way, Maschke's theorem becomes identified with properties of matrices. The choice of a matrix depends upon the basis selected for $\mathbb{G}$, were an ideal choice is a block-diagonal matrix with blocks of smallest possible size. These "smallest blocks" of the $\rho(\sigma)$ matrix correspond to the irreducible representations of $\mathfrak{S}_2$; any representation ρ of a finite group can be written as a direct

---

[8] An example is where the Nash information of the $\mathcal{G}^B$ portion of a game is zero.
[9] See [32] for the proofs and more information about these two theorems.

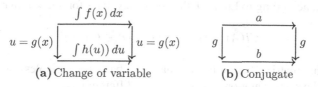

Fig. 7.2 Change of variables

sum of the irreducible representations. Thus these irreducible representations must be found because they determine the irreducible submodules, i.e., the decomposition of $\mathbb{G}$.

Finding the appropriate block representation is what captures the meat of Maschke's theorem, where diagrams such as Fig. 7.2b arise.

7.7.1 Conjugacy Classes

Even a quick perusal of advanced math books discloses Fig. 7.2b type diagrams. The commutativity of the diagram (that is, following either set of arrows arrives at the same place) asserts that the Eq. 7.6 path dependency problem is avoided.

A familiar example (Fig. 7.2a) is to solve $\int f(x)\,dx = \int (x-1)^6\,dx$. A change of variables $u = g(x) = x - 1$ leads to the easier $\int h(u)\,du = \int u^6\,du = \frac{1}{7}u^7 + c$. To convert back to the original variables, use the inverse of the change of variables, $x = g^{-1}(u) = u + 1$, to obtain $\frac{1}{7}(x-1)^7 + c$. The approach's validity is the commutativity of Fig. 7.2a. The operation (top line) followed by the change of variables (line on the right) equals the opposite order of the same change of variables (left line) followed by the operation (bottom line).

Similarly, to determine whether group actions $a, b \in G$ (e.g., permutations) are essentially the same, find whether a "change of variables" (some $g \in G$) makes Fig. 7.2b a commutative diagram. That is, the change of variables g followed by the operation b (or $b \circ g$)[10] agrees with the operation followed by a change of variables $g \circ a$. Avoiding an Eq. 7.6 path dependency concern, as represented by $b \circ g = g \circ a$, means that a and b are *conjugate*.

Solving $b \circ g = g \circ a$ for b leads to $b = g \circ a \circ g^{-1}$. An opportunity offered by this expression is that it shows how to discover *all conjugates* of a—the *conjugacy class of a*. That is, compute everything that can happen with all "change of variables" $g \in G$, which is

$$Cl(a) = \{b \mid \text{there is a } g \in G \text{ so that } b = g \circ a \circ g^{-1}.\} \tag{7.19}$$

---

[10] A standard permutation representation is "from left to right" or $g \circ b$. To avoid changing notion, it is convenient to go from "right to left" where $b \circ g$ has the composition of functions flavor of $b(g)$.

To illustrate, according to Eq. 7.19, the conjugacy class for a $n \times n$ matrix A is

$$Cl(A) = \{B = PAP^{-1} \mid P \in GL(n)\}. \tag{7.20}$$

The invertible matrix P constitutes a change of basis (or variables), so matrix $B = PAP^{-1}$ is the representation of matrix A in a different basis.

For another illustration, the conjugacy classes of the symmetric group $\mathfrak{S}_3$, which consists of all six permutations of the three letters $\{A, B, C\}$, has three conjugacy classes.

1. The first is $Cl(\mathbf{1}) = \{\mathbf{1}\}$, which is the identity map; nothing is permuted. For any $g \in \mathfrak{S}_3, g \circ \mathbf{1} \circ g^{-1} = g \circ g^{-1} = \mathbf{1}$.
2. Next are permutations that exchange two letters; e.g., $ABC \to BAC$ (or $(1, 2)$ that interchanges the first two terms). This defines $Cl((1, 2)) = \{(1, 2), (1, 3), (2, 3)\}$ For instance, $(1, 2, 3)^{-1} = (3, 2, 1)$, so[11] $(1, 2, 3)(1, 2)(3, 2, 1) = (2, 3)$.
3. The third class has the cyclic permutations of all letters, i.e., $ABC \to BCA \to CAB$ (which is $(1, 3, 2)$) and $ABC \to CAB \to BCA$. Here, $Cl((1, 3, 2)) = \{(1, 3, 2), (1, 2, 3)\}$.

A *class function* is a function on group G that is constant valued on each conjugacy class. A linear algebra example is where the eigenvalues of a matrix remain the same after any change of basis; that is, all matrices in the conjugacy class (Eq. 7.20) share the same eigenvalues. As the trace of a matrix (the sum of the terms in its diagonal) equals the sum of the eigenvalues, the trace of A, denoted by $tr(A)$, is another class function.

More equations involving the eigenvalues of A would be welcomed. This is where the symmetries of $\mathbb{G}$ help out. Namely, if σ is a symmetry of $\mathbb{G}$, then $\rho(\sigma)$ is a matrix (Eq. 7.18) with trace $\chi(\sigma) = \operatorname{tr}(\rho(\sigma))$ called the *character of* σ. This collection

$$\chi(\rho) = (\chi(g_1), \cdots, \chi(g_k)), \quad \text{for all } g_i \in G \tag{7.21}$$

is the *character of a representation* ρ. As it must be expected, $\chi(\rho)$ encodes considerable information, such as the number of irreducible representations of a group.

To continue this line of thought, the eigenvalues λ_j and eigenvectors $\mathbf{v}_j$ of matrix $\mathcal{A}$ satisfy $\mathcal{A}(\mathbf{v}_j) = \lambda_j \mathbf{v}_j$. Eigenvector $\mathbf{v}_j$ is a natural direction for $\mathcal{A}$, while λ_j explains whether $\mathbf{v}_j$ is reversed ($\lambda < 0$), enlarged, or even rotated (with imaginary values). Consequently, eigenvalues play critical roles in understanding an object's symmetry structures.

By definition, applying symmetry structure $\mathcal{A}$ to X retains X's shape: Repeatedly applying $\mathcal{A}$ to X keeps preserving X's shape, so $\mathcal{A}^j$ (where $\mathcal{A}$ is applied j times) is another symmetry structure of X. But the eigenvalues can change with $\mathcal{A}^2$, for example,

---

[11]Remember, our operations go from right to left, so with $(3, 2, 1)$, 3 goes to 2. The second operation of $(1, 2)$ converts this 2 to 1. The final operation $(1, 2, 3)$ changes the 1 to a 2, so the total result of these compositions changes 3 to 2. In the same way, after this rash of actions, 2 is interchanged with 3.

$$A^2(\mathbf{v}_j) = A(A(\mathbf{v}_j)) = A(\lambda_j \mathbf{v}_j) = \lambda_j A(\mathbf{v}_j) = \lambda_j^2 \mathbf{v}_j.$$

This expression means that other information about X comes from the eigenvalues of A^j, which are $(\lambda_1^j, \lambda_2^j, \ldots, \lambda_n^j)$. The symmetries of interest to us, such as rotating a square, eventually return to the original orientation after n times, which means that $A^n = \mathcal{I}$ is the identity map where all eigenvalues are unity. Thus, the symmetries of interest have the property $\lambda_j^n = 1$, or that each λ_j is a nth root of unity.

To illustrate, because four rotations of a square by 90° returns to the original position, the eigenvalues of A are fourth roots of unity, or $1, i, -1, -i$. This rotation moves corner 1 to 2, 2 to 3, 3 to 4, and 4 to 1, so matrix A, and its powers, have the representation (where $A^4 = \mathcal{I}$, the identity matrix)

$$A = \begin{pmatrix} 0 & 0 & 0 & 1 \\ 1 & 0 & 0 & 0 \\ 0 & 1 & 0 & 0 \\ 0 & 0 & 1 & 0 \end{pmatrix} \quad A^2 = \begin{pmatrix} 0 & 0 & 1 & 0 \\ 0 & 0 & 0 & 1 \\ 1 & 0 & 0 & 0 \\ 0 & 1 & 0 & 0 \end{pmatrix} \quad A^3 = \begin{pmatrix} 0 & 1 & 0 & 0 \\ 0 & 0 & 1 & 0 \\ 0 & 0 & 0 & 1 \\ 1 & 0 & 0 & 0 \end{pmatrix} \quad A^4 = \begin{pmatrix} 1 & 0 & 0 & 0 \\ 0 & 1 & 0 & 0 \\ 0 & 0 & 1 & 0 \\ 0 & 0 & 0 & 1 \end{pmatrix}$$

So, symmetry information comes from each A^j's eigenvalues as compiled in Eq. 7.22: column entries under A^j are eigenvalues, and the last row specifies the traces.

$$
\begin{array}{c|cccc}
 & \mathcal{I} = A^4 & A & A^2 & A^3 \\
\hline
\chi^{(1)} & 1 & 1 & (1)^2 = 1 & 1 \\
\chi^{(2)} & 1 & i & (i)^2 = -1 & -i \\
\chi^{(3)} & 1 & -1 & (-1)^2 = 1 & -1 \\
\chi^{(4)} & 1 & -i & (-i)^2 = -1 & i \\
\hline
\chi(\rho) & 4 & 0 & 0 & 0 \\
\end{array}
\qquad (7.22)
$$

Here, the sum of each column (the sum of the eigenvalues) equals the trace.

It is standard in mathematics to emphasize classes of behavior rather than special cases. So, replace each A^j with its conjugacy class and the eigenvalue with a "character" $\chi^{(j)}$ term—the *irreducible characters of a group*. This is part of a delightful mathematical theory that, for our purposes, can be brushed over because the needed information comes from more elementary arguments, i.e., although they are not, treat the $\chi^{(j)}$ as eigenvalues. This array reemerges with the *character tables* employed below.

7.7.2   Nash Decomposition

To use this discussion to specify the decomposition of $\mathbb{G}$ with respect to function φ_1, the first step is to determine the matrix representation of $\rho : \mathfrak{S}_2 \to GL(\mathbb{G})$ and compute, the character of this map χ_ρ. This table is then used to find the

Table 7.1 Character table for
$\mathbb{Z}/2\mathbb{Z}$

	ε	$(1, 2)$
$\chi^{(1)}$	1	1
$\chi^{(2)}$	1	-1
χ_ρ	4	0

decomposition. As Player 2's payoffs play no role with φ_1, only the space of Player
1's payoffs $\mathbb{G}^1 \cong \mathbb{R}^4$ is used.

Because $(1, 2) \in \mathfrak{S}_2$ switches the rows of $\mathcal{G} \in G$, if $\{e_1, e_2, e_3, e_4\}$ is the standard
basis of $\mathbb{G}^1 \cong \mathbb{R}^4$, then the action of $\sigma = (1, 2)$ is $\sigma\{e_1, e_2, e_3, e_4\} = \{e_3, e_4, e_1, e_2\}$.[12]
Therefore, the representation $\rho : \mathfrak{S}_2 \to GL_4$ is

$$\rho(\varepsilon) = \begin{pmatrix} 1 & 0 & 0 & 0 \\ 0 & 1 & 0 & 0 \\ 0 & 0 & 1 & 0 \\ 0 & 0 & 0 & 1 \end{pmatrix} \quad \rho((1, 2)) = \begin{pmatrix} 0 & 0 & 1 & 0 \\ 0 & 0 & 0 & 1 \\ 1 & 0 & 0 & 0 \\ 0 & 1 & 0 & 0 \end{pmatrix} \tag{7.23}$$

where ε is the identity permutation. The traces of these two matrices define $\chi_\rho = (4, 0)$.

Next, compile the eigenvalue information in an array similar to that of Eq. 7.22,
but where the $\rho(\sigma)$ matrices are replaced with conjugacy classes. The eigenvalues
for a general $(1, 2)$ permutation are 1 and -1 leading to the following table for $\mathfrak{S}_2$
(Table 7.1).

Where the bottom row contains the character computed for Eq. 7.23.

The sum of column entries do not equal the values of the bottom row. The reason is
that the $\chi^{(i)}$ characters are defined for *all settings*, whether it be an exchange of x and
y (where the $\rho(1, 2)$ permutation matrix is 2×2) or an exchange of two rows of 10
items (where the $\rho(1, 2)$ permutation matrix is 20×20). The structure of the object
being considered determines whether eigenvalues are repeated with multiplicity m_j.
This is a character theory result where if ρ is a representation of G, then

$$\chi_\rho = m_1 \chi^{(1)} + \cdots + m_k \chi^{(k)}, \tag{7.24}$$

where the $\chi^{(i)}$ are the characters of the k inequivalent irreducible representations of
G, and m_i are their multiplicities in ρ. This yields the following equation:

$$(4, 0) = m_1(1, 1) + m_2(1, -1),$$

which has the unique solution $m_1 = m_2 = 2$. So a purpose of Eq. 7.24 and the precise
definition of $\chi^{(j)}$ terms is to find how often an eigenvalue is repeated.

The eigenvalues are known; it remains to find the associated basis for $\mathbb{C}$, or the
irreducible decomposition of $\mathbb{G}$ as an $\mathfrak{S}_2$-module. Taking some liberties with the

---

[12]Recall the standard ordering defined previously to determine the rows in $\mathcal{G}$.

mathematics, think of this as finding the eigenvectors for the eigenvalues. After all, if σ is a symmetry, then the eigenvectors of the invertible matrix $\rho(\sigma)$ are what capture natural directions of the symmetry. So, for eigenvalue λ, find $\mathbf{v}$ so that $\rho(\sigma)(\mathbf{v}) = \lambda\mathbf{v}$.

More to the point, $\rho(1,2)$ is a matrix, which (according to Table 7.1 and $m_1 = m_2 = 2$) has two distinct eigenvectors with eigenvalue 1. This leads to two expressions for vector $\mathbf{c} = (c_1, c_2, c_3, c_4)$ (so $\mathbf{c} = c_1 e_1 + c_2 e_2 + c_3 e_3 + c_4 e_4$): First, matrix $\rho(1,2)$ interchanges the first two coordinates with the second two to have the image (c_3, c_4, c_1, c_2). But, if $\mathbf{c}$ is an eigenvector with eigenvalue 1, then $\rho(1,2)\mathbf{c} = \mathbf{c}$. For both expressions to be true, it must be that $(c_1, c_2, c_3, c_4) = (c_3, c_4, c_1, c_2)$, or $c_1 = c_3$ and $c_2 = c_4$. Consequently, the two basis vectors are

$$\mathbf{b}_1 = (1, 1, 1, 1) \text{ and } \mathbf{b}_2 = (1, -1, 1, -1, 1). \tag{7.25}$$

It also follows from Table 7.1 that matrix $\rho(1,2)$ has two eigenvectors with eigenvalue -1, or $\rho(1,2)\mathbf{c} = -\mathbf{c}$. These eigenvectors must satisfy $(c_3, c_4, c_1, c_2) = -(c_1, c_2, c_3, c_4)$, or $c_3 = -c_1$ and $c_4 = -c_2$. Thus, the last two basis vectors (eigenvectors) are

$$\mathbf{b}_3 = (1, 0, -1, 0) \text{ and } \mathbf{b}_4 = (0, 1, 0, -1). \tag{7.26}$$

To convert these basis vectors into components of G (where e_j is Row's entry in a specified cell), the payoffs for Player 1 have the decomposition

$$\mathbf{b}_1 = \begin{array}{|cc|cc|} \hline 1 & 0 & 1 & 0 \\ 1 & 0 & 1 & 0 \\ \hline \end{array} \quad \mathbf{b}_2 = \begin{array}{|cc|cc|} \hline 1 & 0 & -1 & 0 \\ 1 & 0 & -1 & 0 \\ \hline \end{array} \quad \mathbf{b}_3 = \begin{array}{|cc|cc|} \hline 1 & 0 & 0 & 0 \\ -1 & 0 & 0 & 0 \\ \hline \end{array} \quad \mathbf{b}_4 = \begin{array}{|cc|cc|} \hline 0 & 0 & 1 & 0 \\ 0 & 0 & -1 & 0 \\ \hline \end{array}$$

$$\tag{7.27}$$

From Sect. 3.4, $\mathbf{n}_{1,1} = \mathbf{b}_3$, $\mathbf{n}_{2,1} = \mathbf{b}_4$, and $\mathbf{k}_1 = \mathbf{b}_1$, so with the Chap. 2 notation, this is

$$\eta_{1,1}\mathbf{b}_3 + \eta_{2,1}\mathbf{b}_4 + \beta_1\mathbf{b}_2 + \kappa_1\mathbf{b}_1.$$

A similar analysis holds for Player 2's payoffs yielding the basis vectors

$$\mathbf{b}_5 = \begin{array}{|cc|cc|} \hline 0 & 1 & 0 & 1 \\ 0 & 1 & 0 & 1 \\ \hline \end{array} \quad \mathbf{b}_6 = \begin{array}{|cc|cc|} \hline 0 & 1 & 0 & 1 \\ 0 & -1 & 0 & -1 \\ \hline \end{array} \quad \mathbf{b}_7 = \begin{array}{|cc|cc|} \hline 0 & 1 & 0 & -1 \\ 0 & 0 & 0 & 0 \\ \hline \end{array} \quad \mathbf{b}_8 = \begin{array}{|cc|cc|} \hline 0 & 0 & 0 & 0 \\ 0 & 1 & 0 & -1 \\ \hline \end{array}$$

$$\tag{7.28}$$

and the decomposition

$$\eta_{1,2}\mathbf{b}_7 + \eta_{2,2}\mathbf{b}_8 + \beta_2\mathbf{b}_6 + \kappa_2\mathbf{b}_5.$$

The decomposition of G with respect to the Nash information is given by the basis vectors in Eqs. 7.27 and 7.28. From Schur's lemma, this is, up to isomorphism, the unique decomposition that completely separates the Nash information from the nonstrategic information.

7.7.3   Row-and-Column Decomposition–Potential Games

The Nash decomposition involved two different actions of $\mathfrak{S}_2$ on $\mathbb{G}$, which were selected because they are preserved by the Nash best response functions φ_1 and φ_2. But $\mathbb{G}$ also can be viewed as a $\mathfrak{S}_2 \times \mathfrak{S}_2$ module by combining these two actions: the resulting decomposition defines a basis for potential games.

As before, the first step is computing the representation $\rho : \mathfrak{S}_2 \times \mathfrak{S}_2 \to GL(\mathbb{G})$. In terms of the standard basis on $\mathbb{R}^8$, the σ_{row} symmetry of switching rows is

$$\sigma_{row}\left(\begin{array}{cccc} e_1 & e_5 & e_2 & e_6 \\ e_3 & e_7 & e_4 & e_8 \end{array}\right) = \begin{array}{cccc} e_3 & e_7 & e_4 & e_8 \\ e_1 & e_5 & e_2 & e_6 \end{array}$$

or $\sigma_{row}(e_1, e_2, e_3, e_4; e_5, e_6, e_7, e_8) = (e_3, e_4, e_1, e_2; e_7, e_8, e_5, e_6)$.

Similarly the σ_{col} symmetry of switching columns yields

$$\sigma_{col}(e_1, e_2, e_3, e_4; e_5, e_6, e_7, e_8) = (e_2, e_1, e_4, e_3; e_6, e_5, e_8, e_7).$$

This leads to the respective representations

$$\left(\begin{array}{cccc|cccc} 0 & 0 & 1 & 0 & 0 & 0 & 0 & 0 \\ 0 & 0 & 0 & 1 & 0 & 0 & 0 & 0 \\ 1 & 0 & 0 & 0 & 0 & 0 & 0 & 0 \\ 0 & 1 & 0 & 0 & 0 & 0 & 0 & 0 \\ \hline 0 & 0 & 0 & 0 & 0 & 0 & 1 & 0 \\ 0 & 0 & 0 & 0 & 0 & 0 & 0 & 1 \\ 0 & 0 & 0 & 0 & 1 & 0 & 0 & 0 \\ 0 & 0 & 0 & 0 & 0 & 1 & 0 & 0 \end{array}\right) \left(\begin{array}{cccc|cccc} 0 & 1 & 0 & 0 & 0 & 0 & 0 & 0 \\ 1 & 0 & 0 & 0 & 0 & 0 & 0 & 0 \\ 0 & 0 & 0 & 1 & 0 & 0 & 0 & 0 \\ 0 & 0 & 1 & 0 & 0 & 0 & 0 & 0 \\ \hline 0 & 0 & 0 & 0 & 0 & 1 & 0 & 0 \\ 0 & 0 & 0 & 0 & 1 & 0 & 0 & 0 \\ 0 & 0 & 0 & 0 & 0 & 0 & 0 & 1 \\ 0 & 0 & 0 & 0 & 0 & 0 & 1 & 0 \end{array}\right) \qquad (7.29)$$

A remaining action is the $\sigma_{row}\sigma_{col}$ product where

$$\sigma_{row}\sigma_{col}\left(\begin{array}{cccc} e_1 & e_5 & e_2 & e_6 \\ e_3 & e_7 & e_4 & e_8 \end{array}\right) = \sigma_{row}\left(\begin{array}{cccc} e_2 & e_6 & e_1 & e_5 \\ e_4 & e_8 & e_3 & e_7 \end{array}\right) = \begin{array}{cccc} e_4 & e_8 & e_3 & e_7 \\ e_2 & e_6 & e_1 & e_5 \end{array},$$

so the image of $(e_1, e_2, e_3, e_4; e_5, e_6, e_7, e_8)$ is $(e_4, e_3, e_2, e_1; e_8, e_7, e_6, e_5)$ with the Eq. 7.30 permutation matrix.

Indeed, because ρ is a homomorphism, $\rho(\sigma_{row}\sigma_{col}) = \rho(\sigma_{row})\rho(\sigma_{col})$, which means that the product of the above two matrices must equal the Eq. 7.30 matrix, which is true.

Table 7.2 Character table for $\mathbb{Z}/2\mathbb{Z} \times \mathbb{Z}/2\mathbb{Z}$

	ε	σ_{row}	σ_{col}	$\sigma_{row}\sigma_{col}$
$\chi^{(1)}$	1	1	1	1
$\chi^{(2)}$	1	1	-1	-1
$\chi^{(3)}$	1	-1	1	-1
$\chi^{(4)}$	1	-1	-1	1
χ_ρ	8	0	0	0

$$\rho(\sigma_{row})\rho(\sigma_{col}) = \left(\begin{array}{cccc|cccc}
0 & 0 & 0 & 1 & 0 & 0 & 0 & 0 \\
0 & 0 & 1 & 0 & 0 & 0 & 0 & 0 \\
0 & 1 & 0 & 0 & 0 & 0 & 0 & 0 \\
1 & 0 & 0 & 0 & 0 & 0 & 0 & 0 \\
0 & 0 & 0 & 0 & 0 & 0 & 0 & 1 \\
0 & 0 & 0 & 0 & 0 & 0 & 1 & 0 \\
0 & 0 & 0 & 0 & 0 & 1 & 0 & 0 \\
0 & 0 & 0 & 0 & 1 & 0 & 0 & 0
\end{array}\right). \tag{7.30}$$

Having computed the complete representation ρ for this action of $\mathfrak{S}_2 \times \mathfrak{S}_2$, the next step is to find the character table to determine the irreducibles; the character table for V_4 is

Finding the irreducibles means solving Eq. 7.24, or

$$(8, 0, 0, 0) = m_1(1, 1, 1, 1) + m_2(1, 1, -1, -1) + m_3(1, -1, 1, -1) + m_4(1, -1, -1, 1)$$

with its unique solution $m_1 = m_2 = m_3 = m_4 = 2$. Therefore ρ consists of 2 copies each of the 4 distinct one-dimensional subspaces (Table 7.2).

A basis is found via the "eigenvector" technique from the Nash decomposition. Here, a basis vector (i.e., eigenvector) must either be mapped to zero by one of the symmetries, or it is ± 1 of its image. For instance, the above $\mathbf{b}_3$ would not qualify because a change in columns creates a new vector. To illustrate with one computation

$$\rho(\sigma_{row}\sigma_{col})(c_1, c_2, c_3, c_4; c_5, c_6, c_7, c_8) = (c_4, c_3, c_2, c_1; c_8, c_7, c_6, c_5).$$

If $\mathbf{c}$ is an eigenvector with eigenvalue -1, it must be that

$$(c_4, c_3, c_2, c_1; c_8, c_7, c_6, c_5) = -(c_1, c_2, c_3, c_4; c_5, c_6, c_7, c_8)$$

requiring $c_1 = -c_4, c_2 = -c_3, c_5 = -c_8, c_6 = -c_7$, which, with the symmetry conditions (and orthogonality of basis), leads to $\mathbf{b}_3'$ and $\mathbf{b}_7'$.

The resulting decomposition is

$$\mathbf{b}_1' = \begin{array}{|cc|cc|} 1 & 0 & 1 & 0 \\ \hline 1 & 0 & 1 & 0 \end{array} \quad \mathbf{b}_2' = \begin{array}{|cc|cc|} 1 & 0 & -1 & 0 \\ \hline 1 & 0 & -1 & 0 \end{array} \quad \mathbf{b}_3' = \begin{array}{|cc|cc|} 1 & 0 & 1 & 0 \\ \hline -1 & 0 & -1 & 0 \end{array} \quad \mathbf{b}_4' = \begin{array}{|cc|cc|} 1 & 0 & -1 & 0 \\ \hline -1 & 0 & 1 & 0 \end{array}$$

$$\mathbf{b}_5' = \begin{array}{|cc|cc|} 0 & 1 & 0 & 1 \\ \hline 0 & 1 & 0 & 1 \end{array} \quad \mathbf{b}_6' = \begin{array}{|cc|cc|} 0 & 1 & 0 & 1 \\ \hline 0 & -1 & 0 & -1 \end{array} \quad \mathbf{b}_7' = \begin{array}{|cc|cc|} 0 & 1 & 0 & -1 \\ \hline 0 & 1 & 0 & -1 \end{array} \quad \mathbf{b}_8' = \begin{array}{|cc|cc|} 0 & 1 & 0 & -1 \\ \hline 0 & -1 & 0 & 1 \end{array}$$
$$(7.31)$$

To relate this basis to the Sect. 3.4.4 choice used to create utility functions and discuss the potential game (Sect. 3.4.5), notice that $\mathbf{a}_1 = \mathbf{b}_3'$, $\mathbf{a}_2 = \mathbf{b}_7'$ (from Eq. 3.26) and $\mathbf{d}_1 = \mathbf{b}_4'$, $\mathbf{d}_2 = \mathbf{b}_8'$ (from Eq. 3.27). This basis is a natural one to study potential games, which also arises by viewing a game as a flow on a graph. The symmetry approach, however, does not rely on a game as a flow on a graph, and it provides a basis for $\mathbb{G}$ as a vector space.

7.7.4 Pareto Decomposition–CoCo Solutions

A final example involves the action of $\mathfrak{S}_2$ on $\mathbb{G}$ of switching the players' names.

$$\sigma\mathcal{G} = \langle I = \{\sigma(1), \sigma(2)\}; S_{\sigma(1)}, S_{\sigma(2)}; \pi_{\sigma(1)}, \pi_{\sigma(2)}\rangle,$$

i.e., σ interchanges the labels of the players. Again, a decomposition of $\mathbb{G}$ follows by examining the representation ρ' of $\mathfrak{S}_2$.

The calculation of ρ' is simple: $\rho'(\varepsilon) = \mathbf{1} \in GL_8$; and $\rho'((1, 2))$ is a matrix with 0's down the diagonal since $\rho'((1, 2))$ is a permutation matrix without fixed points. These facts suffice to compute the irreducible characters, which relies only on the trace of ρ'. In this manner, $\chi(\varepsilon) = 8$ and $\chi((1, 2)) = 0$ yielding the character table (Table 7.3)

Finding the character decomposition is again simply solving a system of liner equations

$$(8, 0) = m_1(1, 1) + m_2(1, -1),$$

where $m_1 = m_2 = 4$. Thus there are four copies each of the two different irreducible submodules.

To find a basis corresponding to these irreducible characters, let

$$\mathbf{c} = (c_1, c_2, c_3, c_4; c_5, c_6, c_7, c_8)$$

Table 7.3 Character table for the Pareto decomposition

	ε	(1 2)
$\chi^{(1)}$	1	1
$\chi^{(2)}$	1	-1
χ_ρ	8	0

be a basis vector for the one-dimensional submodule corresponding to $\chi^{(2)}$. Under the action of σ, this corresponds to multiplication by -1. Since the "change of names" $\sigma(\{e_1, e_2, e_3, e_4; e_5, e_6, e_7, e_8\}) = \{e_5, e_6, e_7, e_8; e_1, e_2, e_3, e_4\}$, we have

$$c_1 = -c_5, \quad c_2 = -c_6, \quad c_3 = -c_7, \quad c_4 = -c_8$$

These equalities define the four copies of $\chi^{(2)}$, represented by the basis vectors $\mathbf{d}_i$ below. A similar analysis gives four copies of $\chi^{(1)}$, represented by $\mathbf{p}_i$.

$$
\begin{aligned}
\mathbf{d}_1 &= (1, 0, 0, 0; -1, 0, 0, 0), & \mathbf{p}_1 &= (1, 0, 0, 0; 1, 0, 0, 0) \\
\mathbf{d}_2 &= (0, 1, 0, 0; 0, -1, 0, 0), & \mathbf{p}_2 &= (0, 1, 0, 0; 0, 1, 0, 0) \\
\mathbf{d}_3 &= (0, 0, 1, 0; 0, 0, -1, 0), & \mathbf{p}_3 &= (0, 0, 1, 0; 0, 0, 1, 0) \\
\mathbf{d}_4 &= (0, 0, 0, 1; 0, 0, 0, -1), & \mathbf{p}_4 &= (0, 0, 0, 1; 0, 0, 0, 1)
\end{aligned}
$$

In game form, these basis vectors are

$$
\mathbf{d}_1 = \begin{array}{|cc|cc|} \hline 1 & -1 & 0 & 0 \\ 0 & 0 & 0 & 0 \\ \hline \end{array} \quad
\mathbf{d}_2 = \begin{array}{|cc|cc|} \hline 0 & 0 & 1 & -1 \\ 0 & 0 & 0 & 0 \\ \hline \end{array} \quad
\mathbf{d}_3 = \begin{array}{|cc|cc|} \hline 0 & 0 & 0 & 0 \\ 1 & -1 & 0 & 0 \\ \hline \end{array} \quad
\mathbf{d}_4 = \begin{array}{|cc|cc|} \hline 0 & 0 & 0 & 0 \\ 0 & 0 & 1 & -1 \\ \hline \end{array}
$$

$$
\mathbf{p}_1 = \begin{array}{|cc|cc|} \hline 1 & 1 & 0 & 0 \\ 0 & 0 & 0 & 0 \\ \hline \end{array} \quad
\mathbf{p}_2 = \begin{array}{|cc|cc|} \hline 0 & 0 & 1 & 1 \\ 0 & 0 & 0 & 0 \\ \hline \end{array} \quad
\mathbf{p}_3 = \begin{array}{|cc|cc|} \hline 0 & 0 & 0 & 0 \\ 1 & 1 & 0 & 0 \\ \hline \end{array} \quad
\mathbf{p}_4 = \begin{array}{|cc|cc|} \hline 0 & 0 & 0 & 0 \\ 0 & 0 & 1 & 1 \\ \hline \end{array}
$$

$$(7.32)$$

Vectors $\mathbf{p}_i$ determines which outcome is Pareto superior, while the $\mathbf{d}_i$ vectors determine how the value is divided between players. The $\mathbf{d}_i$ values are zero-sum, so adding a $\mathbf{d}_i$ component preserves a cell's total payoff., which make this decomposition the natural one to use to examine bargaining problems. This is what Kalai and Kalai [15] did.

7.8 Conclusion

The value added by a decomposition is that it simplifies the analysis and provides more general information about a model of interest. This is accomplished by use of abstract algebra, which is new tool for the social sciences. This approach should be more widely used if only because theories in nearly all of the social and behavioral sciences are accompanied by all sorts of symmetries that have yet to be fully exploited. Doing so allows a more sophisticated and complete analysis. The new types of questions that can be addressed are some of the benefits of the decomposition, which is the content of this book.

Chapter 8
Summary

This skinny chapter is positioned off by itself because not all readers will wade through the previous chapter to appreciate homomorphisms, modules, character tables, and the contributions of Maschke and Schur. Yet, expect many to return because fields of rewards involving other social, behavioral, and managerial issues are waiting to be harvested. Think of it this way; if much of what is being done in an area involves, as true with voting and game theories, symmetries with linear processes (equations, algebra, comparisons, etc.), then it is worth learning the Chap. 7 material. And if anyone utters the phrase, "Here are some symmetries that are satisfied, but they are useless, so forget them," then definitely review Chap. 7!

By being separate from Chap. 7, hopefully all readers will glance at this summary. As a quick review, two of this book's basic themes are to concentrate on *classes* of games and to decompose their complex structures into coordinate systems. Of course, when games gain complexity, by adding strategies, players, needs (e.g., potential games), and even formats (e.g., normal to extensive), details of the corresponding decompositions and coordinates change. But the fundamental features remain constant. They are the game's components favoring individual actions (Nash), group activity whether coordination, cooperation or imposed externalities (Behavioral), and payoff inflation or deflation (Kernel).

Among the offered benefits is a new way to analyze games. Rather than requiring a standard but often difficult ad hoc approach, which can differ from one class of games to the next, the decomposition applies the same methodology to all of them: Express games in terms of the coordinates coming from the decomposition. In doing so, new interpretations about games are forthcoming, a systematic way to analyze all of them is possible, creating new examples now becomes routine, and finding new properties is facilitated.

These coordinate systems even exhibit why ad hoc approaches can be so complicated. To illustrate, suppose two students are deciding whether to take class $\mathcal{A}$ or $\mathcal{B}$; here the sign of an a_j variable determines a student's top choice. (So, with $a_j > 0$ the jth student favors $\mathcal{A}$ over $\mathcal{B}$, while $a_j < 0$ has the reversed ranking). Now, whether

© Springer Nature Switzerland AG 2019
D. T. Jessie and D. G. Saari, *Coordinate Systems for Games*, Static & Dynamic Game Theory: Foundations & Applications, https://doi.org/10.1007/978-3-030-35847-1_8

each student wants to be in the same or a different class can be represented by the sign of a c_j variable. Each student can impose an externality on the other, given by β_j; for instance, depending on the instructor, Row might ask questions that add a spirit of humor to the classroom. Any discussion of courses must be accompanied by inflation, or κ_j terms.

Suppose all of this information is aggregated into the game (where TL is $(\mathcal{A}, \mathcal{A})$)

$$
\begin{array}{|cc|cc|}
\hline
6 & 10 & 10 & 4 \\
\hline
8 & 4 & 4 & 6 \\
\hline
\end{array}
\tag{8.1}
$$

What do these payoffs mean; e.g., what information can come from Row's TR payoff of 10? Does Row prefer class $\mathcal{A}$ or $\mathcal{B}$? What is Column's attitude about being in the same class as Row? What kind of dynamic can be expected from this setting? Finding answers can be difficult, which underscore the complexity of an ad hoc analysis.

This is where the associated coordinate system (e.g., see Eq. 3.38) provides significant help; it explicitly separates each payoff into contributions being made to each feature.

$$
\mathcal{G} = \begin{array}{|cc|cc|}
\hline
a_1+c_1+\beta_1+\kappa_1 & a_2+c_2+\beta_2+\kappa_2 & a_1-c_1-\beta_1+\kappa_1 & -a_2-c_2+\beta_2+\kappa_2 \\
\hline
-a_1-c_1+\beta_1+\kappa_1 & a_2-c_2-\beta_2+\kappa_2 & -a_1+c_1-\beta_1+\kappa_1 & -a_2+c_2-\beta_2+\kappa_2 \\
\hline
\end{array}.
$$

Even more, using nothing beyond arithmetic, the decomposition shows how to compute each of a payoff's components, something that cannot normally be done in game theory. (The $a_1 = 1$ value, for instance, means that Row prefers course $\mathcal{A}$.)

This representation of $\mathcal{G}$ carries a powerful message about complexity in even standard games. To see it, track how the different and competing kinds of twisting and weaving of these features, where the appropriate sign of $\pm a_j$, $\pm c_j$, $\pm \beta_j$ change in different manners with moves to different cells, create a game's complicated tapestry! Thus, an ad hoc analysis starts with a given game, say Eq. 8.1, and, in some manner without the guidance of a game's coordinates (e.g., in $\mathcal{G}$), must unravel at least some features to uncover consequences. While not what is done in practice, an ad hoc analysis essentially starts with the game's eight payoffs and explores how to partition each value into portions constituting a game's different properties. The difficulty of doing so, as manifested by the intricate structure of $\mathcal{G}$, underscores why ad hoc approaches can be so complicated!

The analysis need not be complicated! The decompositions and associated coordinate systems for games introduce simple tools to quickly unravel a game's fascinating but bewildering tapestry, to more easily extract essential features, and to discover new results and consequences.

What is given here is a start. Much more could be covered than what would be appropriate if the book is to be kept to a reasonable size, so we adopted the emphases of introducing the basic principles of the decomposition and coordinates and showing how to use them to analyze games. We welcome others to join us!

References

1. Aumann, R.: Acceptable points in general cooperative n-person games. In: Contributions to the theory of games, vol. IV. Princeton University Press, Princeton (1959)
2. Battalio, R., Samuelson, L., Van Huyck, J.: Optimization incentives and coordination failure in laboratory stag hunt games. Econometrica **69**, 749–764 (2001)
3. Bernheim, D., Peleg, B., Whinston, M.: Coalition-Proof nash equilibria. J. Econ. Theory **42**, 1–12 (1987)
4. Borel, E.: La théorie du jeu et les équations intégrales à noyau symétrique gauche. Comptes Rendus del'Acadmie **173**, 13041308 (1921). An English translated version is Savage, L.: The theory of play and integral equations with skew symmetric kernels. Econometrica **21**, 97–100 (1953)
5. Candogan, O., Menache, I., Ozdaglar, A., Parillo, P.: Flows and decompositions of games: harmonic and potential games. Math. Oper. Res. **36**(3) (2011)
6. Dummit, D.S., Foote, R.E.: Abstract Algebra, 3rd edn. Wiley (2004)
7. Fudenberg, D., Tirole, J.: Game Theory. MIT Press, Boston (1991)
8. Harsanyi, J.C., Selten, R.: A General Theory of Equilibrium Selection in Games. MIT Press, Boston (1988)
9. Henrick, J., Boyd, R., Bowles, S., Camerer, C., Fehr, E., Gintis, H. (eds.): Foundations of Human Sociality: Economic Experiments and Ethnographic Evidence from Fifteen Small-Scale Societies. Oxford University Press, Oxford (2004)
10. Hopkins, B.: Expanding the Robinson-Goforth system for 2x2 games. In: The Mathematics of Decisions, Elections, and Games, Jones, M. (ed.), Contemporary Mathematics Series, vol. 624, pp. 177–187. American Mathematical Society (2014)
11. Jessie, D.T., Kendall, R.: Decomposing Models of Bounded Rationality, IMBS Technical Report 15–06. University of California, Irvine (2015)
12. Jessie, D.T., Saari, D.G.: Strategic and Behavioral Decomposition of Games, IMBS Technical Report 15–05. University of California, Irvine (2015)
13. Jessie, D.T., Saari, D.G.: Cooperation in n-player repeated games. In: Jones, M. (ed.), The Mathematics of Decisions, Elections, and Games. AMS Contemporary Mathematics Series, vol. 624, pp. 189–206 (2014)
14. Jessie, D.T., Saari, D.G.: From the luce choice axiom to the quantal response equilibrium. J. Math. Psychol. **75**, 1–9 (2016)
15. Kalai, A., Kalai, E.: Cooperation in strategic games revisited. Q. J. Econ. **128**(2) (2013)
16. Monderer, D., Shapley, L.: Monderer, Fictious play property for games with identical interests. J. Econ. Theory **68**, 258–265 (1996)
17. Monderer, D., Shapley, L.: Potential games. Games Econ Behav. **14**, 124–143 (1996)
18. Nash, J.: Equilibrium points in n-person games. PNAS **36**(1), 48–49 (1950)
19. Nash, J.: Non-cooperative games. Ann. Math. **54**(2), 286–295 (1951)

© Springer Nature Switzerland AG 2019
D. T. Jessie and D. G. Saari, *Coordinate Systems for Games*, Static & Dynamic Game
Theory: Foundations & Applications, https://doi.org/10.1007/978-3-030-35847-1

20. Robinson, D., Goforth, D.: The Topology of the 2×2 Games: a new Periodic Table. Routledge, New York (2005)
21. Rosenthal, R.: A class of games possessing pure-strategy Nash equilibria. Int. J. Game Theory **2**, 65–67 (1973)
22. Rosenthal, R.: Games of perfect information, predatory pricing, and the chain store. J. Econ. Theory **25**(1), 92–100 (1981)
23. Saari, D.G.: A dictionary for voting paradoxes. J. Econ. Theory **48**, 443–475 (1989)
24. Saari, D.G.: Basic Geometry of Voting. Springer, New York (1995)
25. Saari, D.G.: Explaining all three-alternative voting outcomes. J. Econ. Theory **87**, 313–355 (1999)
26. Saari, D.G.: The profile structure for Luce's choice axiom. J. Math. Psychol. **49**, 226–253 (2005)
27. Saari, D.G.: Demystifying Voting Paradoxes: Disposing Dictators. Cambridge University Press, New York (2008)
28. Saari, D.G.: A new way to analyze paired comparison rules. Math. Oper. Res. **39**, 647–655 (2014)
29. Saari, D.G.: Unifying voting theory from Nakamura's to Greenberg's theorems. Math. Soc. Sci. **69**, 1–11 (2014)
30. Saari, D.G.: Mathematics Motivated by the Social and Behavioral Sciences. SIAM, Philadelphia (2018)
31. Saari, D.G.: Arrow, and unexpected consequences of his theorem. Public Choice **179**, 133–144 (2019)
32. Sagan, B.: The Symmetric Group: Representations, Combinatorial Algorithms, and Symmetric Functions, 2nd edn. Springer, New York (2001)
33. Smith, J.M., Parker, G.A.: The logic of animal conflict. Nature **246**(1973), 15–18 (1976)
34. Skyrms, B.: Evolution of the Social Contract. Cambridge University Press (1996)
35. Skyrms, B.: The Stag Hunt and the Evolution of Social Structure. Cambridge University Press (2006)
36. Thorpe, E.: Beat the Dealer. Vintage (1962)

Index

© Springer Nature Switzerland AG 2019
D. T. Jessie and D. G. Saari, *Coordinate Systems for Games*, Static & Dynamic Game
Theory: Foundations & Applications, https://doi.org/10.1007/978-3-030-35847-1

Printed in the United States
By Bookmasters